Corrosion and Corrosion Control

an introduction to corrosion science and engineering

Herbert H. Uhlig

Professor of Metallurgy In Charge of Corrosion Laboratory
Massachusetts Institute of Technology

John Wiley & Sons Inc.,
New York • London • Sydney • Toronto

Corrosion and Corrosion Control

an introduction to
corrosion science
and engineering

Second Edition

Library of Congress Catalog Card Number: 71-162425

ISBN 0-471-89563-6

Printed in the United States of America.

10 9 8 7 6 5 4 3 2 1

Preface

Rapid advances in corrosion science and engineering since the publication of the first edition have made it necessary to update most chapters of the present book. The recently appreciated concept of a *critical pitting potential* and its useful application to practical situations are now described in detail. Similarly, appropriate space is given to the recognized importance of an analogous critical potential below which *stress corrosion cracking* does not occur, and a more detailed account of recent trends in proposed mechanisms of stress corrosion cracking is provided. The section on *corrosion fatigue* is rewritten in accord with new data and their corresponding interpretation; the chapter on *passivity*, which has benefited from several new and exciting experiments conducted in more than one laboratory, is also rewritten. The *intergranular corrosion of nonsensitized stainless steels* and related alloys is a new subject of special interest to the nuclear power industry. An elementary description of *Pourbaix diagrams* is included here in line with the availability of the extensive atlas published by Professor Pourbaix in 1966.

It has become plainly (and sometimes painfully) apparent by events of the past decade that all materials scientists as well as engineers in a broad cross section of professional activities should know something about environmental effects on properties of materials, including the corrosion behavior of metals and alloys. Where safety and economy of performance are vital considerations, the design of critical structures and their components employing modern materials requires more than the traditional attention to mechanical properties alone.

I am gratified that this book has been useful both to stimulate the teaching of corrosion to students in the universities and also to serve as a concise reference for experienced materials scientists and engineers who missed a chance to learn something about the subject or who want to be brought up to date. To help in the necessary understanding of basic principles, additional problems have been added at the end of the book, and, in response to many requests, more answers to such problems are now provided.

Acknowledgment is due several persons who helped to clarify some of the discussions in the first edition, in particular to my former student Iwao Matsushima who, with Seigo Matsuda, also a former student of the Corrosion Laboratory, prepared the Japanese translation. I am indebted to Walter Hamer of the National Bureau of Standards for up-to-date values of reference electrode potentials, to J. Robinson of the Dow Chemical Company for statistical data on magnesium and aluminum production, to M. Vucich of the National Steel Company for production figures on chromium-plated container steel, and to H. Spähn, G. Cheever, and L. Brockway for various photographs.

HERBERT H. UHLIG

Cambridge, Massachusetts
June 1971

Preface to the First Edition

Corrosion research and the teaching of corrosion have a tradition of long standing at M.I.T. Professor W. R. Whitney, who later became the first Director of Research for the General Electric Company, initiated this tradition with his classic paper on "The Corrosion of Iron" published in 1903 in the *Journal of the American Chemical Society*. This paper stimulated the important researches on corrosion by Professor W. H. Walker, and later by Professor Walter Whitman and his associates in the Laboratory of Applied Chemistry, which Walker founded. Formal instruction in corrosion was apparently first presented in a subject entitled "Chemical Resistance of Materials," offered to students by Professor Walker in 1903. The title was changed in 1912 to "Materials of Construction" and was taught with the assistance of Professor W. K. Lewis, and later by Professor Lewis alone. The first M.I.T. subject under the name of "Corrosion" appeared in 1922 and was taught by Professor Whitman. It was subsequently offered by several other members of the Department of Chemical Engineering, including Professor Lewis. In 1931, Professor R. S. Williams of the Department of Metallurgy offered for the first time a subject entitled "Corrosion and Heat Resisting Alloys," which he taught until 1942. Beginning in 1938, I continued the organized lectures on corrosion previously presented by Professor Whitman. After the interruption caused by World War II, these lectures were resumed in the Department of Metallurgy and have continued up to the present time.

This book is based on the one-semester subject called "Corrosion," which I present to seniors and graduate students at M.I.T. The purpose is to introduce the student to the underlying science of corrosion and to the fundamentals of what is now known as corrosion engineering. In brief, the lectures are concerned with why metals corrode and what can be done about it. The subject is treated as quantitatively as present knowledge permits, and, along these lines, problems are assigned which illustrate first principles and the application of these principles to practical situations. Some of the assigned problems are assembled in the latter part of this book. Answers to many of these are given, primarily to assist the professional engineer or the scientist who may use the book for background information on corrosion, and who finds that working out problems helps to obtain the necessary understanding. Better understanding is the justification for working out problems in any subject of instruction, and corrosion is no exception. It is assumed that the reader has familiarity with elementary physical chemistry; this subject is a prerequisite for the M.I.T. corrosion course. Background knowledge in physical metallurgy is also helpful, but is not required.

Treatment of the subject is concise, consistent with time limitations for presentation of all the material within one semester. For this reason, only those literature references are cited which bear directly on the point under discussion. General references are added at the end of the various chapters. These references usually contain additional references if the reader is interested, but in no case is an attempt made to cite all the literature that might justifiably be mentioned. For a more complete guide to the literature, the reader should consult reference books such as *The Corrosion Handbook*, edited by the author, and *The Corrosion and Oxidation of Metals*, by U. R. Evans.

When more than one interpretation of the facts or more than one theory is current, I have tried to present both sides. At the same time, I have made my own preference clear on the premise that the reader is entitled to know my opinion. Corrosion rates are expressed as milligrams per square decimeter per day (mdd) and inches penetration per year (ipy), which are both in common use in the United States. The sign of electrode potential has always been a difficult matter to resolve in view of the opposite conventions followed by major groups in the United States and abroad. It is hoped that by defining *the potential* as opposite in sign to the oxidation potential, consistent with the use recommended by Lewis, Randall, Pitzer, and Brewer in *Thermodynamics*, 2nd edition, McGraw-Hill (1961), some degree of harmony is provided, or at least confusion is avoided.

I am indebted to many friends and colleagues for much of the information which has entered my lectures and is now contained here. Enlightened understanding of corrosion mechanisms in several instances was contributed through the patient effort of many students who sojourned in the Corrosion Laboratory and who searched with me for a critical experiment, or who laboriously accumulated decisive data.

Appreciation is expressed to those who read portions of the manuscript and made valuable suggestions, including Carl Wagner, Milton Stern, W. D. Robertson, Bruce Chalmers, G. T. Paul, G. Schikorr, H. S. Campbell, C. P. Larrabee, W. W. Bradley, and J. P. Pemsler; and also to those who supplied photographs or sketches, in particular, F. L. LaQue, M. VanLoo, J. F. Sebald, and H. M. Bendler. A large part of the manuscript was written during a sabbatical semester spent in Göttingen, West Germany. This opportunity was made possible by financial support provided by the John Simon Guggenheim Memorial Foundation and by the hospitality of Dr. Carl Wagner, Director of the Max Planck Institut für Physikalische Chemie.

If this book stimulates young minds to accept the challenge of continuing corrosion problems, and to help reduce the huge economic losses and dismaying wastage of natural resources caused by metal deterioration, it will have fulfilled the author's major objective.

HERBERT H. UHLIG

Cambridge, Massachusetts
November 1962

Contents

chapter 1
Definition and importance

DEFINITION

Corrosion is the destructive attack of a metal by chemical or electrochemical reaction with its environment. Deterioration by physical causes is not called corrosion, but is described as erosion, galling, or wear. In some instances, chemical attack accompanies physical deterioration as described by the terms: corrosion-erosion, corrosive wear, or fretting corrosion. Nonmetals are not included in the present definition. Plastics may swell or crack, wood may split or decay, granite may erode, and Portland cement may leach away, but the term corrosion is presently restricted to chemical attack of metals.

"Rusting" applies to the corrosion of iron or iron-base alloys with formation of corrosion products consisting largely of hydrous ferric oxides. Nonferrous metals, therefore, corrode but do not rust.

Corrosion Science

Since corrosion involves chemical change, it is obvious that the student must be familiar with principles of chemistry in order to understand corrosion reactions. In particular, because corrosion processes are mostly electrochemical, an understanding of electrochemistry is important. Furthermore, since structure and composition of a metal often determine corrosion behavior, the student should also be familiar with fundamentals of physical metallurgy. Hence, both chemistry and metallurgy are basic to corrosion studies in the same sense as biology and chemistry are basic to the study of medicine.

The *corrosion scientist* is concerned with the study of corrosion mechanisms through which a better understanding is obtained of the causes of corrosion and the available means for preventing or minimizing resulting damage. The *corrosion engineer* applies accumulated scientific knowledge to the abatement of corrosion damage by practical and economical means. For example, the corrosion engineer employs cathodic protection on a large scale to prevent corrosion of buried pipelines, or he tests and develops new and better paints, prescribes proper dosage of corrosion inhibitors, or recommends the correct metal coating. The corrosion scientist, in turn, develops better criteria of cathodic protection, outlines the molecular structure of chemical compounds that behave best as inhibitors, synthesizes corrosion-resistant alloys, and recommends heat treatment and compositional variations of alloys that will improve their performance. Both the scientific and engineering viewpoints supplement each other in the diagnosis of corrosion damage and in the prescription of proper remedies.

Importance of Corrosion

The importance of corrosion studies is threefold. The first area of significance is *economic* including the objective of reducing material losses resulting from the corrosion of piping, tanks, metal components of machines, ships, bridges, marine structures, etc. The second area is *improved safety* of operating equipment which, through corrosion, may fail with catastrophic consequences. Examples are pressure vessels, boilers, metallic containers for radioactive materials, turbine blades and rotors, bridge cables, airplane components, automotive steering mechanisms. Third is *conservation,* applied primarily to metal resources—the world's supply of these is limited, and the wastage of them includes corresponding losses of energy and water reserves associated with the production and fabrication of metal structures. Not least important is the accompanying conservation of human effort entering the design and rebuilding of corroded metal equipment, otherwise available for socially useful purposes.

Currently the prime motive for research in corrosion is provided by the economic factor. Losses sustained by industry, by the military, and by municipalities amount to many billions of dollars annually.

Economic losses are divided into (1) direct losses and (2) indirect losses. By *direct* losses are meant the costs of replacing corroded structures and machinery or their components, such as condenser tubes, mufflers, pipelines, and metal roofing, including necessary labor. Other examples are repainting of structures where prevention of rusting is the prime objective and the capital costs plus upkeep of cathodically protected pipelines. Sizeable direct losses are illustrated by the necessity to replace several million domestic hot water tanks each year because of failure by corrosion or, similarly, the need for annual replacement of millions of corroded automobile mufflers. Direct losses include the extra cost of using corrosion-resistant metals and alloys instead of carbon steel where the latter has adequate mechanical properties but not sufficient corrosion resistance; there are also the costs of galvanizing or nickel plating of steel, of adding corrosion inhibitors to water, or of dehumidifying storage rooms for metal equipment. The total combined losses of this kind to the United States alone is estimated conservatively to be about $5.5 billion annually.

Indirect losses are more difficult to assess, but a brief survey of typical losses of this kind compels the conclusion that they add several billion dollars to the direct losses already outlined. Examples of indirect losses are as follows:

1. *Shutdown.* The replacement of a corroded tube in an oil refinery may cost a few hundred dollars, but shutdown of the unit while repairs are underway may cost $10,000 per hour in lost production. Similarly, replacement of corroded boiler or condenser tubes in a large power plant may necessitate $25,000 per day for power purchased from interconnected electric systems to supply customers while the boiler is down. Losses of this kind cost the electrical utilities in the United States tens of millions of dollars annually.[1]

2. *Loss of Product.* Losses of oil, gas, or water occur through a corroded pipe system until repairs are made. Antifreeze may be lost through a corroded auto radiator, or gas leaking from a corroded pipe may enter the basement of a building causing an explosion.

3. *Loss of Efficiency.* This may occur because of diminished heat transfer through accumulated corrosion products, or because of the clogging of pipes with rust necessitating increased pumping capacity. (It is estimated that in the United States, for example, increased pumping capacity, made necessary by partial clogging of water mains with rust, costs approximately $40 million per year.[2])

A further example is provided by internal combustion engines of automobiles where piston rings and cylinder walls are continuously corroded by combustion gases and condensates. Loss of critical dimensions leading to excess gasoline and oil consumption is caused by corrosion to an extent often equal to or greater than that caused by wear.

4. *Contamination of Product.* A small amount of copper picked up by slight corrosion of copper piping or of brass equipment that is otherwise durable may damage an entire batch of soap. Copper salts acceler-

[1] H. Klein and J. Rice, *Trans. Amer. Soc. Mech. Eng., Ser. A*, **88**, 232 (1966).
[2] Private communication, H. Jordan, Amer. Water Works Assoc.

ate rancidity of soaps and shorten the time they can be stored before use. Traces of metals may similarly alter the color of dyes.

Lead equipment, otherwise durable, is not permitted in the preparation of foods and beverages because of the toxic properties imparted by very small quantities of lead salts.[3] Similarly, soft waters conducted through lead piping are not safe for drinking purposes.*

Included in the category of contamination is spoilage of food in corroded metal containers. One cannery of fruits and vegetables lost more than $1 million in one year before the metallurgical factors causing localized corrosion were analyzed and remedied. Another company using metal caps on glass food jars lost $0.5 million in one year because the caps soon perforated by a pitting type of corrosion, thereby allowing bacterial contamination of the contents.

5. *Overdesign.* This factor is common in the design of reaction vessels, boilers, condenser tubes, oil-well sucker rods, buried pipelines, water tanks, and marine structures. Because corrosion rates are unknown or methods for corrosion control are uncertain, equipment is often designed many times heavier than normal operating pressures or applied stresses would require in order to insure reasonable life. With adequate knowledge of corrosion, more reliable estimates of equipment life can be made, and design can be simplified in terms of materials and labor.

A typical example of overdesign, currently less common than formerly, is in the laying of buried oil pipes. A line of 8-in. diameter pipe 225 miles long was specified to have a wall thickness of 0.322 in. to allow for corrosion from the soil side. With adequate corrosion protection, a wall thickness of only 0.250 in. could have been used, saving 3700 tons of steel, as well as increasing internal capacity by 5%.[5]

Similarly, oil-well sucker rods are normally overdesigned to increase their service life before ultimate failure occurs by corrosion fatigue. Were the corrosion factor eliminated, losses would be cut at least in half. There would be further savings because less power would be re-

---

* The poisonous effects of small amounts of lead have been known for a long time. In a letter to Benjamin Vaughn dated July 31, 1786, Benjamin Franklin[4] warned against possible ill effects of drinking rain water collected from lead roofs or consuming alcoholic beverages exposed to lead. The symptoms were called in his time "dry bellyache" and were accompanied by paralysis of the limbs. The disease originated because New England rum distillers used lead coil condensers. On recognizing the cause, the Massachusetts Legislature passed an act outlawing use of lead for this purpose.

[3] Bureau of Foods and Drugs permit not over 1 ppm of lead in food.

[4] *Benjamin Franklin's Autobiographical Writings,* edited by Carl Van Doren, p. 671, Viking Press, 1945.

[5] J. Stirling, *Corrosion,* **1,** 17 (1945) (March).

quired to operate a light-weight rod and the expense for recovering a light-weight rod after breakage would be lower.

Obviously, indirect losses are a substantial part of the economic tax imposed by corrosion, although it is difficult to arrive at a reasonable estimate of total losses from this source, even within one industry. In the event of loss of health or life through explosion, unpredictable failure of chemical equipment, or wreckage of airplanes, trains, or automobiles through sudden failure by corrosion of critical parts, the indirect losses are still more difficult to assess and are beyond interpretation in terms of dollars.

GENERAL REFERENCES

H. H. Uhlig, "The Cost of Corrosion to the United States," *Chem. Eng. News,* **27,** 2764 (1949) ; or *Corrosion,* **6,** 29 (1950).

W. Vernon, "Metallic Corrosion and Conservation," excerpt from *The Conservation of Natural Resources,* Institute of Civil Engineers (London), 1957.

chapter 2

Electrochemical mechanisms

THE DRY CELL ANALOGY, FARADAY'S LAW

As mentioned in Chapter 1, corrosion processes are most often electrochemical. In aqueous media, the action is similar to that taking place in a flashlight cell made up of a center carbon electrode and a zinc cup electrode separated by an electrolyte consisting essentially of NH_4Cl solution* (Fig. 1). An incandescent light bulb connected to both electrodes glows continuously, the electrical energy being supplied by chemical reactions at both electrodes. At the carbon electrode (positive pole),$(+)$ chemical reduction occurs, and at the zinc electrode (negative pole)$-)$ oxidation occurs, metallic zinc being converted into hydrated zinc ions, $Zn^{++} \cdot nH_2O$.† The greater the flow of electricity through the cell, the greater is the amount of zinc that corrodes. The relationship is quantitative, as Michael Faraday showed in the early nineteenth century (Faraday's law):

$$\text{weight of metal reacting} = kIt \tag{1}$$

where I is in amperes, t in seconds, and k is a constant called the *electrochemical equivalent*. The value of k in the case of zinc is 3.39×10^{-4} g/C, (gm per coulomb), the coulomb being defined as the amount of electricity represented by 1 A (ampere) flowing for 1

---

*The function of carbon granules for conduction and manganese dioxide as depolarizer, both surrounding the carbon electrode, need not concern us at this point.

† Ions in aqueous solution attach to themselves water molecules, but their number is not well defined. They differ in this way from gaseous ions, which are not hydrated. It is common practice, however, to omit mention of the appended H_2O molecules and to designate hydrated zinc ions, for example, as Zn^{++}.

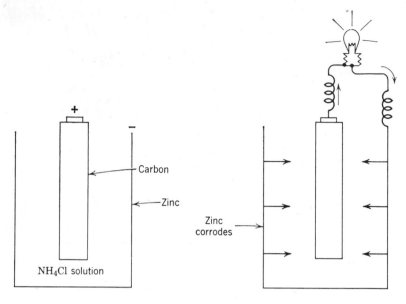

Fig. 1. Dry cell.

sec.* On short circuiting the cell with a low-resistance metallic connec-
tor, the zinc cup perforates by corrosion within a matter of hours; but
when the cell is left disconnected (open circuit), the zinc may remain
intact for years. The slow consumption of zinc occurring on open circuit
is accounted for largely by activity of minute impurities, like iron, em-
bedded in the surface of zinc; these impurities assume the same role
as carbon and allow the flow of electricity accompanied by corrosion
of zinc. Current of this kind is called *local-action current,* and the
corresponding cells are called *local-action cells.* Local-action current,
of course, produces no useful energy, but acts only to heat up the
surroundings.

Any metal surface, similar to the situation for zinc, is a composite
of electrodes electrically short-circuited through the body of the metal
itself (Fig. 2). So long as the metal remains dry, local-action current
and corrosion are not observed. But on exposure of the metal to water
or aqueous solutions, local-action cells are able to function and are accompanied by chemical conversion of the metal to corrosion products.
Local-action currents, in other words, may account for the corrosion
of metals exposed to water, salt solutions, acids, or alkalies.

* Similarly, electrochemical equivalents are obtainable for other metals (see e.g.,
Corrosion Handbook, p. 1133).

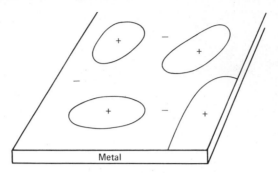

Fig. 2. Metal surface enlarged, showing schematic arrangement of local-action cells.

Whenever impurities in a metal constitute the electrodes of local-action cells, their removal, as might be expected, appreciably improves corrosion resistance. Accordingly, purified aluminum and magnesium are much more resistant to corrosion in seawater or in acids than are the commercial varieties of these metals, and high-purity zinc resists dilute hydrochloric acid much better than does commercial zinc. However, it is not correct to assume, as was done many years ago when the electrochemical theory was first proposed, that pure metals do not corrode at all. As we see later, local-action cells are also set up when there are variations in the environment or in temperature. With iron or steel in aerated water, for example, the negative electrodes are commonly portions of the iron surface itself covered perhaps by porous rust (iron oxides), and positive electrodes are areas exposed to oxygen, the positive and negative electrode areas interchanging and shifting from place to place as the corrosion reaction proceeds. Accordingly, high-purity iron in air-saturated water corrodes at essentially the same rate as impure or commercial iron. A difference in rates is observed in acids, however, because impurities now enter predominantly as electrodes of local-action cells. This matter is discussed again later.

DEFINITION OF ANODE AND CATHODE

A combination of two electrical conductors (electrodes) immersed in an electrolyte is called a *galvanic cell* in honor of Luigi Galvani, a physician in Bologna, Italy, who published his studies of electrochemical action in 1791. A galvanic cell converts chemical energy into electrical energy. On short-circuiting such a cell (attaching a low-resistance wire to each electrode), positive current flows through the metallic path from positive electrode to negative electrode. This direction of current flow follows an arbitrary convention, established before anything was known about the nature of electricity, and is employed today despite contemporary knowledge that only negative carriers or electrons move in a metal. Electrons, of course, go from negative to positive pole, opposite to the imaginary flow of positive carriers. Whenever current is said to flow, however, without designating the sign of the carrier, positive current is always implied.

Within the electrolyte, current is carried by both negative and positive carriers known as ions (electrically charged atoms or groups of atoms). The current carried by each ion depends on its respective mobility and electric charge. The total of positive and negative current in the electrolyte of a cell is always exactly equivalent to the total current carried in the metallic path by electrons alone. Ohm's law, i.e., $I = E/R$ where I is the current in amperes, E the potential difference in volts, and R the resistance in ohms, applies precisely, under conditions with which we are presently concerned, to current flow in electrolytes as well as in metals.

The electrode at which chemical reduction occurs (or $+$ current enters the electrode from the electrolyte) is called the *cathode*. Examples of cathodic reactions are:

$$H^+ \rightarrow \tfrac{1}{2}H_2 - e^-$$

$$Cu^{++} \rightarrow Cu - 2e^-$$

$$Fe^{+3} \rightarrow Fe^{++} - e^-$$

all of which represent reduction in the chemical sense.

The electrode at which chemical oxidation occurs (or $+$ electricity leaves the electrode and enters the electrolyte) is called the *anode*. Examples of anodic reactions are:

$$Zn \rightarrow Zn^{++} + 2e^-$$

$$Al \rightarrow Al^{+3} + 3e^-$$

$$Fe^{++} \rightarrow Fe^{+3} + e^-$$

These represent oxidation in the chemical sense. Corrosion of metals usually occurs at the anode.*

In galvanic cells, the cathode is the positive pole and the anode is the negative pole. However, when current is impressed on a cell from a generator or an external battery, for example as in electroplating, reduction occurs at the electrode connected to the negative pole of the external current source, and this electrode, consequently, is the cathode.

---

* Alkaline reaction products forming at the cathode can sometimes cause secondary corrosion of amphoteric metals, e.g., Al, Zn, Pb, or Sn. These are metals that corrode rapidly on exposure to either acids or alkalies.

Similarly, the electrode connected to the positive pole of the generator is the anode. It is perhaps best, therefore, not to remember anode and cathode as negative and positive electrodes, or vice versa, but instead to remember the cathode as the electrode at which current enters from the electrolyte, and the anode as the electrode at which current leaves to return to the electrolyte. This situation is true whether current is impressed on or drawn from the cell.

Cations are ions that migrate toward the cathode when electricity flows through the cell (e.g., H^+, Fe^{++}) and are always positively charged whether current is drawn from or supplied to the cell. Similarly, *anions* are always negatively charged (e.g., Cl^-, OH^-, SO_4^{--}).

TYPES OF CELLS

There are three main types of cells that take part in corrosion reactions.

1. *Dissimilar Electrode Cells.* These are illustrated by the dry cell, discussed earlier. A metal containing electrically conducting impurities on the surface as a separate phase, a copper pipe connected to an iron pipe, and a bronze propeller in contact with the steel hull of a ship are examples of this type of corrosion cell. These cells also include cold-worked metal in contact with the same metal annealed, grain-boundary metal in contact with grains, and a single metal crystal of definite orientation in contact with another crystal of differing orientation.*

2. *Concentration Cells.* These are cells having two identical elec-

---

* The various crystal faces of a metal, although initially exhibiting different potentials (tendencies to corrode), all tend to achieve the same potential in time when exposed to an environment capable of reacting with it.[1, 2] This results from the most corrodible planes of atoms reacting first, leaving behind the least corrodible planes; hence, the latter eventually are the only faces exposed regardless of the original orientation. The corrosion rates continue to differ, however, because of differing absolute surface areas of what were once differing crystal faces.

The least corrodible crystal face of any metal is not always the same but varies with environment. The (110) face of copper corrodes most rapidly in 0.3 N HCl–0.1 N H_2O_2, but in 0.3 N HNO_3–0.1 N H_2O_2 the (111) and (110) faces are most reactive.[3] In dilute nitric acid, the (100) face of iron is least reactive.[4, 5]

[1] W. Tragert and W. D. Robertson, *J. Electrochem. Soc.,* **102,** 86 (1955).

[2] C. Walton, *Trans. Electrochem. Soc.,* **85,** 239 (1944).

[3] R. Glauner and R. Glocker, *Z. Kristallogr.,* **80,** 377 (1931).

[4] H. M. Howe, *Metallography of Steel and Cast Iron,* p. 269, McGraw-Hill, New York, 1916.

[5] C. M. Barrett, *Structure of Metals,* 2nd ed., p. 194, McGraw-Hill, New York, 1952.

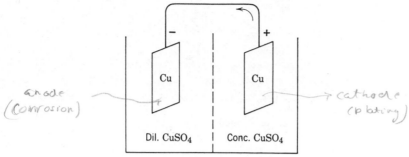

anode
(Corrosion)

cathode
(plating)

Fig. 3.   Salt concentration cell.

trodes each in contact with a solution of differing composition, and there are two kinds. The first is called a *salt concentration cell.* For example, if one copper electrode is exposed to a concentrated copper sulfate solution, and another to a dilute copper sulfate solution (Fig. 3), on short-circuiting such a cell, copper dissolves from the electrode in contact with the dilute solution (anode) and plates out on the other electrode (cathode). Both reactions tend to bring the solutions to the same concentration.

The second kind of concentration cell, which in practice is the more important, is called a *differential aeration cell.* This may include two iron electrodes in dilute sodium chloride solution, the electrolyte around one electrode being thoroughly aerated (cathode), and the other de-aerated (anode), brought about, for example, by bubbling through nitrogen. The difference in oxygen concentration produces a potential differ-

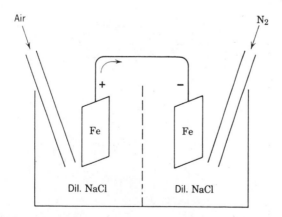

Fig. 4.   Differential aeration cell.

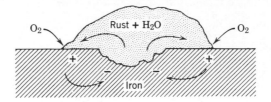

Fig. 5. Differential aeration cell formed by rust on iron.

ence and causes current to flow (Fig. 4). This type of cell accounts for pronounced damage at crevices (*crevice corrosion*) such as are formed at the interface of two coupled pipes or at threaded connections, because the oxygen concentration is lower within the crevice or at the threads than elsewhere. It also accounts for pitting damage under rust (Fig. 5) or at the water line (water-air interface) (Fig. 6). The amount of oxygen reaching the metal that is covered by rust or other insoluble reaction products is less than the amount that contacts other portions where the permeable coating is thinner or absent.

Differential aeration cells also usually initiate pits in the stainless steels, aluminum, nickel, and other so-called passive metals when they are exposed to aqueous environments, such as seawater.

3. *Differential Temperature Cells.* Components of these cells are electrodes of the same metal, each of which is at a different temperature, immersed in an electrolyte of the same initial composition. Less is known about the practical importance and fundamental theory of differential temperature cells than for the cells previously described. They are found in heat exchangers, boilers, immersion heaters, and similar equipment.

In copper sulfate solution the copper electrode at the higher tempera-

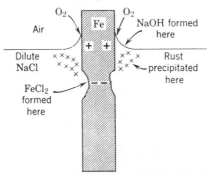

Fig. 6. Differential aeration cell illustrated by water-line corrosion.

ture is cathode and the copper electrode at the lower temperature is anode.[6] On short-circuiting the cell, copper deposits on the hot electrode and dissolves from the cold electrode. Lead acts similarly, but for silver the polarity is reversed.

For iron immersed in dilute aerated sodium chloride solutions, the hot electrode is anodic to colder metal of the same composition; but after a matter of hours, depending on aeration, stirring rate, and whether the two metals are short-circuited, the polarity may reverse.[7, 8]

In practice, cells responsible for corrosion may be a combination of these three types.

TYPES OF CORROSION DAMAGE

Corrosion is often thought of only in terms of rusting and tarnishing. However, corrosion damage also occurs in other ways, resulting, for example, in failure by cracking or in loss of strength or ductility. In general, most types, with some few exceptions, occur by an electrochemical mechanism, but corrosion products are not necessarily observable and metal weight loss need not be appreciable. The five main types of corrosion classified with respect to outward appearance or altered physical properties are:

1. *Uniform Attack*. This includes the commonly recognized rusting of iron or tarnishing of silver. "Fogging" of nickel and high-temperature oxidation of metals are also examples of uniform attack.

Rate of uniform attack is reported in various units, the usual terminology in the United States being *inches penetration per year* (ipy), and *milligrams per square decimeter per day* (mdd). These units refer to metal penetration or to weight loss of metal, excluding any adherent or nonadherent corrosion products on the surface. Steel, for example, corrodes at a relatively uniform rate in seawater equal to about 25 mdd or 0.005 ipy. These represent time-averaged values, the initial rate of attack usually being greater than the final rate. Hence duration of exposure should always be given when corrosion rates are reported, for it is often not safe to extrapolate a reported rate to times of exposure far exceeding the test period.

Conversion of ipy to mdd or vice versa requires knowledge of the metal density. A given weight loss per unit area for a light metal (e.g., aluminum) represents a greater actual loss of metal thickness than the

[6] N. Berry, *Corrosion*, **2**, 261 (1946).

[7] H. H. Uhlig and O. Noss, *Corrosion*, **6**, 140 (1950).

[8] V. Simpson, Jr., S.B. thesis, Department of Chemical Engineering, M.I.T., 1950.

same weight loss for a heavy metal (e.g., lead). Conversion tables are given in the Appendix, p. 407.

For handling chemical media whenever attack is uniform, metals are classified into three groups according to their corrosion rates and intended application. These classifications are:

1. <0.005 ipy (<0.015 cm/year)
   Metals in this category have good corrosion resistance to the extent that they are suitable for critical parts, e.g., valve seats, pump shafts and impellors, springs, etc.
2. 0.005 to 0.05 ipy (0.015–0.15 cm/year)
   Metals in this group are satisfactory if a higher rate of corrosion can be tolerated, e.g., for tanks, piping, valve bodies, bolt heads, etc.
3. >0.05 ipy (>0.15 cm/year)
   Usually not satisfactory.

2. *Pitting*. This is a localized type of attack, the rate of corrosion being greater at some areas than at others. If appreciable attack is confined to a relatively small fixed area of metal, acting as anode, the resultant pits are described as deep. If the area of attack is relatively larger and not so deep, the pits are called shallow. Depth of pitting is sometimes expressed by the term *pitting factor*. This is the ratio of deepest metal penetration to average metal penetration as determined by weight loss of the specimen. A pitting factor of unity represents uniform attack (Fig. 7).

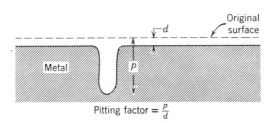

Pitting factor = $\frac{p}{d}$

Fig. 7. Sketch of deepest pit with relation to average metal penetration and the pitting factor.

Iron buried in the soil corrodes with formation of shallow pits, whereas stainless steels immersed in seawater characteristically corrode with formation of deep pits. Many metals, when subjected to high-velocity liquids, undergo a pitting type of corrosion called *impingement attack,* or sometimes *corrosion-erosion* (Fig. 1, p. 323; Fig. 9B, p. 1107, *Corrosion Handbook*). Copper and brass condenser tubes, for example, are subject to this type of attack.

Fretting corrosion, which results from slight relative motion (as in vibration) of two substances in contact, one or both being metals, usually leads to a series of pits at the metal interface. Metal-oxide debris usually fills the pits so that only after the corrosion products are removed do the pits become visible.

Cavitation-erosion resulting from formation and collapse of vapor bubbles at a dynamic metal-liquid interface (as in rotors of pumps or on trailing faces of propellers) causes a sequence of pits, sometimes appearing as a honeycomb of small relatively deep fissures (Fig. 12, p. 110; Figs. 1, 2, and 3, pp. 598–9, *Corrosion Handbook*).

3. *Dezincification and Parting.* Dezincification is a type of attack occurring with zinc alloys (e.g., yellow brass) in which zinc corrodes preferentially, leaving a porous residue of copper and corrosion products (Fig. 4, p. 328). The alloy so corroded often retains its original shape, and may appear undamaged except for surface tarnish, but its tensile strength and especially ductility are seriously reduced. Dezincified brass pipe may retain sufficient strength to resist internal water pressures until an attempt is made to uncouple the pipe, or a water hammer occurs, causing the pipe to split open.

Parting is similar to dezincification in that one or more reactive components of the alloy corrode preferentially, leaving a porous residue that may retain the original shape of the alloy. Parting is usually restricted to such noble metal alloys as gold-copper or gold-silver and is made use of practically in refining of gold. For example, an alloy of Au-Ag containing more than 65% gold resists concentrated nitric acid as well as does gold itself. However, on addition of silver to form an alloy of approximately 25% Au-75% Ag, reaction with concentrated HNO_3 forms silver nitrate and a porous residue or powder of pure gold.

Copper-base alloys that contain aluminum are subject to a form of corrosion resembling dezincification, with aluminum corroding preferentially.

4. *Intergranular Corrosion.* This is a localized type of attack at the grain boundaries of a metal, resulting in loss of strength and ductility. Grain-boundary material of limited area, acting as anode, is in contact with large areas of grains acting as cathode. The attack is often rapid, penetrating deeply into the metal, and sometimes causing catastrophic failures. Improperly heat-treated 18-8 stainless steels or Duralumin-type alloys (4% Cu-Al) are among the alloys subject to intergranular corrosion. An example of nonelectrochemical grain-boundary attack is

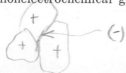

provided by nickel heated in a sulfur-bearing atmosphere which fails because sulfur penetrates along the grain boundaries (Fig. 14, p. 1109, *Corrosion Handbook*).

5. *Cracking*. If a metal cracks when subjected to *repeated* or *alternate* tensile stresses in a corrosive environment, it is said to fail by *corrosion fatigue*. In absence of a corrosive environment, the metal stressed similarly but at values below a critical stress, called the *fatigue limit* or *endurance limit*, will not fail by fatigue even after a very large or infinite number of cycles. A true endurance limit does not commonly exist in a corrosive environment—the metal fails after a prescribed number of stress cycles no matter how low the stress. The types of environment causing corrosion fatigue are many and are not specific.

If a metal, subject to a *constant* tensile stress and exposed simultaneously to a *specific* corrosive environment, cracks immediately or after a given time, the failure is called *stress corrosion cracking*.* The stress may be residual in the metal, as from cold working or heat treatment, or it may be externally applied. The observed cracks are intergranular or transgranular, depending on the metal and the damaging environment. Failures of this kind differ basically from intergranular corrosion defined earlier, which proceeds without regard to whether the metal is stressed.

Almost all structural metals (e.g., carbon- and low alloy-steels, brass, stainless steels, Duralumin, magnesium alloys, titanium alloys, nickel alloys, and many others) are subject to stress corrosion cracking in some environments. Fortunately, either the damaging environments are often restricted to a few chemical species, or the necessary stresses are sufficiently high to limit observed failures of this kind in practice. As knowledge accumulates regarding the specific media that cause cracking and regarding the limiting stresses necessary to avoid failure within a given time period, it will be possible to design metal structures without incidence of stress corrosion cracking. Unfortunately not all highly stressed metal structures are designed today with adequate assurance that stress-corrosion cracking will not occur.

---

* Both stress corrosion cracking, and cracking caused by absorption of hydrogen generated by a corrosion reaction follow this definition. Distinguishing differences between the two types of cracking are discussed on p. 142.

chapter 3

Corrosion tendency and electrode potentials

FREE-ENERGY CHANGE

The tendency for any chemical reaction to go, including the reaction of a metal with its environment, is measured by the Gibbs free energy change ΔG. The more negative the value of ΔG, the greater is the tendency for the reaction to go. For example, consider the following reaction at 25°C:

$$Mg + H_2O\ (l) + \tfrac{1}{2}O_2\ (g) \rightarrow Mg(OH)_2\ (s) \qquad \Delta G° = -142,600\ \text{cal}$$

The large negative value for $\Delta G°$ (reactants and products in standard states) indicates a pronounced tendency for magnesium to react with water and oxygen. On the other hand, for the reaction

$$Cu + H_2O\ (l) + \tfrac{1}{2}O_2\ (g) \rightarrow Cu(OH)_2\ (s) \qquad \Delta G° = -28,600\ \text{cal}$$

the reaction tendency is less. Or we can say that the corrosion tendency of copper in aerated water is not as pronounced as for magnesium.

For the reaction

$$Au + \tfrac{3}{2}H_2O\ (l) + \tfrac{3}{4}O_2\ (g) \rightarrow Au(OH)_3\ (s) \qquad \Delta G° = +15,700\ \text{cal}$$

the free energy is positive, indicating that the reaction has no tendency to go at all, and gold, correspondingly, does not corrode in aqueous media to form $Au(OH)_3$.

It should be emphasized that the tendency to corrode is not a measure of reaction rate. A large negative ΔG may or may not be accompanied by a high corrosion rate, but when ΔG is positive it can be stated with certainty that the reaction will not go at all under the particular conditions described. If ΔG is negative, the reaction rate may be rapid or slow, depending on various factors described in detail later.

In view of the electrochemical mechanism of corrosion, the tendency for a metal to corrode can also be expressed in terms of the electromotive force (emf) of the corrosion cells that are an integral part of the corrosion process. Since electrical energy is expressed as the product of volts by coulombs (joules, J), the relation between ΔG in joules and emf in volts, E is defined by $\Delta G = -EnF$ where n is the number of electrons (or chemical equivalents) taking part in the reaction, and F is the Faraday (96,500 C/eq). The term ΔG can be converted from calories to joules by making use of the factor 1 cal = 4.184 absolute joules.

Accordingly, the greater the value of E for any cell, the greater is the tendency for the overall reaction of the cell to go. This applies to any of the types of cells described earlier.

THE POTENTIOMETER

The emf of a cell, as set up in the laboratory or in the field, is measured by opposing it with a known emf until no current flows through a sensitive galvanometer in series with the cell. The known emf is obtained by means of a convenient circuit known as a potentiometer, a simplified diagram of which is given in Fig. 1.

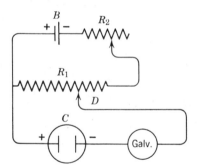

Fig. 1. Potentiometer circuit.

A calibrated uniform resistance R_1 is connected to a battery B of 1.5 to 4 V through a variable resistance, R_2. Each position of D corresponds to a labeled voltage varying from zero at the extreme left to the maximum value at the extreme right. A standard Weston cell of known emf is first substituted for the cell C, and R_2 is adjusted until no current flows through the galvanometer when the contact D

is placed at the position corresponding to the voltage of the standard cell. Because the standard cell would soon be exhausted or damaged if used continuously, it is employed only to calibrate the battery *B*. The unknown cell is then placed in the circuit as shown and *D* is again adjusted until the galvanometer shows zero current. The corresponding reading in volts of the position *D* indicates the exact emf of the cell.

At exact balance, no current flows through the cell, but since balance is never realized perfectly, some current is apt to be drawn from the cell during measurement (hence, the key is depressed only momentarily). This is not as important to large-size cells, with so-called slightly polarizable electrodes as with small-size cells or those with high internal resistance. For the latter it is essential to use a galvanometer capable of operating with only minute flow of current. Electronic galvanometers used for glass electrode pH measurements, for example, have an input resistance of about $10^{12}\Omega$ (ohms) or higher, meaning that if the potentiometer and cell are unbalanced to the extent of 1 V, a current of only 10^{-12} A flows. This current is not sufficient to polarize (temporarily alter the emf of) the cell. Sensitive galvanometers of high input resistance are conveniently used, therefore, for measurement of emf, since the correct value is obtained even if the potentiometer has been set previously at an off-balance position. The disadvantages of such sensitive galvanometers are (1) the need to be very careful of all insulation used for lead wires, particularly on humid days and (2) the need for shielding all leads to guard against external electrical disturbances caused, for example, by adjacent high-frequency generators, commutator-type motors, switches, and similar equipment.

THE NERNST EQUATION, CALCULATION OF THE HALF-CELL POTENTIAL

Based on thermodynamic reasoning,[1] an equation can be derived to express the emf of a cell in terms of the concentrations of reactants and reaction products. The general reaction for a galvanic cell can be assumed to be:

$$lL + mM + \cdots \rightarrow qQ + rR + \cdots$$

meaning that *l* moles of substance *L* plus *m* moles of substance *M*, etc., react to form *q* moles of substance *Q*, *r* moles of substance *R*, etc. The corresponding change of Gibbs free energy, ΔG, for this reaction

---

[1] See, for example, G. Lewis, M. Randall, K. Pitzer, and L. Brewer, *Thermodynamics,* McGraw-Hill, New York, 1961.

is given by the difference in molal free energy of products and reactants, where G_Q represents the molal free energy of substance Q, etc.

$$\Delta G = (qG_Q + rG_R + \cdots) - (lG_L + mG_M + \cdots) \qquad (1)$$

A similar expression is obtained for each substance in the standard state or arbitrary reference state, the symbol G° indicating standard molal free energy.

$$\Delta G^\circ = (qG_Q^\circ + rG_R^\circ + \cdots) - (lG_L^\circ + mG_M^\circ + \cdots) \qquad (2)$$

If a_L is the corrected concentration or pressure of substance L, called its activity, the difference of free energy for L in any given state and in the standard state is related to a_L by the expression

$$l(G_L - G_L^\circ) = lRT \ln a_L = RT \ln a_L^l \qquad (3)$$

where R is the gas constant (8.314 J/deg-mole) and T is the absolute temperature (deg C + 273.16). Subtracting (2) from (1) and equating to corresponding activities, we have the expression

$$\Delta G - \Delta G^\circ = RT \ln \frac{a_Q^q \cdot a_R^r \cdots}{a_L^l \cdot a_M^m \cdots} \qquad (4)$$

When the reaction is at equilibrium, there is no tendency for it to go, $\Delta G = 0$, and

$$\frac{a_Q^q \cdot a_R^r \cdots}{a_L^l \cdot a_M^m \cdots} = K$$

where K is the equilibrium constant for the reaction. Hence

$$at\ equilibrium\quad \Delta G^\circ = -RT \ln K \qquad (5)$$

On the other hand, when all the activities of reactants and products are equal to unity, the logarithm term becomes zero ($\ln 1 = 0$) and $\Delta G = \Delta G^\circ$.

Since $\Delta G = -EnF$, it follows that $\Delta G^\circ = -E^\circ nF$, where E° is the emf when all reactants and products are in their standard states (activities equal to unity). Corresponding to (4), we have

$$E = E^\circ - \frac{RT}{nF} \ln \frac{a_Q^q \cdot a_R^r \cdots}{a_L^l \cdot a_M^m \cdots} \qquad (6)$$

This is the *Nernst equation*, which expresses the exact emf of a cell in terms of activities of products and reactants of the cell. The activity a_L of a dissolved substance L is equal to its concentration in *moles per thousand grams of water* (*molality*) multiplied by a correction factor, γ, called the activity coefficient. The activity coefficient is a function of temperature and concentration and, except for very dilute solutions, must be determined experimentally. If L is a gas, its activity is equal to its fugacity, approximated at ordinary pressures by the pressure in atmospheres. The activity of a pure solid is arbitrarily set equal to unity. Similarly, for a substance like water whose concentration is essentially constant throughout the reaction, the activity is set equal to unity.

Since the emf of a cell is always the algebraic sum of two electrode potentials or of two half-cell potentials, it is convenient to calculate each electrode potential separately. For example, for the electrode reaction:

$$Zn \rightarrow Zn^{++} + 2e^- \tag{7}$$

$$E_{Zn} = E_{Zn}^{\circ} - \frac{RT}{2F} \ln \frac{(Zn^{++})}{(Zn)} \tag{8}$$

where (Zn^{++}) represents the activity of zinc ions (molality $\times$ activity coefficient), (Zn) is the activity of metallic zinc, the latter being a pure solid and equal therefore to unity, and E_{Zn}° is the so-called standard oxidation potential of zinc (equilibrium potential of zinc in contact with Zn^{++} at unit activity).

Since it is more convenient to work with logarithms to the base 10, the value of the coefficient RT/F is multiplied by the conversion factor 2.303. Then from the value of $R = 8.314$ J/deg-mole, $T = 298.2°$K, and $F = 96,500$ C/eq, the coefficient 2.303 RT/F at 25°C becomes 0.0592 V. This coefficient appears frequently in expressions representing potentials or emf.

Measured or calculated values of standard oxidation potentials E° at 25°C are found in several reference books (e.g., *Oxidation Potentials*, by W. M. Latimer, Prentice-Hall, Englewood Cliffs, N. J., 1952) and in various handbooks (e.g., *Corrosion Handbook*, edited by H. H. Uhlig, p. 1134, Wiley, New York, 1948, or any of the chemical handbooks). Some values are listed on p. 29 and in the Appendix, p. 409. Values of activity coefficients for various electrolytes are given in the Appendix, p. 388, and definitions and rules applying to the use of activity coefficients are outlined on p. 386.

HYDROGEN ELECTRODE, STANDARD HYDROGEN SCALE

Since absolute potentials of electrodes are not known with certainty and their meaning is still being discussed, it is convenient to assume arbitrarily that the standard potential for the reaction

$$H_2 \rightarrow 2H^+ + 2e^- \tag{9}$$

is equal to zero at all temperatures. Hence

$$E_{H_2} = 0 - \frac{RT}{2F} \ln \frac{(H^+)^2}{p_{H_2}} \tag{10}$$

don't forget pressure term

where p_{H_2} is the fugacity of hydrogen in atmospheres and (H^+) is the activity of hydrogen ions. All values of electrode potentials, therefore, are with reference to the hydrogen electrode. By measuring the emf of a cell made up, for example, of a zinc and hydrogen electrode in a zinc salt solution of known activity of Zn^{++} and H^+, the standard oxidation potential, E°, for zinc can be calculated; the accepted value is 0.763 V.

The hydrogen electrode potential is measured by immersing a piece of platinized platinum in a solution saturated with hydrogen gas at 1 atm (Fig. 2), or more conveniently by the glass electrode, whose poten-

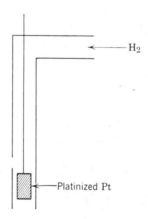

Fig. 2. Hydrogen electrode.

tial is also reversible to (in equilibrium with) hydrogen ions. Note that the potential of the electrode equals zero if the hydrogen ion activity and the pressure of hydrogen gas in atmospheres are both unity. This is the standard hydrogen potential. Hence the half-cell potential for any electrode is equal to the emf of a cell with the standard hydrogen electrode as the other electrode. The half-cell potential for any electrode expressed on this basis is said to be on the *normal hydrogen* or *standard hydrogen scale*, sometimes expressed as ϕ_H or ϕ S.H.E.

Note: Standard hyd. electrode otherwise must consider pH and press.

CONVENTION OF SIGNS AND CALCULATION OF EMF

In accord with the foregoing discussion, the standard oxidation potential of zinc, which is not separately measurable, refers to the emf of a cell whose other electrode is the standard hydrogen electrode:

$$Zn; \ Zn^{++}, H^+, H_2; \ Pt \tag{11}$$

The corresponding reaction, somewhat simplified, is written arbitrarily subtracting right-hand oxidation reaction from the left-hand oxidation reaction, or

$$Zn + 2H^+ \rightarrow Zn^{++} + H_2 \qquad \text{standard emf} = \underline{0.763 \ V} \tag{12}$$

The free energy $\Delta G°$ equals $-0.763 \times 2F$ joules, the negative value indicating that the reaction is thermodynamically possible as written, for products and reactants in their standard states. On the other hand for the cell

$$Pt; H_2, H^+, Zn^{++}; \ Zn \qquad \text{Standard emf} = -0.763 \ V \tag{13}$$

the corresponding reaction is $Zn^{++} + H_2 \rightarrow Zn + 2H^+$, the standard emf is negative and $\Delta G°$ is positive.

Obviously, the standard reduction potential for zinc has a sign opposite to that for the oxidation potential or

$$Zn^{++} \rightarrow Zn - 2e^- \qquad \phi° = -0.763 \ V \tag{14}$$

where $\phi°$ is given as the standard reduction potential.* It follows that $\phi° = -E°$.

It was agreed at the 1953 meeting of the International Union of Pure and Applied Chemistry that the reduction potential for any half-cell electrode reaction would be called *the potential*. This designation of sign has the advantage of conforming with the physicist's concept of potential defined as the work necessary to bring unit positive charge to the point at which the potential is given. It also has the advantage of corresponding in sign to the polarity of a voltmeter or potentiometer to which an electrode may be connected. Thus zinc has a negative reduction potential and is also the negative pole of a galvanic cell of which the standard hydrogen electrode is the other electrode. It is said to be negative to the hydrogen electrode.

---

* For the so-called European sign convention, $\phi°$ values apply to half-cell *oxidation* reactions, and the sign of the logarithm term in the Nernst equation is made positive instead of negative. The sign convention used in this book follows *Thermodynamics*, G. Lewis, M. Randall, K. Pitzer, and L. Brewer, McGraw-Hill, New York, 1961.

In setting up the emf of a cell, the foregoing conventions of sign dictate the direction regarding spontaneous flow of electricity. If on short-circuiting a cell, positive current through the electrolyte within the cell flows from *left* to *right*, then the emf is positive, and correspondingly the left electrode is anode and the right electrode is cathode. If current flows within the cell from right to left, the emf is negative.

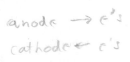

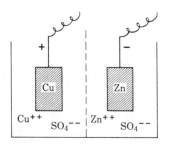

Fig. 3. Copper-zinc cell.

If, therefore, we wish to calculate the emf of the cell shown in Fig. 3: Cu; Cu⁺⁺, Zn⁺⁺; Zn, we can first write the reaction of the left electrode (Cu) as if it were anode (whether it is or not is clarified later).

$$Cu \rightarrow Cu^{++} + 2e^- \qquad E° = -0.337 \text{ V} \qquad (15)$$

$$E_{Cu} = -0.337 - \frac{0.0592}{2} \log (Cu^{++}) \qquad (16)$$

The reaction for the right electrode is

$$Zn \rightarrow Zn^{++} + 2e^- \qquad E° = 0.763 \text{ V} \qquad (17)$$

$$E_{Zn} = 0.763 - \frac{0.0592}{2} \log (Zn^{++}) \qquad (18)$$

Reaction 17 is then subtracted from (15) in such a way as to cancel out the total electrons (multiplying, if necessary, by a numerical factor). This results in the tentative reaction for the cell:

$$Cu + Zn^{++} \rightarrow Cu^{++} + Zn \qquad (19)$$

The emf is obtained by adding algebraically the corresponding half-cell oxidation potentials (16) and (18).* Note that although reversing

---

* Reaction 19 is a simplification of the actual reaction. A more exact approach would include calculation of the liquid junction potential between CuSO₄ and ZnSO₄ and elimination of single-ion activities for which values are not measurable.

a reaction changes the sign of potential, <u>multiplying by any factor has no effect on either emf or $E°$ values</u>, since the tendency for a reaction to go is independent of the amount of substance reacting (in contrast to the total free-energy change, which does depend on amount of substance reacting).

$$\text{emf} = E_{Cu} - E_{Zn} = -1.100 - \frac{0.0592}{2} \log \frac{(Cu^{++})}{(Zn^{++})} \qquad (20)$$

If the activities of Cu^{++} and Zn^{++} are chosen to be the same, the emf = -1.100 V. <u>Since the emf is negative, current flows spontaneously</u> from right to left within the cell. This fixes the true polarity of the cell, with the left electrode (Cu) as positive (cathode) and the right electrode (Zn) as negative (anode). From the relation $\Delta G = -EnF$, the free energy of (19) is positive and is therefore not spontaneous as written but goes instead in the opposite direction. In other words, when current is drawn from the cell, Cu^{++} plates out on the copper electrode, and the zinc electrode corrodes. *it real reaction is $Zn + Cu^{++} \rightarrow Cu + Zn^{++}$*

Similarly, the emf, polarity, and spontaneous reaction can be determined for any cell for which half-cell reactions and standard potentials are known.

MEASUREMENT OF pH

Hydrogen ion activity is commonly expressed, for convenience, in terms of pH. This is defined as

$$pH = - \log (H^+)$$

Hence for the half-cell reaction, $H_2 \rightarrow 2H^+ + 2e^-$, with the pressure of hydrogen equal to one atmosphere

$$E_{H_2} = 0.0592 \; pH$$

Since pure water contains equal concentrations of H^+ and OH^- in equilibrium with undissociated water, $H_2O \rightarrow H^+ + OH^-$, it is possible to calculate the activity of either the hydrogen ion or the hydroxyl ion from the ionization constant, the value for which at 25°C is 1.01×10^{-14}. Obviously, therefore, the pH of pure water at 25°C equals $- \log \sqrt{1.01 \times 10^{-14}} = 7.0$. If (H^+) exceeds (OH^-) as in acids, the pH is less than 7, or if the pH is greater than 7, the solution is alkaline. The pH of strong acids can be negative, and for strong alkalies it can be greater than 14.

At temperatures higher than 25°C, the ionization constant of H_2O is larger; therefore above 25°C the pH of pure water is less than 7 (Table 1).

TABLE 1

Ionization Constant, K_W, and pH of Pure Water at Various Temperatures

Temp.	K_W	pH
0°C	0.115×10^{-14}	7.47
10	0.293	7.27
25	1.008	7.00
40	2.916	6.77
60	9.614	6.51

THE OXYGEN ELECTRODE AND DIFFERENTIAL AERATION CELL

The oxygen electrode can be represented by platinized platinum immersed in an electrolyte saturated with oxygen. This electrode is particularly important to corrosion studies because of the part it plays in differential aeration cells that enter the mechanism of crevice corrosion and pitting.

Ideally, the equilibrium for such an electrode is expressed as

$$OH^- \rightarrow \tfrac{1}{2}H_2O + \tfrac{1}{4}O_2 + e^- \qquad E° = -0.401 \text{ V} \tag{21}$$

and

$$E_{O_2} = -0.401 - 0.0592 \log \frac{p_{O_2}^{\frac{1}{4}}}{(OH^-)} \tag{22}$$

This reaction, however, unlike the hydrogen electrode reaction, is not strictly reversible under practical conditions of measurement, and hence the measured potential may vary with time and is not reproducible. The observed potential tends to be less noble than the calculated reversible value. Nevertheless, it is useful to know the direction of expected potential change as, for example, when oxygen pressure is altered. For illustration, consider two oxygen electrodes in an aqueous solution, one in contact with O_2 at 1 atm and the other with O_2 at 0.2 atm. The oxidation potentials of the left- and right-hand electrodes, respectively, are as follows:

$$OH^- \rightarrow \tfrac{1}{2}H_2O + \tfrac{1}{4}O_2(1 \text{ atm}) + e^- \tag{23}$$

$$E_1 = -0.401 - 0.0592 \log \frac{1^{\frac{1}{4}}}{(OH^-)} \qquad \textit{cathode} \tag{24}$$

more neg than

$$OH^- \rightarrow \tfrac{1}{2}H_2O + \tfrac{1}{4}O_2(0.2 \text{ atm}) + e^- \tag{25}$$

$$E_2 = -0.401 - 0.0592 \log \frac{0.2^{\frac{1}{4}}}{(OH^-)} \qquad \textit{anode} \tag{26}$$

(23) − (25)

$$\tfrac{1}{4} O_2(0.2 \text{ atm}) \rightarrow \tfrac{1}{4} O_2(1 \text{ atm}) \tag{27}$$

(24) − (26)

$$E_1 - E_2 = -0.0592 \log \frac{1^{\frac{1}{4}}}{0.2^{\frac{1}{4}}} = \frac{0.0592}{4} \log 0.2 = -0.0103 \text{ V} \tag{28}$$

The negative value of emf indicates that ΔG for (27) is positive and hence the reaction is not spontaneous as written. Instead, positive electricity flows spontaneously within the cell from right to left. Therefore the left-hand electrode—(23)—is positive (cathode) and the right-hand electrode—(25)—is negative (anode). This expresses the fact formulated earlier that in any differential aeration cell the electrode in contact with lower pressure oxygen tends to be anode and the electrode in contact with higher pressure oxygen tends to be cathode.

When a similar cell is made up of iron electrodes instead of platinum, an adherent, electrically conducting oxide of iron forms at cathodic areas; in contact with aerated solutions, this oxide acts as an oxygen electrode. But at anodic areas, Fe^{++} forms and the electrode acts as an iron electrode ($E° = 0.440$ V). The operating emf of such a cell is much larger than for the cell made up of platinum electrodes, the value being given by

$$E = -0.440 - 0.401 - \frac{0.0592}{2} \log \frac{p_{O_2}^{\frac{1}{2}}}{(Fe^{++})(OH^-)^2} \tag{29}$$

If the ferrous ion activity at the anode is assumed equal to 0.1, the pH of water at the cathode equal to 7.0, and the partial pressure of oxygen at the cathode equal to that of air (0.2 atm), the operating emf of the corresponding cell is 1.26 V. This is an appreciable value for resultant flow of current and accompanying corrosion at the anode. In practice, the emf is less than this because of the irreversible nature of the oxygen electrode, especially as approximated by an iron oxide film on iron, but the emf would in general be larger than the small value calculated for two platinum electrodes.

POURBAIX DIAGRAMS

M. Pourbaix devised a compact summary of thermodynamic data in the form of potential-pH diagrams, which relate to the electrochemical and corrosion behavior of any metal in water. These diagrams are now available for most of the common metals.[2] They have the advantage

---

[2] M. Pourbaix, *Atlas of Electrochemical Equilibria in Aqueous Solutions,* Pergamon Press, New York, 1966.

of showing at a glance specific conditions of potential and pH under which the metal either does not react (immunity) or can react to form specific oxides or complex ions. Since the data are thermodynamic, they convey no information on rates of reaction, that is, whether any possible reaction proceeds rapidly or slowly when the energy changes are favorable. The data indicate the conditions for which diffusion barrier films may form on an electrode surface, but they provide no measure of how effective such barrier films may be in presence of specific anions such as SO_4^{--} or Cl^-. Similarly, they do not indicate the detailed conditions under which nonstoichiometric metal compound films are possible; some of these films are important in determining observed corrosion rates (see discussion under "Passivity," Chapter 5). However, the diagrams do plainly outline the nature of the stoichiometric compounds into which any less stable compounds may transform whenever equilibrium is achieved. Employed with these limitations in mind, they usefully describe the equilibrium status of a metal either as immersed in acids or alkalies, or when a given potential is impressed on it. The Pourbaix diagram for iron and a discussion of its use are given in the Appendix, p. 393.

EMF AND GALVANIC SERIES

The Emf Series is an orderly arrangement of the standard oxidation or reduction potentials for all metals. The more positive oxidation values, or the more negative reduction values correspond to the more reactive metals (Table 2). Note that position in the Emf Series is determined by the *equilibrium* potential of a metal in contact with its ions at a concentration equal to unit activity. Of two metals composing a cell, the anode is the more active metal in the Emf Series, provided that the ion activities in equilibrium are both unity. Since unit activity corresponds in some cases to impossible concentrations of metal ions because of restricted solubility of metal salts, it is obvious that the Emf Series has only limited utility for predicting which metal is anodic to another. Also, in practice, the actual activities of ions in equilibrium with a given metal vary greatly with the environment.

For example, tin is noble to iron according to the Emf Series. This is also the normal galvanic relationship of tin to iron in tin plate exposed to aerated aqueous media. But on the inside of tin-plated iron containers ("tin cans") in contact with food, certain constituents of foods combine chemically with Sn^{++} ions to form soluble tin complexes. Reactions of this kind greatly lower the activity of Sn^{++} ions with which the tin is in equilibrium, producing a tin potential that is much more

TABLE 2
Electromotive Force Series

Electrode Reaction	Standard Oxidation Potential, $E°$ (V), 25°C*
$Li = Li^+ + e^-$	3.05
$K = K^+ + e^-$	2.93
$Ca = Ca^{++} + 2e^-$	2.87
$Na = Na^+ + e^-$	2.71
$Mg = Mg^{++} + 2e^-$	2.37
$Be = Be^{++} + 2e^-$	1.85
$U = U^{+3} + 3e^-$	1.80
$Hf = Hf^{+4} + 4e^-$	1.70
$Al = Al^{+3} + 3e^-$	1.66
$Ti = Ti^{++} + 2e^-$	1.63
$Zr = Zr^{+4} + 4e^-$	1.53
$Mn = Mn^{++} + 2e^-$	1.18
$Nb = Nb^{+3} + 3e^-$	ca. 1.1
$Zn = Zn^{++} + 2e^-$	0.763
$Cr = Cr^{+3} + 3e^-$	0.74
$Ga = Ga^{+3} + 3e^-$	0.53
$Fe = Fe^{++} + 2e^-$	0.440
$Cd = Cd^{++} + 2e^-$	0.403
$In = In^{+3} + 3e^-$	0.342
$Tl = Tl^+ + e^-$	0.336
$Co = Co^{++} + 2e^-$	0.277
$Ni = Ni^{++} + 2e^-$	0.250
$Mo = Mo^{+3} + 3e^-$	ca. 0.2
$Sn = Sn^{++} + 2e^-$	0.136
$Pb = Pb^{++} + 2e^-$	0.126
$H_2 = 2H^+ + 2e^-$	0.000
$Cu = Cu^{++} + 2e^-$	−0.337
$Cu = Cu^+ + e^-$	−0.521
$2Hg = Hg_2^{++} + 2e^-$	−0.789
$Ag = Ag^+ + e^-$	−0.800
$Pd = Pd^{++} + 2e^-$	−0.987
$Hg = Hg^{++} + 2e^-$	−0.854
$Pt = Pt^{++} + 2e^-$	ca. −1.2
$Au = Au^{+3} + 3e^-$	−1.50

* Standard reduction potentials, $\phi°$, have the opposite sign.

active and which may actually be less noble than iron. The polarity of the iron-tin couple under these conditions reverses sign. The ratio of Sn^{++} to Fe^{++} within the can must be very small for the reversal of polarity to occur, as can be calculated from $E°$ values for iron and tin in accord with the following reaction:

$$Fe^{++} + Sn \rightarrow Sn^{++} + Fe \tag{30}$$

$$E = 0.136 - 0.440 - \frac{0.0592}{2} \log \frac{(Sn^{++})}{(Fe^{++})} \tag{31}$$

The cell reverses polarity when $E = 0$. Hence,

$$\log \frac{(Sn^{++})}{(Fe^{++})} = -0.304 \times \frac{2}{0.0592} = -10.30$$

or the ratio $(Sn^{++})/(Fe^{++})$ must be less than 5×10^{-11} for tin to become active to iron. This small ratio can occur only through formation of tin complexes. Complexing agents in general, such as EDTA, cyanides, and strong alkalies, tend to increase the corrosion rates of many metals by reducing the metal-ion activity, thereby shifting metal potentials markedly in the active direction.

Another factor that alters the galvanic position of some metals is the tendency, especially in oxidizing environments, to form specific surface films. These films shift the measured potential in the noble direction. In this state the metal is said to be passive (see section under Chapter 5, *Passivity*). Hence, chromium, although normally near zinc in the Emf Series, behaves galvanically more like silver in many air-saturated aqueous solutions because of a passive film that forms over its surface. The metal acts like an oxygen electrode instead of like chromium, and hence, when coupled with iron, chromium becomes the cathode and current flow accelerates corrosion of iron. In the active state (e.g., in hydrochloric acid), the reverse polarity occurs or chromium becomes anodic to iron. Many metals, especially the transition metals of the periodic table, commonly exhibit passivity in aerated aqueous solutions.

Because of the limitations of the Emf Series for predicting galvanic relations, and also because alloys are not included (conditions affecting equilibrium of solid alloys with their environment are not well understood), the so-called *Galvanic Series* has been suggested. This series is an arrangement of metals and alloys in accord with their actual measured potentials in a given environment. The potentials that determine the position of a metal in this series may include steady-state values in addition to truly reversible values, and hence alloys and passive metals

TABLE 3
Galvanic Series in Seawater[3]

Active (Read down)	
Magnesium	18-8 stainless steel, type 304 (active)
Magnesium alloys	18-8, 3% Mo stainless steel, type 316 (active)
Zinc	Lead
	Tin
Aluminum 52SH	Muntz metal
Aluminum 4S	Manganese bronze
Aluminum 3S	Naval brass
Aluminum 2S	
Aluminum 53S-T	Nickel (active)
Alclad	76% Ni-16% Cr-7% Fe (Inconel) (active)
	Yellow brass
Cadmium	Aluminum bronze
	Red brass
Aluminum 17S-T	Copper
Aluminum 17S-T	Silicon bronze
Aluminum 24S-T	5% Zn-20% Ni, Bal. Cu (Ambrac)
	70% Cu-30% Ni
Mild steel	88% Cu-2% Zn-10% Sn (composition G-bronze)
Wrought iron	88% Cu-3% Zn-6.5% Sn-1.5% Pb (comp. M-bronze)
Cast iron	Nickel (passive)
Ni-Resist	76% Ni-16% Cr-7% Fe (Inconel) (passive)
13% Chromium stainless steel, type 410 (active)	70% Ni-30% Cu (Monel)
	18-8 stainless steel, type 304 (passive)
50-50 lead-tin solder	18-8, 3% Mo stainless steel, type 316 (passive)
	Noble (Read up)

[3] F. L. LaQue, in *Corrosion Handbook*, p. 416, Wiley, New York, 1948.

are included. The Galvanic Series for metals in contact with seawater is given in Table 3. Note that some metals occupy two positions in the Galvanic Series, depending on whether they are active or passive, whereas in the Emf Series only the active positions are possible, since only in this state is true equilibrium attained. The passive state, to the contrary, represents a nonequilibrium state in which the metal, because of surface films, is no longer in normal equilibrium with its ions. Although only one Emf Series exists, obviously there can be several Galvanic Series because of differing complexing tendencies of various environments or differences in tendency to form surface films. In general, therefore, a specific Galvanic Series exists for each environment, and the relative positions of metals in such series may vary from environment to environment.

The damage incurred by coupling of two metals depends not only on how far apart they are in the Galvanic Series (open-circuit potential difference) but also on their relative areas and the extent to which they are polarized (see p. 37). The potential difference of the polarized electrodes and the conductivity of the corrosive environment determine how much current flows between them.

LIQUID JUNCTION POTENTIALS

In addition to potential differences between two metals in an electrolyte, potential differences also arise whenever two solutions of different composition or concentration come into contact. The potential difference is called the liquid junction potential, and its sign and magnitude are determined by the relative mobility of ions and their concentration differences across the liquid junction. For example, in a junction formed between dilute and concentrated hydrochloric acid, H^+ ions move with greater velocity than Cl^- ions (mobility at infinite dilution = 36×10^{-4} and 7.9×10^{-4} cm/sec, respectively). Hence the dilute aqueous solution acquires a positive charge with respect to the concentrated solution. For potassium chloride, mobility of K^+ and Cl^- are similar, hence liquid junction potentials between dilute and concentrated KCl are small in comparison with HCl junctions. In fact, if the HCl solutions discussed previously are saturated with KCl, so that most of the current across the boundary is carried by K^+ and Cl^- ions, the liquid junction potential is very much decreased. Use of a saturated KCl solution whenever

TABLE 4
Characteristic Liquid Junction Potentials of Salt Solutions (MacInnes)

Electrolyte	Concentration	
	0.1 N	0.01 N
(−)HCl	35.65 mV	33.87 mV
KCl	8.87	8.20
NH$_4$Cl	6.92	6.89
NaCl	2.57	2.63
(+)LiCl	0.00	0.00

liquid junctions are formed is one practical approach to minimizing liquid junction potentials.

Calculations of liquid junction potentials can be made on the basis of certain assumptions, but the derivation of such calculations is relatively complex even for simple-type junctions. Calculations of this kind are discussed, for example, by MacInnes.[4] Measurements show that potentials of junctions formed between salt solutions of the same concentration having a common ion (e.g., Cl⁻) are additive.[4] Characteristic values are given in Table 4, assigning zero arbitrarily to LiCl. For example, the potential of the junction HCl$(0.1N)$:KCl$(0.1N)$ is equal to $35.65 - 8.87 = 26.78$ mV with KCl solution positive and HCl solution negative. On the other hand, for the junction LiCl$(0.1N)$:NH$_4$Cl $(0.1N)$ the value is $0.00 - 6.92 = -6.92$ mV and the NH$_4$Cl solution is negative and the LiCl solution positive. It is obvious that values at 0.01 and $0.1N$ are nearly alike and that, except perhaps for strongly acidic or basic solutions, the values are small and are not of great concern to most corrosion measurements.

REFERENCE HALF-CELLS

In the measurement of emf, the observed value represents a tendency for simultaneous reactions to occur at both electrodes of the cell. Actually, our interest is usually centered on the reaction that occurs at one electrode only. The criterion of complete cathodic protection based on potential measurements is an example. Measurements of this kind are made by using an electrode having a relatively fixed value of potential, regardless of the environment in which it is used, called *reference half-cell*, or *reference electrode*. Therefore, any change occurring in emf is the result of change in potential of the electrode under observation and not of the reference electrode. The latter makes use of stable reversible electrode systems; examples of these are discussed next.

Calomel Half-Cell

The calomel half-cell has long been a standard reference electrode for use in the laboratory. It consists of mercury in equilibrium with Hg$_2^{++}$, the activity of which is determined by the solubility of Hg$_2$Cl$_2$ (mercurous chloride or calomel). The half-cell reaction is

$$2Hg + 2Cl^- \rightarrow Hg_2Cl_2 + 2e^- \qquad E° = -0.268 \text{ V}$$

[4] D. MacInnes, *Principles of Electrochemistry,* Reinhold, New York, 1939.

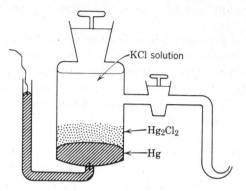

Fig. 4. One type of calomel reference electrode.

One design of electrode is pictured in Fig. 4. Pure mercury covers a platinum wire sealed through the bottom of a glass tube. The mercury is covered with powdered mercurous chloride, which is only slightly soluble in potassium chloride solution, the latter filling the cell. The activity of Hg_2^{++} depends on the concentration of KCl since the solubility product $(Hg_2^{++})(Cl^-)^2$ is a constant. Potentials on the standard hydrogen scale for various KCl concentrations are as follows:

$Hg_2Cl_2 + 2e^- \rightarrow$

reduction

KCl Concentration	ϕ (V)	Temperature Coefficient (V/°C)
0.1 N	0.3338	-0.88×10^{-4}
1.0 N	0.2801	-2.75×10^{-4}
Saturated KCl	0.2416	-6.6×10^{-4}

The saturated KCl electrode is convenient to prepare, but the potential is stated to be somewhat more sluggish in responding to temperature changes than are the unsaturated KCl electrodes. The 0.1N electrode has the lowest temperature coefficient.

The values cited neglect the liquid junction potential at the KCl boundary, which in the case of strong acids, for example, increase the absolute value an average of several millivolts. The potentials of reference electrodes are usually listed in handbooks with positive sign, corresponding to the reduction potential, ϕ.

Silver-Silver Chloride Half-Cell

This electrode is prepared, according to one method, by electroplating with silver a platinum wire sealed into a glass tube, using a silver cyanide electrolyte of high purity.[5] The silver coating is then converted partly into silver chloride by making it anode in dilute hydrochloric acid (Fig. 5). An electrode suitable for corrosion measurements can also be pre-

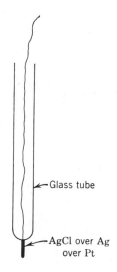

Fig. 5.   Silver-silver chloride reference electrode.

pared from pure annealed silver wire, chloridized as just described. The electrode potential should be checked frequently against either freshly prepared electrodes or against the calomel electrode because of a gradual change on aging. When the electrode is immersed in a chloride solution, the following equilibrium is established:

$$Ag + Cl^- \rightarrow AgCl + e^- \qquad E° = -0.222 \text{ V}$$

Like the calomel electrode, the potential is more active the higher the KCl concentration. In $0.1N$ KCl, the value is 0.288 V, the temperature coefficient for which is -4.3×10^{-4} V/°C. Potentials for other concentrations of KCl can be obtained by substituting the corresponding mean ion activity of Cl^- into the Nernst equation.

[5] A. S. Brown, *J. Amer. Chem. Soc.*, **56**, 646 (1934).

The Saturated Copper-Copper Sulfate Cell

This electrode consists of metallic copper immersed in saturated copper sulfate (Fig. 6). Its main use is for field measurements where the electrode must be resistant to shock and where its usual large size minimizes polarization errors. The accuracy of this electrode is adequate for most

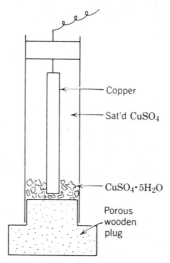

Fig. 6. Copper-saturated copper sulfate reference electrode.

corrosion investigations, even though it falls somewhat below the precision obtainable with the calomel or silver chloride electrodes. The half-cell reaction is

$$Cu \rightarrow Cu^{++} + 2e^- \qquad E° = -0.337 \text{ V}$$

For saturated copper sulfate, the potential is 0.316 V and the temperature coefficient is 7×10^{-4} V/°C.[6]

---

[6] S. Ewing, "The Copper-Copper Sulfate Half-cell for Measuring Potentials in the Earth," p. 10, Committee Report, American Gas Assoc., 1939. G. Scott proposes the value 0.300 V based on calculations employing activity coefficients for $CuSO_4$ extrapolated to saturated solution. [*Corrosion,* 14, 136t (1958).] This lower value can probably be attributed to uncertainty in the extrapolation and to the absence of a liquid junction potential, the latter being included in Ewing's measurement of copper-saturated $CuSO_4$ versus saturated calomel.

chapter 4
Polarization and corrosion rates

POLARIZATION

The foregoing chapters have dealt with equilibria between metals and their environments, which provides a basis for the concept of corrosion tendency. In practice, however, rates of corrosion are our main concern. Some metals with pronounced tendency to react, for example aluminum and magnesium, nevertheless react so slowly that they generally satisfy the requirements of a structural metal and may actually be more resistant in some media than other metals that have inherently less tendency to react.

It must not be concluded at this point, however, that considerations of equilibria are irrelevant to the study of corrosion. Instead, a fundamental approach to nonequilibrium states, and calculation of corrosion rates, begins with the primary consideration that equilibrium has been disturbed. In general, therefore, we must know the equilibrium state of the system before we can appreciate the various factors entering the rate at which the system tends toward equilibrium (i.e., corrodes).

An electrode is no longer at equilibrium when a net current flows to or from its surface. The measured potential of such an electrode is altered to an extent that depends on the magnitude of the external current and its direction. The direction of potential change always opposes the shift from equilibrium and hence opposes the flow of current, whether the current is impressed externally or is of galvanic origin. When current flows in a galvanic cell, for example, the anode always becomes more cathodic in potential and the cathode always becomes more anodic, the difference of potential becoming smaller. The extent of potential change caused by net current to or from an electrode, measured in volts, is called *polarization*.

THE POLARIZED CELL

Consider a cell made up of zinc in $ZnSO_4$ solution and copper in $CuSO_4$ solution (the Daniell cell), the electrodes of which are connected to a variable resistance R, voltmeter V, and ammeter A (Fig. 1). The poten-

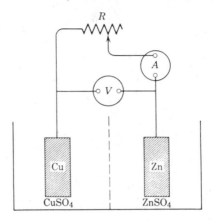

Fig. 1. Polarized copper-zinc cell.

tial difference (emf) of zinc and copper electrodes of the cell without current flow is about 1 V. If a small current is allowed to flow through the external resistance, the measured potential difference falls below 1 V because both electrodes polarize. The voltage continues to fall as the current increases. On complete short-circuit (very small external resistance), maximum current flows and the potential difference of copper and zinc electrodes becomes almost zero. The effect of net current flow on voltage of the Daniell cell can be represented by plotting separate potentials, ϕ, of copper and zinc electrodes (as described next under "How Measured") with total current, I, as shown in Fig. 2. The so-called *open-circuit potentials* (no current through the cell) are given by ϕ_{Zn} and ϕ_{Cu}. The zinc electrode polarizes along curve *abc* and the copper electrode along *def*. At a value of current through the ammeter equal to I_1, the polarization of zinc in volts is given by the difference between the actual potential of zinc at *b* and the open circuit value *a* or ϕ_{Zn}. Similarly, the polarization of copper is given by the difference of potential *e–d*. The potential difference of the polarized electrodes *b–e* is equal to the current I_1 multiplied by the total resistance of both the external metallic resistance, R_m, and the internal electrolytic resistance, R_e, in series, or $I_1(R_e + R_m)$. On short circuit, the current becomes maximum, I_{max}. Then R_m can be neglected, and the potential difference of both electrodes

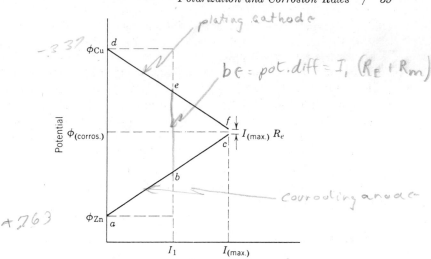

Fig. 2. Polarization diagram for copper-zinc cell.

decreases to a minimum, equal to $I_{max}R_e$. The maximum current is equivalent to $(65.38/2)\,I_{max}/F$ grams zinc corroding per second where I_{max} is in amperes, F is equal to 96,500 C/eq., and 65.38/2 is the equivalent weight of zinc.

The cathodic reaction corresponds to the identical chemical equivalents of copper depositing per second on the cathode. The corrosion rate of zinc can exceed the indicated equivalent corrosion rate, I_{max}, only if some means are introduced for reducing the polarization of zinc or copper, or both, thereby reducing the slopes of *abc* or *def*, causing an approach to intersection at a larger value of I. Similarly, any factor tending to increase polarization will decrease current through the cell and decrease the corresponding corrosion rate of zinc. Obviously the polarization curves can never actually intersect, although they can approach each other very closely if anodes and cathodes are closely spaced in media of moderate to good conductivity. There will always be a finite potential difference accompanying an observed flow of current.

Electrolytic cells that account for the corrosion of metals are analogous to the previously mentioned short-circuited cell. The measured potential of a corroding metal is the compromise potential of both polarized anodes and cathodes known as the *corrosion potential*, ϕ_{corros}. The value I_{max} is known as the corrosion current, I_{corros}. By Faraday's law, the corrosion rate of anodic areas on a metal surface is proportional to I_{corros}, and hence the corrosion rate per unit area can always be expressed as a

<u>current density</u>.   For zinc, a corrosion rate of 1 mdd is equivalent to 3.42×10^{-7} A/cm². For Fe corroding to Fe^{++}, the corresponding value is 4.0×10^{-7} A/cm². Values for other metals are listed in the Appendix.

Referring to Fig. 2, we can obviously càlculate the corrosion rate of a metal if data are available for the <u>corrosion</u> potential and for the <u>polarization</u> behavior and open-circuit potential of either anode *or* <u>cathode</u>. In general, the relative anode-cathode area ratio for the corroding metal must also be known, since polarization data are usually obtained under conditions where the electrode surface is all anode or all cathode. The success of such calculations in comparison with measured rates is one of the strongest supports of the electrochemical theory of corrosion (see p. 53).

HOW MEASURED

Referring to Fig. 3, showing a two-compartment cell separated by a sintered glass disk G, assume electrode B to be polarized by current from electrode D.* The probe L (sometimes called Luggin capillary)

---

* Current density at B must be uniform. This is sometimes achieved using a three-compartment cell with B in the center and two auxiliary electrodes in the outer compartments.

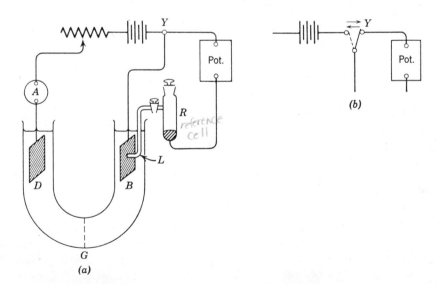

Fig. 3. (*a*) Cell for measuring polarization. (*b*) Oscillating contact for indirect method.

of reference cell R (or of a salt bridge between R and B) is placed close to the surface of B, thereby minimizing extraneous potentials caused by IR drop through the electrolyte. The emf of cell B-R is recorded for each value of current as read on ammeter A, allowing sufficient time for steady-state conditions. Polarization of B, whether anode or cathode, is recorded in volts with reference to half-cell electrode R for various values of current density. The potentials are often converted to the standard hydrogen scale. This is called the direct method for measuring polarization and is the method usually employed in corrosion studies.

The correct distance of the probe L from B has been discussed at great length in the electrochemical literature because of the disturbing effect of the probe on distribution of current to the electrode. One way of overcoming the IR drop without placing the probe near the electrode is to install an oscillating contact at point Y, which interrupts the current during measurement of potential (Fig. 3). Interruptions can be effected by a tuning fork, by a commutator, or electronically. Measurements of potential at various frequencies of the oscillating contact are extrapolated to infinite frequency. The advantage of this so-called *interrupter* or *commutator* or *indirect method* for measuring polarization is that the IR drop is completely eliminated both between probe and electrode and through films on the electrode, therefore the probe can be located well away from the electrode surface. A disadvantage is that rapid transients accompanying decay of polarization may lead to error, the observed polarization being low compared with that measured by the direct method.

Another approach is to measure potentials directly with probe L adjusted at various measured distances from B, and then extrapolate to zero distance. This correction, as shown in the next paragraph, is needed only when the measurements require highest accuracy, when the current densities are unusually high, or when the conductivity of the electrolyte is unusually low as in distilled water. This correction, however, *does not include* any high-resistance reaction product film that may cover the surface of the electrode. A special electrical circuit has been proposed for high-resistivity electrolytes; it provides for convenient measurement of the relevant IR drop, including that of electrode surface films, and introduction of the corresponding correction to the measured potential.[1, 2]

[1] D. Jones, *Corros. Sci.*, **8**, 19 (1968).
[2] B. Wilde, *Corrosion*, **23**, 379 (1967).

Calculation of *IR* Drop in an Electrolyte

$R = \dfrac{\rho l}{A}$

The resistance of an electrolyte solution measuring l cm long and S cm^2 in cross section is equal to $l/\kappa S$ ohms, where κ is the specific conductivity. Hence the *IR* drop in volts equals il/κ where i is the current density. For sea water, $\kappa = 0.05\ \Omega^{-1}$ cm^{-1}; therefore a current density of 1×10^{-5} A/cm^2 (representing the magnitude of current density that might be applied for cathodically protecting steel) produces an *IR* drop correction for a 1-cm separation of probe from cathode equal to $(1 \times 10^{-5}\text{ V})/0.05 = 0.2$ mV. This value is negligible in establishing the critical minimum current density for adequate cathodic protection. In some soft waters, however, where κ may be $10^{-5}\ \Omega^{-1}$ cm^{-1}, the corresponding *IR* drop equals 1 V/cm.

CAUSES OF POLARIZATION

The causes of electrode polarization fall into three different categories: concentration polarization, activation polarization, and *IR* drop.

1. *Concentration Polarization.** For example, if copper is made cathode in a solution of dilute $CuSO_4$ whose activity of cupric ion is represented by (Cu^{++}), then the oxidation potential, E_1, in absence of external current, is given by the Nernst equation

$$E_1 = -0.337 - \frac{0.0592}{2} \log (Cu^{++}) \tag{1}$$

When current flows, copper is deposited on the electrode, thereby decreasing surface concentration of copper ions to an activity $(Cu^{++})_s$. The oxidation potential, E_2, of the electrode now becomes

$$E_2 = -0.337 - \frac{0.0592}{2} \log (Cu^{++})_s \tag{2}$$

Since $(Cu^{++})_s$ is less than (Cu^{++}), the potential of the polarized cathode is less noble or more active than in the absence of external current. The difference of potential, $E_2 - E_1$, is known as *concentration polarization*, equal to

$$E_2 - E_1 = \frac{0.0592}{2} \log \frac{(Cu^{++})}{(Cu^{++})_s} \tag{3}$$

The larger the current, the smaller is the surface concentration of copper ion, or the smaller is $(Cu^{++})_s$; therefore the larger is the corresponding polarization. Infinite concentration polarization is ap-

* Also called *diffusion overpotential*.

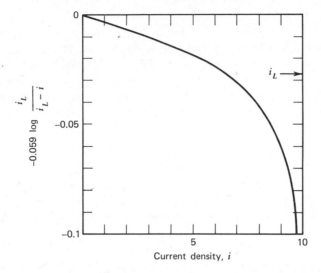

Fig. 4. Dependence of concentration polarization at a cathode on applied current density.

proached when $(Cu^{++})_s$ approaches zero at the electrode surface, the corresponding current density producing this limiting lower value of $(Cu^{++})_s$ being called the *limiting current density*. Obviously in practice polarization can never reach infinity; instead another electrode reaction establishes itself at a more active potential than corresponds to the first reaction. In the case of copper deposition, for example, the potential moves to that for hydrogen evolution: $2H^+ \rightarrow H_2 - 2e^-$ and hydrogen gas is liberated together with the simultaneous plating out of copper.

If i_L is the limiting current density for a cathodic process and i is the applied current density, it can be shown[3] that

$$\phi_2 - \phi_1 = -\frac{RT}{nF} \ln \frac{i_L}{i_L - i} \tag{4}$$

As i approaches i_L, $\phi_2 - \phi_1$ approaches minus infinity. This is shown by the plot of $\phi_2 - \phi_1$ versus i in Fig. 4.

The limiting current density (A/cm^2) can be evaluated from the expression

$$i_L = \frac{DnF}{\delta} c \times 10^{-3} \tag{5}$$

[3] S. Glasstone, *Principles of Electrochemistry*, p. 448, Van Nostrand, New York, 1942; G. Kortüm, *Treatise on Electrochemistry*, pp. 449–453, Elsevier, New York, 1965.

where D is the diffusion coefficient for the ion being reduced, n and F have their usual significance, δ is the thickness of the stagnant layer of electrolyte next to the electrode surface (about 0.05 cm in an unstirred solution), t is the transference number of all ions in solution except the ion being reduced (equal to unity if many other ions are present), and c is the concentration of diffusing ion in moles/liter. Since D for all ions at 25°C in dilute solution, except H^+ and OH^-, average about 1×10^{-5} cm²/sec, the limiting current density is approximated by

$$i_L = 0.02 \, nc \tag{6}$$

For H^+ and OH^-, D equals 9.3×10^{-5} and 5.2×10^{-5} cm²/sec, respectively (infinite dilution), so that the corresponding values of i_L are higher.

Should the copper electrode be polarized anodically, concentration of copper ion at the surface is larger than in the body of solution. The ratio $(Cu^{++})/(Cu^{++})_s$ then becomes less than unity and $E_2 - E_1$ of (3) changes sign. In other words, concentration polarization at an anode polarizes the electrode in the cathodic or noble direction, opposite to the direction of potential change when the electrode is polarized as cathode. For a copper anode the limiting upper value for concentration polarization corresponds to formation of saturated copper salts at the electrode surface. This limiting value is not as large as for cathodic polarization where the Cu^{++} activity approaches zero.

2. *Activation Polarization.* This is polarization caused by a slow electrode reaction. Or stated in another way, the reaction at the electrode requires an activation energy in order to go. The most important example is that of hydrogen ion reduction at a cathode, $H^+ \rightarrow \frac{1}{2}H_2 - e^-$, the corresponding polarization term being called *hydrogen overvoltage.* At a platinum cathode, for example, the following reactions are thought to occur in sequence:

$$H^+ \rightarrow H_{ads} - e^-$$

where H_{ads} represents hydrogen atoms adsorbed on the metal surface. This relatively rapid reaction is followed by a combination of adsorbed hydrogen atoms to form hydrogen molecules and bubbles of gaseous hydrogen.

$$2H_{ads} \rightarrow H_2$$

This reaction is relatively slow, and its rate determines the value of hydrogen overvoltage on platinum. The controlling slow step of H^+ discharge is not always the same, but varies with metal, current density, and environment.

Pronounced activation polarization also occurs with discharge of OH⁻ at an anode accompanied by oxygen evolution.

$$2OH^- \rightarrow \tfrac{1}{2}O_2 + H_2O + 2e^-$$

This is known as oxygen overvoltage. Overvoltage may also occur with Cl⁻ or Br⁻ discharge, but the values at a given current density are much smaller than for O_2 or H_2 evolution.

Activation polarization is also characteristic of metal-ion deposition or dissolution. The value may be small for nontransition metals like silver, copper, and zinc, but it is larger for the transition metals, e.g., iron, cobalt, nickel, chromium (see Table 1). The anion associated with the metal ion influences metal overvoltage values more than in the case of hydrogen overvoltage. The controlling step in the reaction is not known precisely, but in some cases it is probably a slow rate of hydration of the metal ion as it leaves the metal lattice, or dehydration of the hydrated ion as it enters the lattice.

Activation polarization, η, of any kind increases with current density, i, in accord with the Tafel equation:*

$$\eta = \beta \log \frac{i}{i_0}$$

TAFEL EQU.

where β and i_0 are constants for a given metal and environment and are both dependent on temperature. The exchange current, i_0, represents the current density equivalent to the equal forward and reverse reactions at the electrode at equilibrium. The larger the value of i_0 and the smaller the value of β, the smaller is the corresponding overvoltage.

A typical plot of activation polarization or overvoltage for H⁺ discharge is shown in Fig. 5. At the equilibrium potential for the hydrogen electrode (−0.059 pH), for example, overvoltage is zero. At applied current density, i_1, it is given by η, the difference between measured and equilibrium potentials. Although usually listed as positive, hydrogen overvoltage values are negative and, correspondingly, oxygen overvoltage values are positive on the ϕ scale.

3. *IR Drop.* Polarization measurements include a so-called ohmic potential drop through a portion of the electrolyte surrounding the electrode, through a metal–reaction product film on the surface, or both. This contribution to polarization is equal to iR where i is the current density and R equal to l/κ represents the value in ohms of the resistance path of length l cm, and specific conductivity, κ. The product iR decays

---

*Named after J. Tafel [Z. *Physik. Chem.*, **50**, 641 (1904)], who first proposed a similar equation to express hydrogen overvoltage as a function of current density.

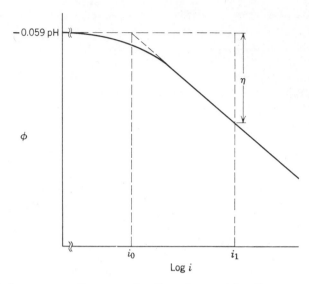

Fig. 5. Hydrogen overvoltage as a function of current density.

simultaneously with shutting off of the current, whereas concentration polarization and activation polarization usually decay at measurable rates. As mentioned previously, values of polarization obtained by the indirect method do not include the iR contribution.

NOTE: Concentration polarization decreases with stirring, whereas activation polarization and iR drop are not affected significantly.

HYDROGEN OVERVOLTAGE

The dominant polarization term controlling the corrosion rate of many metals in deaerated water or in nonoxidizing acids is hydrogen overvoltage at cathodic areas of the metal. In accord with the previously discussed definition of polarization, hydrogen overvoltage is the difference of potential between a cathode at which hydrogen is being evolved, $\phi_{measured}$, and a hydrogen electrode at equilibrium in the same solution, or H_2 overvoltage $= \phi_{measured} - (-0.059 \text{ pH})$. Hydrogen overvoltage, therefore, is measured precisely as is polarization. Ideally, hydrogen overvoltage includes only the activation polarization term corresponding to the reaction: $2H^+ \rightarrow H_2 - 2e^-$, but reported values often include iR drop and sometimes concentration polarization in addition.

$$2H^+ + 2e^- \rightarrow H_2 \uparrow$$

The absolute values of hydrogen overvoltage for a given metal *decrease* with:

1. *Increasing Temperature* (i_0 increases). For metals that corrode by hydrogen evolution, decreasing hydrogen overvoltage is one factor accounting for increase of corrosion as the temperature is raised.

2. *Roughening of the Surface*. A sand-blasted surface has a lower hydrogen overvoltage than a polished surface. Greater area and improved catalytic activity of a rough surface account for the observed effect.

3. *Decreasing Current Density*. The Tafel equation η (overvoltage) $= \beta \log i/i_0$ is obeyed within the region of applied current density, i, below the limiting current density for concentration polarization, and above the exchange current density, i_0. The interpretation of the exchange current density has already been mentioned. The term β is equal to $2.3RT/\alpha F$ where α is a constant and the other terms have their usual significance. The term α is approximately 2 for platinum and palladium, and is approximately 0.4 to 0.6 for Fe, Ni, Cu, Hg, and several other metals. Although hydrogen overvoltage values usually differ in acid compared with alkaline media, values are not greatly sensitive to pH within either range.

Stern[4] showed that for a corroding metal

$$\eta = \beta \log \frac{i + i_r + i_{corros}}{i_0}$$

corroding (handwritten)

where i_r is the reverse current for the reaction $H_2 \rightarrow 2H^+ + 2e^-$, which varies with potential and which is equal at equilibrium to i_0. This equation describes the small observed slope of $d\eta/d \log i$ for small values of impressed current density, i, the slope increasing as i approaches $i_{corros} + i_r$ and reaching the true Tafel slope, β, at $i \gg i_r + i_{corros}$. Similarly, the overvoltage for a noncorroding metal can be represented by a *non corrode* (handwritten) modified Tafel equation equal to $\beta \log (i + i_r)/i_0$ which holds from zero to finite values of i (Fig. 5). Values of i_r determined from measured values of η also follow the Tafel equation, but with a slope of opposite sign.

The slow step in the discharge of hydrogen ions on platinum or palladium as described earlier, appears to be the recombination of adsorbed hydrogen atoms. This assumption is consistent, from kinetic considerations, with an observed value of $\alpha = 2$. For iron, the value of α is approximately 0.5 and, correspondingly, $\beta = 0.1$. To account for these

[4] M. Stern, *J. Electrochem. Soc.*, **102**, 609, 663 (1955); (with Geary), *Ibid.*, **104**, 56 (1957).

values, the proposal has been made that the slow step in the hydrogen evolution reaction on iron is probably

$$H^+ + H_{ads} \rightarrow H_2 - e^-$$

The same slow step may apply to various metals having intermediate values of hydrogen overvoltage (e.g., iron, nickel, copper, etc.).

For metals of high hydrogen overvoltage (e.g., mercury or lead), the slow discharge of the hydrated hydrogen ion is apparently the slow step:

$$H^+ \rightarrow H_{ads} - e^-$$

For many metals at high current densities, this slow discharge step is also the controlling reaction. Or, instead, it may be the reduction of H_2O in accord with

$$H_2O \rightarrow OH^- + \tfrac{1}{2}H_2 - e^-$$

The reduction of water as the controlling reaction applies generally to metals in alkaline solutions at high or low current densities.

The rapidity with which H_{ads} combines to form H_2, either by combination with itself or with H^+, is affected by the catalytic properties of the electrode surface. A good catalyst (e.g., platinum or iron) leads to a low value of hydrogen overvoltage, whereas a poor catalyst (e.g., mercury or lead) accounts for a high value of overvoltage. If a catalyst poison like hydrogen sulfide or certain arsenic or phosphorus compounds are added to the electrolyte, these retard the rate of formation of molecular H_2 and increase the accumulation of adsorbed hydrogen atoms on the electrode surface.* The increased concentration of surface hydrogen favors entrance of hydrogen atoms into the metal lattice, causing *hydro-*

---

* Increase of hydrogen overvoltage normally decreases the corrosion rate of steel in acids, but presence of sulfur or phosphorus in steels is observed instead to increase the rate. This increase probably results from the low hydrogen overvoltage of ferrous sulfide or phosphide existing in the steel as separate phases in the metal, or which are formed as a surface compound by reaction of iron with H_2S or phosphorus compounds in solution. It is also possible[5] that the latter compounds in addition stimulate the anodic dissolution reaction $Fe \rightarrow Fe^{++} + 2e^-$ (reduce activation polarization), or alter the anode-cathode area ratio. The whole subject requires further study.

Similarly, arsenious oxide in small amount accelerates corrosion of steel in acids (e.g., H_2SO_4), perhaps forming arsenides, but when present in larger amount (e.g., 0.05% As_2O_3 in 72% H_2SO_4) it is an effective corrosion inhibitor, probably because elementary arsenic having a high hydrogen overvoltage deposits out at cathodic areas. Tin salts have the same inhibiting effect and are sometimes used to protect steels from attack by pickling acids during descaling operations.

[5] M. Stern, *J. Electrochem. Soc.,* **102**, 663 (1955).

TABLE 1
Overvoltage Values

$$\eta = \beta \log \frac{i}{i_0}$$

Hydrogen Overvoltage

Metal	Temperature (°C)	Solution	β (V)	i_0 (A/cm^2)	η [1 mA/cm^2 (V)]
Pt (smooth)	20	1 N HCl	0.03	10^{-3}	0.00
	25	0.1 N NaOH	0.11	6.8×10^{-5}	0.13
Pd	20	0.6 N HCl	0.03	2×10^{-4}	0.02
Mo	20	1 N HCl	0.04	10^{-6}	0.12
Au	20	1 N HCl	0.05	10^{-6}	0.15
Ta	20	1 N HCl	0.08	10^{-5}	0.16
W	20	5 N HCl	0.11	10^{-5}	0.22
Ag	20	0.1 N HCl	0.09	5×10^{-7}	0.30
Ni	20	0.1 N HCl	0.10	8×10^{-7}	0.31
	20	0.12 N NaOH	0.10	4×10^{-7}	0.34
Bi	20	1 N HCl	0.10	10^{-7}	0.40
Nb	20	1 N HCl	0.10	10^{-7}	0.40
Fe	16	1 N HCl	0.15	10^{-6}	0.45
	25	4% NaCl pH 1–4	0.10	10^{-7}	0.40 (Stern)
Cu	20	0.1 N HCl	0.12	2×10^{-7}	0.44
	20	0.15 N NaOH	0.12	1×10^{-6}	0.36
Sb	20	2 N H$_2$SO$_4$	0.10	10^{-9}	0.60
Al	20	2 N H$_2$SO$_4$	0.10	10^{-10}	0.70
Be	20	1 N HCl	0.12	10^{-9}	0.72
Sn	20	1 N HCl	0.15	10^{-8}	0.75
Cd	16	1 N HCl	0.20	10^{-7}	0.80
Zn	20	1 N H$_2$SO$_4$	0.12	1.6×10^{-11}	0.94
Hg	20	0.1 N HCl	0.12	7×10^{-13}	1.10
	20	0.1 N H$_2$SO$_4$	0.12	2×10^{-13}	1.16
	20	0.1 N NaOH	0.10	3×10^{-15}	1.15
Pb	20	0.01–8 N HCl	0.12	2×10^{-13}	1.16

Oxygen Overvoltage

Metal	Temperature (°C)	Solution	β (V)	i_0 (A/cm^2)	η [1 mA/cm^2 (V)]
Pt (smooth)	20	0.1 N H$_2$SO$_4$	0.10	9×10^{-12}	0.81
	20	0.1 N NaOH	0.05	4×10^{-13}	0.47
Au	20	0.1 N NaOH	0.05	5×10^{-13}	0.47

Metal Overvoltage (deposition)

Metal	Temperature (°C)	Solution	β (V)	i_0 (A/cm^2)	η [1 mA/cm^2 (V)]
Zn	25	1 M ZnSO$_4$	0.12	2×10^{-5}	0.20 (Bockris)
Cu	25	1 M CuSO$_4$	0.12	2×10^{-5}	0.20 (Bockris)
Fe	25	1 M FeSO$_4$	0.12	10^{-8}	0.60 (Bockris)
Ni	25	1 M NiSO$_4$	0.12	2×10^{-9}	0.68 (Bockris)

Data from:

B. E. Conway, *Electrochemical Data*, Elsevier, New York, 1952. *Modern Aspects of Electrochemistry*, edited by J. Bockris, Academic Press, New York, 1954. M. Stern, *J. Electrochem. Soc.*, **102**, 609 (1955). H. Kita, *J. Electrochem. Soc.*, **113**, 1095 (1966).

gen embrittlement (loss of ductility) and in some stressed high-strength ferrous alloys may induce spontaneous cracking (hydrogen cracking) (see p. 142). Catalyst poisons increase absorption of hydrogen whether the metal is polarized by externally applied current or by a corrosion reaction with accompanying hydrogen evolution. For this reason, some oil-well brines containing H_2S are difficult to handle in low-alloy steel tubing subject to the usual high stresses of a structure extending several thousand feet underground. Slight general corrosion of the tubing produces hydrogen, a portion of which enters the stressed steel to cause hydrogen cracking. In the absence of hydrogen sulfide, general corrosion occurs but without hydrogen cracking. High-strength steels, because of their limited ductility, are more susceptible to hydrogen cracking than are low-strength steels, but hydrogen enters the lattice in either case, tending to blister low-strength steels instead.*

Values of β, i_0, and of η at 1 mA/cm^2 for H^+ discharge are given in Table 1. Note that values of hydrogen overvoltage vary greatly with the metal. Values also change with concentration of electrolyte, but the effect comparatively is not large.

INFLUENCE OF POLARIZATION ON CORROSION RATE

Both resistance of the electrolyte and polarization of the electrodes limit the magnitude of current produced by a galvanic cell. For local-action cells on the surface of a metal, electrodes are in close proximity to each other, consequently resistance of the electrolyte is usually a secondary factor compared to the more important factor of polarization. When polarization occurs mostly at the anodes, it is said that the corrosion reaction is *anodically* controlled. Typical polarization curves are shown in Fig. 6. Note that the corrosion potential is close to the open-circuit potential of the cathode. A practical example is impure lead immersed in sulfuric acid, where a lead-sulfate film covers the anodic areas and exposes cathodic impurities (e.g., copper). Other examples are magnesium exposed to natural waters and iron immersed in a chromate solution.

When polarization occurs mostly at the cathode, the corrosion rate is said to be *cathodically controlled*. The corrosion potential is then

---

* Austenitic steels (e.g., 18-8 stainless steels) are ordinarily not susceptible to hydrogen cracking because the diffusion rate of hydrogen into ferrous alloys is much lower for the face-centered cubic structure than for the body-centered cubic structure, and hydrogen that enters has less effect on ductility and is therefore less damaging.

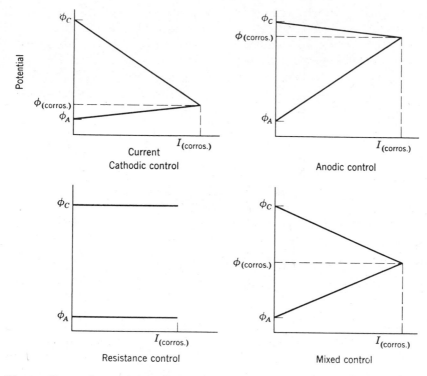

Fig. 6. Types of corrosion control.

near the open-circuit anode potential. Examples are zinc corroding in sulfuric acid and iron exposed to natural waters.

Resistance control occurs when the electrolyte resistance is so high that the resultant current is not sufficient to appreciably polarize anodes or cathodes. An example occurs with a porous insulating coating covering a metal surface. The corrosion current is then controlled by the IR drop through electrolyte in pores of the coating.

It is common for polarization to occur in some degree at both anodes and cathodes. This situation is described as *mixed control*.

The extent of polarization, it should be noted, depends not only on the nature of the metal and electrolyte, but also on the actual exposed area of the electrode. If the anodic area of a corroding metal is very small, caused, for example, by porous surface films, there may be considerable anodic polarization accompanying corrosion, even though measurements show that unit area of the bare anode polarizes only slightly at a given current density. Consequently, the anode-cathode area ratio

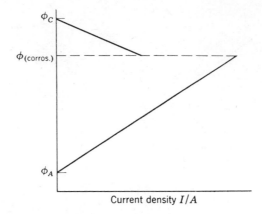

Fig. 7. Polarization diagram for corroding metal when anode area = $\frac{1}{2}$ cathode area.

is also an important factor in determining the corrosion rate. If current density is plotted instead of total corrosion current, as for example when the anode area is half the cathode area, corresponding polarization curves are described by Fig. 7.

A significant experiment was performed by Wagner and Traud[6] bearing on the electrochemical mechanism of corrosion. They measured the cor-

[6] C. Wagner and W. Traud, *Z. Elektrochem.*, **44**, 391 (1938).

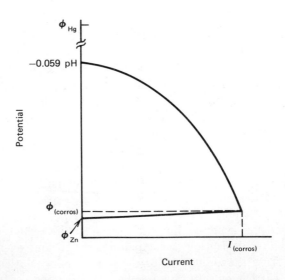

Fig. 8. Polarization diagram for zinc amalgam in deaerated HCl.

rosion rate of a dilute zinc amalgam in an acid calcium chloride mixture and the cathodic polarization of mercury in the same electrolyte. The current density equivalent to the corrosion rate was found to correspond to the current density necessary to polarize mercury to the corrosion potential of the zinc amalgam (Fig. 8). In other words, mercury atoms of the amalgam composing the majority of the surface apparently act as cathodes (hydrogen electrodes)* and zinc atoms act as anodes of corrosion cells. The amalgam polarizes anodically very little, and the corrosion reaction is controlled almost entirely by rate of hydrogen evolution at cathodic areas. It is, consequently, the high hydrogen overvoltage of mercury that limits the corrosion rate of amalgams in nonoxidizing acids. A piece of platinum coupled to the amalgam considerably increases the rate of corrosion, because hydrogen is more readily evolved from a low-overvoltage cathode at the operating emf of the zinc-hydrogen electrode cell.

The very low corrosion rate and the absence of appreciable anodic polarization explain why amalgams in corresponding metal salt solutions exhibit corrosion potentials closely approaching the reversible potential of the alloyed component. For example, the corrosion potential of cadmium amalgam in $CdSO_4$ solution is nearer the thermodynamic value for $Cd \rightarrow Cd^{++} + 2e^-$ than is observed for pure cadmium in the same solution. The steady-state corrosion rate of pure cadmium is appreciably higher than that of cadmium amalgam, leading to greater deviation of the measured corrosion potential from the corresponding thermodynamic value. By and large, the steady-state potential of any metal more active than hydrogen (e.g., iron, nickel, zinc, cadmium) in an aqueous solution of its ions deviates from the true thermodynamic value by an amount that depends on the prevailing corrosion rate accompanied by H^+ discharge.[7] The measured value tends to be more noble than the true value. This situation also holds for the steady-state potentials of more noble metals (e.g., copper, mercury) which undergo corrosion in the presence of dissolved oxygen.

CALCULATION OF CORROSION RATES FROM POLARIZATION DATA

As mentioned earlier, the corrosion current can be calculated from the corrosion potential and the open-circuit potential if the equation expres-

---

* Mercury in aqueous solutions acts first as a mercury electrode, but when cathodically polarized, all mercury ions in solution are deposited before H^+ is discharged. Any conducting surface on which H^+ ions are discharged acts as a polarized hydrogen electrode and can be so considered in evaluating a corrosion cell.

[7] H. H. Uhlig, *Proc. Nat. Acad. Sci. (U. S.)*, **40, 276** (1954).

sing polarization of the anode or cathode is known, and the anode-cathode area ratio can be estimated. For corrosion of active metals in deaerated acids, for example, the surface of the metal is probably covered largely with adsorbed H atoms and can be assumed therefore to be mostly cathode. The open-circuit potential is -0.059 pH, and if i_{corros} is sufficiently larger than i_0 for $H^+ \leftrightarrows \frac{1}{2}H_2 + e^-$, the Tafel equation expresses cathodic polarization behavior. Then

$$\phi_{corros} = -\left[0.059 \text{ pH} + \beta \log \frac{i_{corros}}{i_0} \right] \qquad (7)$$

from which i_{corros} and the equivalent corrosion rate can be calculated. Stern[4] showed that calculated corrosion rates for iron, using (7) and employing empirical values for β and i_0, were in excellent agreement with observed rates. Typical values are given in Table 2.

TABLE 2
Comparison of Calculated and Observed Corrosion Currents for Pure Iron in Various Deaerated Acids (Stern)

	$\phi_{H (corros.)}$ + 0.059 pH (Volt)	β (volt)	i_0 (μamp/ cm^2)	$i_{(corros.)}$ (μamp/cm^2) calc.	obs.
0.1M Citric Acid pH = 2.06	-0.172	0.084	0.093	10.4	11.5
0.1M Malic Acid pH = 2.24	-0.158	0.083	0.015	1.2	1.2
4% NaCl + HCl					
pH = 1	-0.203	0.100	0.10	10.5	11.1
pH = 2	-0.201	0.100	0.11	11.0	11.3

Subsequently, Stern and Geary[8] derived the equation

$$I_{corros} = \frac{I_{applied}}{2.3 \, \Delta\phi} \left(\frac{\beta_c \beta_a}{\beta_c + \beta_a} \right) \qquad (8)$$

where β_c and β_a refer to Tafel constants for the cathodic and anodic reactions, respectively, and $I_{applied}/\Delta\phi$ is the polarization slope in the region

[8] M. Stern and A. Geary, *J. Electrochem. Soc.*, **104**, 56 (1957); M. Stern, *Corrosion*, **14**, 440t (1958); M. Stern and E. Weisert, *Proc. Amer. Soc. Testing Mat.*, **59**, 1280 (1959).

near the corrosion potential for which the change of $\Delta\phi$ with $I_{applied}$ is essentially linear. (For derivation, see Appendix, p. 387.) Under conditions of slight polarization, for which $\Delta\phi$ is not more than about 10 mV, the anode-cathode area ratio, which need not be known, remains essentially constant and conditions otherwise at the surface of the corroding metal are largely undisturbed.

In the event corrosion is controlled by concentration polarization at the cathode, as when oxygen depolarization is controlling, the equation simplifies to

$$I_{corros} = \frac{\beta_a}{2.3}\frac{I_{applied}}{\Delta\phi} \tag{9}$$

Although β values are relatively well known for H^+ discharge, they are not as generally available for other electrode reactions. Stern showed, however, that the majority of reported β values fall between 0.06 and 0.12 V. Then, if β_c is known to be 0.06 V, for example, and β_a falls within 0.06–0.12 V, the calculated corrosion current is within at least 20% of the correct value. Under other assumptions, the corrosion rate can be calculated to at least a factor of 2.

The general validity of (8) and (9) is shown by data summarized in Fig. 9. The observed corrosion current, corresponding to data on corrosion of nickel in HCl, and steels and cast iron in acids and in natural waters, extends over six orders of magnitude. Some exchange current densities for $Fe^{+++} \rightleftharpoons Fe^{++} - e^-$ on passive surfaces are included because the same principle applies in calculating i_0 for a noncorroding electrode as in calculating I_{corros} for a corroding electrode. Also, straight lines are shown representing values calculated on the basis of several assumed β values within which most of the empirical data lie. Equations 7 and 8 have been used successfully for determining the corrosion rates of various metals in several aqueous environments at ordinary or elevated temperatures.[9–12]

The linear polarization method has obvious advantages in calculating instantaneous corrosion rates for many metals in a wide variety of environments and under various conditions of velocity and temperature. It can also be used to evaluate inhibitors and protective coatings, as well as for detecting changes of corrosion with time. A calibration is required if an IR drop is involved in the measurement. Also, the equations are not applicable for calculating corrosion rates in presence of interfer-

[9] J. Evans and E. Koehler, *J. Electrochem. Soc.*, **108**, 509 (1961).
[10] A. Cohen and R. Jellinek, *Corrosion*, **22**, 39 (1966).
[11] D. Jones and N. Greene, *Ibid.*, p. 198.
[12] B. Wilde, *Ibid.*, **23**, 379 (1967).

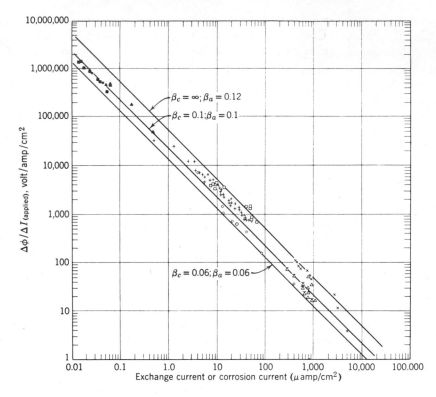

Fig. 9. Relation between polarization slope at low current densities and observed corrosion or exchange current densities (Stern and Weisert).

ing oxidation-reduction systems that are not involved directly in the corrosion reaction.

ANODE–CATHODE AREA RATIO

The reduction of H^+ at a platinum cathode at a given rate is accompanied by the simultaneous reverse reaction at a lower rate for H_2 oxidizing to H^+. The oxidation and reduction reactions are assumed to take place on the same surface sites. At equilibrium, the forward and reverse reactions are equal and the equivalent current density is called the exchange current density. For a corroding metal, the anode and cathode reactions are distinct and different; one is not merely the reverse reaction of the other. Hence, the oxidation reaction can take place only on sites of the metal surface that are different and distinct from those on which reduction takes place. For this situation, the distance between anode and cathode may vary from atomic dimensions to a separation of many meters. Correspondingly, an anode-cathode area ratio exists, and since the observed polarization of anodic or cathodic sites depends in part on the area over which oxidation or reduction occurs, the anode-cathode area ratio enters as an important factor in the observed corrosion rate.

Stern,[13] treated the general relation for which anodic and cathodic polarization follows Tafel behavior and IR drop is negligible. Employing the polarization diagram illustrated on p. 391:

$$\phi_c{}' = \phi_c - \beta_c \frac{\log I_c}{A_c i_{0c}} \tag{10}$$

$$\phi_a{}' = \phi_a + \beta_a \frac{\log I_a}{A_a i_{0a}} \tag{11}$$

where subscripts a and c refer to anode and cathode respectively, ϕ' is the polarized potential, ϕ is the open-circuit potential, β is the Tafel slope, I is the current per unit metal surface area, A_a is the fraction of surface that is anode such that $A_a + A_c = 1$ and i_{0a} and i_{0c} are the exchange current densities for the anode and cathode reactions, respectively. At steady state $\phi_c{}' = \phi_a{}' = \phi_{corros}$ and $I_a = I_c = I_{corros}$. Then

$$\beta_c \log I_{corros} = \phi_c - \phi_{corros} + \beta_c \log A_c i_{0c} \tag{12}$$

$$\phi_{corros} = \phi_a + \beta_a \log I_{corros} - \beta_a \log A_a i_{0a} \tag{13}$$

Substituting (13) into (12):

$$\log I_{corros} = \frac{\phi_c - \phi_a}{\beta_c + \beta_a} + \frac{\beta_a}{\beta_c + \beta_a} \log A_a i_{0a} + \frac{\beta_c}{\beta_c + \beta_a} \log A_c i_{0c} \tag{14}$$

Since $A_a = 1 - A_c$

$$\frac{d \log I_{corros}}{dA_c} = - \frac{\beta_a}{2.3(\beta_c + \beta_a)} \frac{1}{(1 - A_c)} + \frac{\beta_c}{2.3(\beta_c + \beta_a)} \frac{1}{A_c} \tag{15}$$

$$= \frac{\beta_c(1 - A_c) - \beta_a A_c}{2.3(\beta_c + \beta_a) A_c A_a} \tag{16}$$

Maximum I_{corros} occurs at $d \log I_{corros}/dA_c = 0$ or when the numerator of (16) equals zero. This occurs at $A_c = \beta_c/(\beta_c + \beta_a)$. If $\beta_c = \beta_a$, which is frequently observed, the maximum corrosion rate occurs at $A_c = \frac{1}{2}$ or at an anode-cathode area ratio of unity. At any other anode-cathode

[13] M. Stern, *Corrosion*, **14**, 329t (1958).

area ratio, the corrosion rate is less, reaching zero at a ratio equal to zero or infinity.

Although these conclusions are logical consequences of the electrochemical theory of corrosion, there are some investigators who believe that typical simultaneous oxidation and reduction reactions (e.g., corrosion of zinc amalgam in HCl) occur on the same surface sites and that anode-cathode area ratios for these situations have no meaning. An experiment could conceivably be devised in which the conclusions of Stern's model could be tested, but such an experiment remains to be done.

THEORY OF CATHODIC PROTECTION *(making cathode anodic i.e. same potential as anod)*

Observing the polarization diagram for the copper-zinc cell in Fig. 2, it is obvious that if polarization of the cathode is continued by use of external current beyond the corrosion potential to the open-circuit potential of the anode, both electrodes attain the same potential and no corrosion of zinc can occur. This is the basis for cathodic protection of metals. It provides one of the most effective practical means for reducing the corrosion rate to zero (this subject is discussed further in Chapter 12). Cathodic protection is accomplished by supplying an external current to the corroding metal, on the surface of which local-action cells operate as illustrated schematically in Fig. 10. Current leaves the auxiliary anode (composed of any metallic or nonmetallic conductor) and enters both the cathodic and anodic areas of the corrosion cells, returning to the source of d-c current, B. When the cathodic areas are polarized by external current to the open-circuit potential of the anodes, all the metal surface is at the same potential and local-

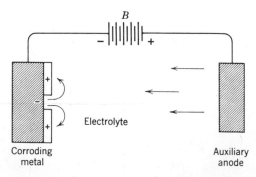

Fig. 10. Cathodic protection by superposition of impressed current on local-action current.

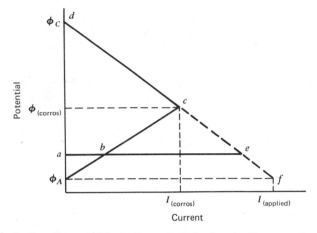

Fig. 11. Polarization diagram illustrating principle of cathodic protection.

action current no longer flows. The metal, therefore, cannot corrode so long as the external current is maintained. The corresponding polarization diagram is given in Fig. 11 where $I_{applied}$ is the current necessary for complete protection.

If the metal is polarized slightly beyond the open-circuit potential, ϕ_A, of the anode, the corrosion rate remains zero. Net current to the anode areas is from electrolyte to metal, and hence metal ions cannot enter into solution. Current used in excess of that required does no good, however, and may do harm to amphoteric metals or to coatings. In practice, therefore, the impressed current is kept close to the theoretical minimum. Should the applied current fall below that required for complete protection, some degree of protection is afforded nevertheless. For example, if the corrosion potential is moved from ϕ_{corros} to a in Fig. 11, by applied current b–e, the corrosion current decreases from I_{corros} to a–b. Total cathodic current (corrosion current plus applied current) is a–e. As the applied current b–e is increased, the potential a moves to more active values, and the corrosion current a–b becomes smaller. At the limit, when a coincides with ϕ_A, corrosion current a–b becomes zero and applied current for complete cathodic protection equals $I_{applied}$.

chapter 5
Passivity

A passive metal is one that is active in the Emf Series but which corrodes nevertheless at a very low rate. Passivity is the property underlying the useful natural corrosion resistance of many structural metals, including aluminum, chromium, and the stainless steels. Some metals and alloys can be made passive by exposure to passivating environments (e.g., iron in chromate or nitrite solutions) or by anodic polarization at sufficiently high current densities (e.g., in H_2SO_4).

Near the beginning of the nineteenth century, it was observed that iron reacts rapidly in dilute HNO_3 but is visibly unattacked by concentrated HNO_3. Upon removing iron from the concentrated acid and immersing it into the dilute acid, a temporary state of corrosion resistance persists. Schönbein[1] in 1836 defined iron in the corrosion-resistant state as "passive." He also showed that iron could be made passive by anodic polarization. Faraday[2] at the same time performed several experiments on passivity showing, among other things, that a cell made up of passive iron coupled to platinum in concentrated nitric acid produced little or no current, in contrast to the high current produced by amalgamated zinc coupled to platinum in dilute sulfuric acid. Although passive iron corrodes only slightly in concentrated HNO_3, and similarly amalgamated zinc corrodes only slightly in dilute H_2SO_4, Faraday emphasized that a low corrosion rate is not alone a measure of passivity. Instead, he stated, the magnitude of current produced in the cell versus platinum is a better criterion, for on this basis iron is passive but not zinc. This in essence defined a passive metal, as we do today, as one that is ap-

[1] C. Schönbein, *Pogg. Ann.*, **37**, 590 (1836).
[2] M. Faraday, *Experimental Researches in Electricity,* vol. II, University of London, 1844.

preciably polarized by a small anodic current. Later investigators deviated from this definition, however, and also called metals passive that corrode only slightly despite their pronounced tendency to react in a given environment. This usage brought about two definitions of passivity which are still in force today.[3]

Definition 1. A metal active in the Emf Series, or an alloy composed of such metals, is considered passive when its electrochemical behavior approaches that of an appreciably less active or noble metal.

Definition 2. A metal or alloy is passive if it substantially resists corrosion in an environment where thermodynamically there is a large free-energy decrease associated with its passage from the metallic state to appropriate corrosion products.

C. Wagner[4] offered an improvement of Def. 1, the essence of which is the following: A metal is passive if, on increasing the electrode potential toward more noble values, the rate of anodic dissolution in a given environment under steady-state conditions becomes less than the rate at some less noble potential. Alternatively, a metal is passive if, on increasing the concentration of an oxidizing agent in an adjacent solution or gas phase, the rate of oxidation, in absence of external current, is less than the rate at some lower concentration of the oxidizing agent. These alternative definitions are equivalent under conditions where the electrochemical theory of corrosion applies.

Thus lead immersed in sulfuric acid, or magnesium in water, or iron in inhibited pickling acid, would be called passive by Def. 2 based on low corrosion rates, despite pronounced corrosion tendencies; but they are not passive by Def. 1. The corrosion potentials of these metals are relatively active and polarization is not pronounced when they are made the anode of a cell.

Examples of metals under Def. 1, on the other hand, include chromium, nickel, molybdenum, titanium, zirconium, the stainless steels, 70% Ni–30% Cu alloys (Monel), and several other metals and alloys. Also included are metals that become passive in passivator solutions, like iron in dissolved chromates. Metals and alloys in this category show a marked tendency to polarize anodically. Pronounced anodic polarization reduces observed reaction rates, so that metals under Def. 1 usually conform as well to Def. 2 based on low corrosion rates. The corrosion potentials of metals passive by Def. 1 approach the open-circuit cathode potentials (for example, the oxygen electrode) and

[3] *Corrosion Handbook,* edited by H. H. Uhlig, p. 21, Wiley, New York, 1948.

[4] C. Wagner, Discussions at 1st Int. Symp. Passivity, Heiligenberg, West Germany, 1957; *Corros. Sci.,* **5,** 751 (1965).

hence as components of galvanic cells they exhibit potentials near those of the noble metals.

CHARACTERISTICS OF PASSIVATION, THE FLADE POTENTIAL

Suppose iron is made anode in $1N$ H_2SO_4 arranged so that as the potential is gradually increased, the corresponding polarizing current reaches whatever value is required, but no more, to maintain the prevailing potential with respect to some reference electrode. This can be done manually, or better still by using a device called the *potentiostat,* which automatically adjusts current to potential—making use, in most instances, of an appropriate electronic circuit. The resulting polariza-

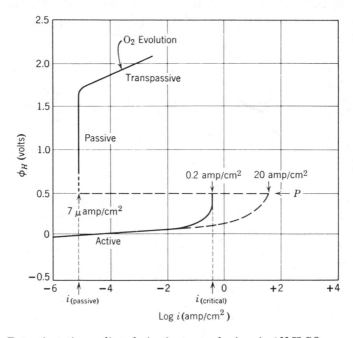

Fig. 1. Potentiostatic anodic polarization curve for iron in $1N$ H_2SO_4.

tion curve is shown in Fig. 1.[5] It is called a potentiostatic polarization curve in contrast to a galvanostatic curve using, for example, the circuit shown in Fig. 3, p. 40, for which current is maintained constant and potential is allowed to vary at will (Fig. 2). The potentiostatic polari-

[5] K. Bonhoeffer, *Z. Metallk.,* **44, 77** (1953).

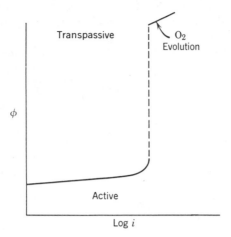

Fig. 2. Galvanostatic anodic polarization curve for iron in $1N$ H_2SO_4.

zation curve provides more information than the galvanostatic curve because it corresponds more closely to the actual behavior of passive metals as components of galvanic cells. Referring to Fig. 1, iron is active at small current densities and corrodes anodically as Fe^{++} in accord with Faraday's law. As current increases, an <u>insulating film</u> forms over the electrode surface, composed probably of $FeSO_4$. At a critical current density, $i_{critical}$, of about 200 mA/cm² (higher on stirring or on lowering pH of the environment), the current suddenly drops to a value orders of magnitude lower called the passive current density [$i_{passive}$]. At this point, <u>the thick insulating film dissolves, being</u> replaced by a much thinner film, and iron becomes passive. The value of $i_{passive}$ decreases with time, diminishing to about 7 μA/cm² at steady-state conditions in $1N$ H_2SO_4. In other electrolytes $i_{passive}$ may be higher or lower than in $1N$ H_2SO_4.

On further gradual change of the potential, the current density remains at the above low value, and the corrosion product is now $\underline{Fe^{+3}}$. At about 1.2 V, the equilibrium oxygen electrode potential is reached, but oxygen is not evolved appreciably until the potential exceeds the equilibrium value by several tenths volt (oxygen overvoltage). The true critical current density for achieving passivity of iron in absence of an insulating reaction product layer is estimated to be about 10 to 20 A/cm² as was shown by short-time current-pulse measurements.

When the anodic current is interrupted, <u>passivity decays within a short time in the manner shown by Fig. 3.</u> The potential first changes quickly to a value still noble on the hydrogen scale, then changes slowly for a matter of seconds to several minutes. Finally, it decays rapidly

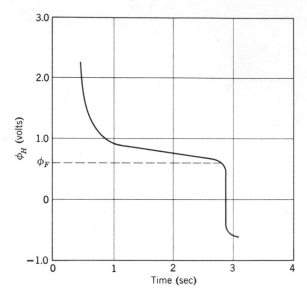

Fig. 3. Decay of passivity of iron in $1N$ H_2SO_4 showing Flade potential, ϕ_F.

to the normal active potential of iron. The noble potential arrived at just before rapid decay to the active value was found by Flade[6] to be more noble the more acid the solution in which passivity decayed. This characteristic potential, ϕ_F, was later called the Flade potential, and Franck[7] found that it was a linear function of pH. His measurements, combined with later data by others, provide the relation at 25°C:

$$\phi_F(V), \text{ s.h.e.} = +0.63 - 0.059 \text{ pH} \tag{1}$$

or

$$E_F = -0.63 + 0.059 \text{ pH}$$

This reproducible Flade potential and its 0.059 pH dependence is characteristic of the passive film on iron. A similar potential-pH relation is found for the passive film on chromium, Cr-Fe alloys,* and for nickel, for which the standard Flade potentials (pH = 0) are less noble than for iron, in accord with more stable passivity.

Stability of passivity is related to the Flade potential, assuming the

---

* When activated cathodically, the Flade potential of chromium and stainless steels follows the relation $n(0.059 \text{ pH})$ where n may be as high as 2. For self-activation, n is 1 [P. King and H. Uhlig, *J. Phys. Chem.*, **63**, 2026 (1959)].

[6] F. Flade, *Z. Physik. Chem.*, **76**, 513 (1911).

[7] U. F. Franck, *Z. Naturforschung.*, **4A**, 378 (1949).

following schematic reaction to take place during anodic passivation:

$$M + H_2O \rightarrow O \cdot M + 2H^+ + 2e^- \qquad \text{(2)}$$

the oxidation potential for which is E_F, and $O \cdot M$ refers to oxygen in the passive film on metal M whatever the passive film composition and structure may be. The amount of oxygen assumed to be combined with M has no effect on present considerations. It follows, as observed, that

$$E_F = E_F^\circ - \frac{0.059}{2} \log (H^+)^2 = E_F^\circ + 0.059 \text{ pH} \qquad \text{(3)}$$

The negative value of E_F° for iron (-0.63 V) indicates considerable tendency for the passive film to decay—reverse reaction of (2)—whereas an observed positive value of $E_F^\circ = 0.2$ V for chromium indicates conditions more favorable to passive film formation and hence greater stability of passivity. The value of E_F° for nickel is -0.2 V. For chromium-iron alloys, values range from -0.63 V for pure iron to increasingly positive values as chromium is alloyed, changing most rapidly in the range 10–15% chromium, and reaching about 0.1 V at 25% chromium.[8] The corresponding ϕ_F° values are shown in Fig. 4.

The potential P in Fig. 1 at which passivity of iron initiates (passivating potential) approximates but is not the same as the Flade potential because of IR drop through the insulating layer first formed and because the pH of the electrolyte at the base of pores in this layer differs from that in the bulk of solution (concentration polarization). These effects are absent on decay of passivity.

BEHAVIOR OF PASSIVATORS (See also pp. 257–265)

The same Flade potential is obtained whether iron is passivated by concentrated nitric acid or is anodically polarized in sulfuric acid, indicating that the passive film is essentially the same in both instances. In fact, when iron is passivated by immersion in solutions of chromates (CrO_4^{--}), nitrites (NO_2^-), molybdates (MoO_4^{--}), tungstates (WO_4^{--}), ferrates (FeO_4^{--}), or pertechnetates (TcO_4^-),[9, 10] (called passivators) the corresponding Flade potentials are also close to values obtained other-

---

[8] H. Rocha and G. Lennartz, *Arch. Eisenhüttenw.* **26**, 117 (1955). A longer decay time for passivity of Cr-Ni-Fe alloys with increasing chromium content and less positive Flade potential, ϕ_F, was demonstrated by H. Feller and H. Uhlig, *J. Electrochem. Soc.*, **107**, 864 (1960).

[9] G. Cartledge and R. Simpson, *J. Phys. Chem.*, **61**, 973 (1957).

[10] H. Uhlig and P. King, *J. Electrochem. Soc.*, **106**, 1 (1959).

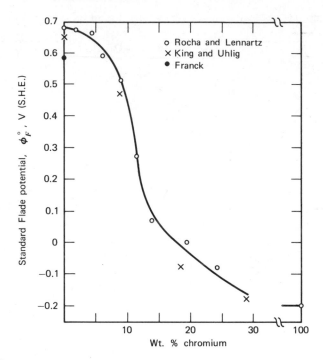

Fig. 4. Standard Flade potentials for chromium-iron alloys and for chromium.

wise.* Hence it can be concluded that the passive film on iron is essentially the same regardless of the passivation process. The amount of passive film substance as determined by coulometric and other type measurements is approximately 0.01 C/cm² in all cases, and it also appears that passivation proceeds (with some exceptions discussed later) by an electrochemical mechanism. For example, passivators are reduced at cathodic areas at a current density equivalent to a true current density at anodic areas equaling or exceeding i_{critical} (10–20 A/cm²) for passivation of iron. The passivator is reduced over a large cathodic area of the metal surface to an extent not less than that necessary to form a chemically equivalent passive film at small residual anodic areas. The small passive areas, in turn, adsorb passivator thereby becoming noble

* The standard Flade potential of iron passivated by chromates ($\phi_F^\circ = 0.54$ V) is less noble than for iron passivated by HNO₃ ($\phi_F^\circ = 0.63$ V). The explanation is proposed[10] that chromates adsorb on the passive film more strongly than do nitrates, thereby reducing the overall free energy of the system and increasing the stability of the passive film. Other passivators presumably adsorb similarly, but with differing energies of adsorption.

to adjoining passive or nonpassive areas, causing passivity to spread. When the passive film is complete, it acts as cathode over its entirety and further reduction of passivator proceeds at a much slower rate, equivalent to the rate of continuous passive film breakdown, or to $i_{passive}$. Since breakdown is accelerated by presence of chlorides and by elevated temperatures, consumption of passivator is also increased correspondingly.

Passivators as a group are inorganic oxidizing agents that have the unique property of reacting only slowly when in direct contact with iron, but which are reduced more rapidly by cathodic currents. For this reason they can adsorb first on the metal surface, each site of adsorption adding to the cathodic area. The higher the concentration of passivator, the more readily it adsorbs, and the smaller become the residual anodic areas; this situation obviously favors increased anodic polarization and ultimately passivation. It apparently requires about 0.5 to 2 hr for the passive film to form completely when iron is immersed in 0.1% K_2CrO_4, the shorter time being characteristic of aerated solution and the longer time of deaerated solution.†

Passivation of Iron by HNO₃

In nitric acid, the cathodic depolarizer (passivator) is nitrous acid (HNO_2).[5] This must form first in sufficient quantity by an initial rapid reaction of iron with HNO_3. As nitrous acid accumulates, anodic current densities increase, eventually reaching $i_{critical}$. Passivity thereupon is achieved and the corrosion rate falls to the comparatively low value of about 20 mdd.[13]

If urea is added to concentrated HNO_3, passivation is interrupted because urea reacts with nitrous acid in accord with

$$(NH_2)_2CO + 2HNO_2 \rightarrow 2N_2 + CO_2 + 3H_2O \qquad (4)$$

thereby decreasing the nitrous acid concentration. The rate of reaction is nevertheless sufficiently below that of HNO_2 production so that passivity can still occur, although periodic breakdown and formation of the passive film usually result.

Hydrogen peroxide added to concentrated nitric acid also causes periodic breakdown and formation of passivity, probably by oxidizing

† Based on potential-time measurements[11] and on reaction-rate studies of iron, previously exposed to chromate solution, with concentrated HNO_3.[12]

[11] R. M. Bruns, *J. Appl. Phys.*, **8**, 398 (1937).
[12] H. Gatos and H. Uhlig, *J. Electrochem. Soc.*, **99**, 250 (1952).
[13] H. Uhlig and T. O'Connor, *J. Electrochem. Soc.*, **102**, 562 (1955).

HNO_2 to HNO_3.[13]   The peroxide of itself is not as efficient a cathodic depolarizer as HNO_2, and hence the passive film in its presence can repair itself only when the momentary surface concentration of HNO_2 formed by reaction of iron with HNO_3 is sufficiently high.   After passivity is achieved, the surface concentration of HNO_2 is diminished by reaction with peroxide below that needed to maintain passivity, whereupon the cycle is repeated.

If iron is first immersed in dilute chromate solution for a sufficient time (several minutes), it remains passive in concentrated nitric acid without the initial reaction to form HNO_2.   The passive film is preformed by chromate, and nitrous acid is no longer necessary as depolarizer to reach $i_{critical}$, but is needed only in concentration sufficient to maintain the film already present.

ANODIC PROTECTION AND TRANSPASSIVITY

The electrochemical nature of the passivation process explains why anodic polarization using applied current, or increase of either the cathodic area or cathodic reaction rate (which increases polarization of remaining anodic areas) favors both formation and retention of passivity.   High-carbon steels, for example, which contain areas of cementite (Fe_3C)

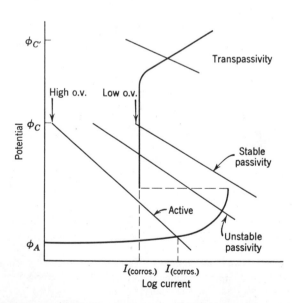

Fig. 5.   Polarization diagram for metal that is either active or passive, depending on overvoltage of cathodic areas (differing cathodic reaction rates).

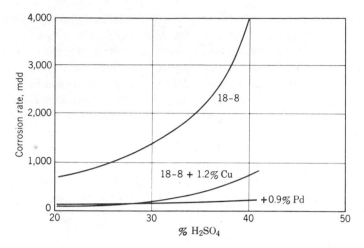

Fig. 6. Corrosion rates in sulfuric acid of 18-8 stainless steel alloyed with copper or palladium; 360-hr test, 20°C (Tomashov).

acting as cathodes, are more readily passivated by concentrated nitric acid than is pure iron. For this reason, nitrating mixtures of sulfuric and nitric acids are best stored or shipped in drums of steel of the highest carbon content consistent with required mechanical properties. Similarly, stainless steels that may lose passivity in dilute sulfuric acid retain their corrosion resistance if alloyed with small amounts of more noble constituents of low hydrogen overvoltage, or of low overvoltage for cathodic reduction of dissolved oxygen (e.g., Pd, Pt, or Cu).[14] The polarization diagram corresponding to increased passivity by low overvoltage cathodes, and the corresponding improved corrosion resistance to sulfuric acid are shown in Figs. 5 and 6.

Because titanium, unlike 18-8 stainless steel, has a low critical current density for passivity in chlorides as well as in sulfates, passivity in boiling 10% HCl is made possible by alloying titanium with 0.1% Pd or Pt.[15] The pure metal, on the other hand, corrodes in the same acid at very high rates (see Fig. 1, p. 363).

Employing the same underlying principle, the corrosion resistance of many metals and alloys (e.g., Cr-Fe alloys in H_2SO_4) can be greatly improved by an impressed *anodic* current initially equal to or greater than the critical current for passivity. The potential of the metal moves into the passive region (Fig. 1) and the final current density and accom-

[14] N. Tomashov, *Corrosion*, **14**, 299*t* (1958).
[15] M. Stern and H. Wissenberg, *J. Electrochem. Soc.*, **106**, 759 (1959).

panying corrosion rate correspond to the low value of $i_{passive}$. This process is called *anodic protection* (see p. 227), because the current flow is in opposite direction to that which is used in cathodic protection. However, whereas cathodic protection can, in principle, be applied to both passive and nonpassive metals, anodic protection is applicable only to metals that can be passivated when anodically polarized (Def. 1).

If the cathodic polarization curves of Fig. 5 intersect the anodic curve at a still more noble potential, the corrosion rate of, for example stainless steel, is greatly increased over the corrosion rate at less noble potentials within the passive region, and the corrosion products are $Cr_2O_7^{--}$ and Fe^{+3}. This situation is called *transpassivity*. It occurs not only with stainless steels, but also with chromium for which the potential for the reaction

$$2Cr + 7H_2O \rightarrow Cr_2O_7^{--} + 14H^+ + 12e^- \qquad E^\circ = -0.30 \text{ V} \qquad (5)$$

is less noble than for the oxygen evolution reaction

$$2H_2O \rightarrow O_2 + 4H^+ + 4e^- \qquad E^\circ = -1.23 \text{ V} \qquad (6)$$

Appreciable corrosion in the transpassive region does not occur for iron in sulfuric acid, (oxygen evolution is the primary reaction), but increased corrosion does occur in alkaline solutions that favor formation of ferrate (FeO_4^{--}). Transpassivity accounts for an observed increase of corrosion rate with time for 18-8 stainless steels in boiling concentrated nitric acid in which corrosion products accumulate, in particular $Cr_2O_7^{--}$, which move the corrosion potential into the transpassive region.

THEORIES OF PASSIVITY

There are two commonly expressed points of view regarding the nature of the passive film. The first holds that the passive film (Def. 1 or 2) is always a diffusion-barrier layer of reaction products, e.g., metal oxide or other compound, which separates metal from its environment and which slows down the rate of reaction. This is sometimes referred to as the oxide-film theory.

The second holds that metals passive by Def. 1 are covered by a chemisorbed film, e.g., of oxygen. Such a layer displaces the normally adsorbed H_2O molecules and slows down the rate of anodic dissolution involving hydration of metal ions. Or expressed in another way, adsorbed oxygen decreases the exchange current density (increases anodic overvoltage) corresponding to the reaction $M \rightleftarrows M^{+z} + ze^-$. Even less than a monolayer on the surface is observed to have a passivating effect[16, 17], hence it is suggested that the film cannot act primarily as a diffusion barrier layer. This second point of view is called the "adsorption theory of passivity."

There is no question on either viewpoint that a diffusion barrier film accounts for passivity of many metals passive only by Def. 2. A visible lead sulfate film on lead immersed in H_2SO_4, or an iron fluoride film on steel immersed in aqueous HF are examples of protective films that successfully isolate the metal from its environment. But for metals within Def. 1, based on marked anodic polarization, the films are usually invisible and are even too thin on stainless steels or on chromium, for example, to be detected by high-energy electron diffraction.* Metals and alloys in this category have been the source of extended debate and discussion on the mechanism of passivity over the past 125 years. Main support of the viewpoint that a reaction-product film is always the source of passivity has come from isolating thin oxide films from passive iron accomplished, for example, by immersing the metal in iodine-aqueous potassium iodide, or iodine-methyl alcohol solutions[18, 19] and floating the detached film to the surface of the solution for examination and analysis. Electron diffraction patterns of films isolated from iron previously passivated in aerated sodium hydroxide solution or in chromates have disclosed the presence of γFe_2O_3.[20, 21]

The adsorption theory, on the other hand, derives support from the fact that most of the metals falling within Def. 1 are transition metals of the periodic table (i.e., they contain electron vacancies or uncoupled electrons in the d shells of the atom). The uncoupled electrons account for strong bond formation with components of the environment, especially O_2, which also contains uncoupled electrons (hence its slight paramagnetic susceptibility) resulting in electron-pair or covalent bonding supplementary to ionic bonding. Transition metals, furthermore, have high heats of sublimation compared to nontransition metals, a property which favors adsorption of the environment because metal atoms tend

---

* This is true when the metals are air-exposed after etching. If they are abraded, local high temperatures generated at the surface produce a detectable oxide, but this is not the passive film. Low-energy electron diffraction techniques can be used to detect adsorbed films and hence show greater promise in the study of surface films responsible for passivity.

[16] B. Kabanov, R. Burstein, and A. Frumkin, *Discussions Faraday Soc.*, **1**, 259 (1947).

[17] R. Frankenthal, *J. Electrochem. Soc.*, **114**, 542 (1967).

[18] U. R. Evans, *J. Chem. Soc.*, **1927**, p. 1024.

[19] W. Vernon, F. Wormwell, and T. Nurse, *J. Chem. Soc.*, **1939**, p. 621.

[20] J. Mayne and M. Pryor, *J. Chem. Soc.*, **1949**, p. 1831.

[21] J. Mayne, J. Menter, and M. Pryor, *Ibid.*, **1950**, p. 3229.

to remain in their lattice, whereas oxide formation requires metal atoms to leave their lattice. The characteristic high energies for adsorption of oxygen on transition metals correspond to chemical bond formation, and hence such films are called *chemisorbed* in contrast to lower energy films, which are called *physically adsorbed*. On the nontransition metals (e.g., copper and zinc), oxides tend to form immediately, and any chemisorbed films on the metal surface are short lived. On transition metals, the life of chemisorbed species is much longer. Multilayer chemisorbed passive films, however, react in time with the underlying metal to form compounds such as oxides, but such oxides are less important in accounting for passivity than the chemisorbed films that form initially and continue to form on metal exposed at pores in the oxide.

The adsorption theory emphasizes that the observed Flade potential of passive iron is too noble by about 0.6 V to be explained by any of the known iron oxides in equilibrium with iron. The noble value has been explained by some investigators on the assumption that a high potential gradient (10^6 V/cm) exists in the supposed oxide,[22] or that a film of γFe_2O_3 is in equilibrium with an underlying Fe_3O_4 layer.[23]

On the other hand, observed values of the Flade potential are consistent with a chemisorbed film of oxygen on the surface of iron, the corresponding potential of which is calculated (see Prob. 2, Ch. 5, p. 380) making use of the observed heat and estimated entropy of adsorption of oxygen on iron in accord with the reaction.[24]

$$3H_2O + Fe \text{ surface} \rightarrow O_2{\cdot}O_{ads. \, on \, Fe} + 6H^+ + 6e^-,$$
$$E^{\circ}_{calc'd} = -0.56 \text{ V} \quad (7)$$

The measured chemical equivalents of passive film substance (about 0.01 C/cm²) correspond (roughness factor = 4) to one atomic layer of oxygen atoms ($r = 0.7$ Å) over which one layer of oxygen molecules ($r = 1.2$ Å) is chemisorbed—hence the adsorbed passive film is represented above by $O_2{\cdot}O_{ads. \, on \, Fe}$. The value of the Flade potential, as cited in (1), corresponds to $E^{\circ}_{observed} = -0.63$ V, which is in reasonable accord with the calculated value, -0.56 V. The agreement is still better if account is taken of adsorbed water displaced from the metal surface by the passive film during the passivation reaction, this displacement presumably involving a greater free-energy change than the adsorption of water on the passive film itself.[25]

Further confirming evidence that the passive film on iron contains

[22] K. Vetter, *Z. Elehtrochem.,* **62**, 642 (1958).
[23] H. Göhr and E. Lange, *Z. Elektrochem.,* **61**, 1291 (1957).
[24] H. H. Uhlig, *Z. Elektrochem.,* **62**, 626 (1958).
[25] P. King and H. Uhlig, *J. Phys. Chem.,* 63, 2026 (1959).

oxygen in a higher energy state than corresponds to any iron oxide is obtained from the ability of the passive film to oxidize chromite (CrO_2^-) to chromate (CrO_4^{--}) in $NaOH$ solution.[13] This oxidation does not occur with active iron. The maximum oxidizing capacity corresponds to 0.012 C/cm^2 of passive film substance in accord with the reaction

$$O_2 \cdot O_{ads.\,on\,Fe} + Fe + OH^- + CrO_2^- + H_2O \rightarrow$$
$$Fe(OH)_3 + CrO_4^{--} \qquad \Delta G° = -33,000 \text{ cal.} \quad (8)$$

Calculations show that the same reaction will not go for chromium or the stainless steels on which the passive film is more stable.

On breakdown of passivity, 0.01 C/cm^2 of adsorbed oxygen on iron reacts with underlying metal in accord with

$$O_2 \cdot O_{ads.\,on\,Fe} + 2Fe + 3H_2O \rightarrow 2Fe(OH)_3 \text{ or } Fe_2O_3 \cdot 3H_2O \qquad (9)$$

A layer of Fe_2O_3 (mol. wt 159.7; d. = 5.12) is formed equal to a minimum of $(0.01 \times 159.7)/(6 \times 96,500 \times 5.12) = 54$ Å thick (based on apparent area). The hydrated oxide would be thicker. This value compares in magnitude with measured values for the thickness of the decomposed passive film (25–100 Å). It is the decomposed passive film that is presumably isolated in experiments designed to remove the passive film from iron.

According to the adsorption theory, passivity of chromium and the stainless steels, because of their pronounced affinity for oxygen, can occur by direct chemisorption of oxygen from the air or from aqueous solutions, and the equivalents of oxygen so adsorbed are found[26] to be of the same order of magnitude as the equivalents of passive film formed on iron when passivated either anodically, by concentrated nitric acid, or by exposure to chromates. Similarly, oxygen in air can adsorb directly on iron and passivate it in aerated alkaline solutions, or in near-neutral solutions if the partial pressure of oxygen is increased sufficiently.

More Stable Passive Films with Time

It was observed by Flade[6] that the passive film on iron remains stable in sulfuric acid for a longer time the longer the iron remains beforehand in concentrated nitric acid. In other words, the film is stabilized by continued exposure to the passivating environment. Frankenthal[17] noted similarly that although less than monolayer quantities of oxygen (measured coulometrically) suffice to fully passivate 24% Cr-Fe in $1N$

[26] H. Uhlig and S. Lord, *J. Electrochem. Soc.*, **100**, 216 (1953).

H_2SO_4, the film thickened and became more resistant to cathodic reduction when the alloy was maintained for some time at potentials noble to the passivating potential (P in Fig. 1). It is probable that the observed stabilizing effect is the result of positively charged metal ions entering the adsorbed layers of negatively charged oxygen ions and molecules, the coexisting opposite charges tending to stabilize the adsorbed film. Low-energy electron diffraction data for nickel single crystals,[27] for example, indicate that the first-formed adsorbed film consists of a regular array of oxygen and nickel ions located in the same approximate plane of the surface. This initial adsorbed layer is found to be more stable thermally than is the oxide, NiO. At increasing oxygen pressures, several adsorbed layers, probably consisting of O_2, form on top of the first layer and result in an amorphous film. It is likely that additional metal ions in time succeed in entering such a film, particularly in the noble potential range, becoming a mobile species within the adsorbed oxygen layer. Protons from the aqueous environment are probably also incorporated. Okamoto and Shibata,[28] for example, showed that the passive film on 18-8 stainless steel contains H_2O; similar results have been reported for passive iron.[29] Eventually the stoichiometric oxide is nucleated at favorable sites of the metal surface and such nuclei then grow laterally to form a uniform oxide film; but the adsorbed (passive) film remains intact at pores in the oxide. A schematic structure of the first-formed adsorbed passive film is shown in

O Metal

● Oxygen

Fig. 7. Schematic illustration of chemisorbed atomic and molecular oxygen making up the passive film.

Fig. 7.[30] This initial passive film, therefore, presumably grows into a multilayer adsorbed structure, which can be considered to be an amorphous nonstoichiometric oxide. It differs markedly in properties from the stoichiometric oxide into which it eventually converts.

[27] A. MacRae, *Surface Sci.,* **1,** 319 (1964); *Science,* **139,** 379 (1963).

[28] G. Okamoto and T. Shibata, *Nature,* **206,** 1350 (1965); G. Okamoto et al., *Proc. 2nd Congr. Metallic Corrosion,* p. 558, Nat. Assoc. Corros. Engrs., Houston, Texas, 1966.

[29] K. Kudo, et al., *Corros. Sci.,* **8,** 809 (1968); P. Yolken et al., *Ibid.,* **8,** 103 (1968).

[30] H. Uhlig, *Corros. Sci.,* **7,** 325 (1967).

Action of Chloride Ions, Passive-Active Cells

*[handwritten: increasing exchange i
decreasing overvoltage]*

Chloride ions, and to a lesser extent other halogen ions, break down passivity or prevent its formation in iron, chromium, nickel, cobalt, and the stainless steels. From the point of view of the oxide film theory, Cl^- penetrates the oxide film through pores or defects easier than do other ions (e.g., SO_4^{--}). Or Cl^- may colloidally disperse the oxide film and increase its permeability.

On the other hand, according to the adsorption theory, Cl^- adsorbs on the metal surface in competition with dissolved O_2 or OH^-.[16] Once in contact with the metal surface, Cl^- favors hydration of metal ions and increases the ease with which metal ions enter into solution, opposite to the effect of adsorbed oxygen, which decreases the rate of metal dissolution. In other words, adsorbed chloride ions increase the exchange current (decrease overvoltage) for anodic dissolution of the above-mentioned metals over the value prevailing when oxygen covers the surface. The effect in this regard is so pronounced that iron and the stainless steels are not readily passivated anodically in solutions containing an appreciable concentration of Cl^-. Instead, the metal continues to dissolve at high rates both in the active and passive potential ranges.

Breakdown of passivity by Cl^- occurs locally rather than generally over the passive surface, the preferred sites being determined perhaps by small variations in the passive film structure and thickness. Minute anodes of active metal are formed surrounded by large cathodic areas of passive metal. The potential difference between such areas is 0.5 V or more, and the resulting cell is called a *passive-active cell*. High current densities at the anode cause high rates of metal penetration, accompanied by cathodic protection of the metal area immediately surrounding the anode. This fixes the anode in place, and results in corrosion of the pitting type. Also, the greater the current flow and cathodic protection at any pit, the less likely it is that another pit will initiate in its neighborhood. Hence the observed number of deep pits per unit area is usually less than the density of smaller shallow pits. It is evident, because of the possibility of passive-active cell formation, that deep pitting is much more common with passive metals than with nonpassive metals.

Halogen ions have less effect on the anodic behavior of titanium, tantalum, molybdenum, tungsten, and zirconium, and their passivity may continue in media of high chloride concentration in contrast to iron, chromium, and Fe-Cr alloys which lose passivity. This is sometimes explained by formation of insoluble protective Ti-, Ta-, Mo-, etc., basic chloride films. However, the true situation is probably related to the high affinity of these metals for oxygen, making it more difficult for Cl⁻ to displace oxygen of the passive film, in accord with the noble critical potentials of these metals above which, but not below, pitting is initiated, if pitting occurs at all.

CRITICAL PITTING POTENTIAL

When iron is anodically polarized potentiostatically in $1N$ H_2SO_4 to which sodium chloride is added ($>3 \times 10^{-4}$ mole/liter), the apparent transpassive region is shifted to lower potentials. However, instead of oxygen being evolved, the metal corrodes locally with formation of visible pits.[31] Similarly, if 18-8 stainless steel is anodically polarized in $0.1N$ NaCl within the lower passive region, the alloy remains passive, analogous to its behavior in Na_2SO_4 solution, no matter how long the time. But above a critical potential, there is a rapid increase in the current, accompanied by random formation of pits (Fig. 8). The critical potential for 18-8 moves toward less noble values as the Cl⁻ concentration increases, and toward more noble values with increasing pH and with lower temperatures.[32] The critical potential is also moved in the noble direction by adding extraneous salts to the NaCl solution, e.g., Na_2SO_4,

[31] H. Engell and N. Stolica, *Arch. Eisenhüttenw.*, **30**, 239 (1959).
[32] H. Leckie and H. Uhlig, *J. Electrochem. Soc.*, **113**, 1262 (1966).

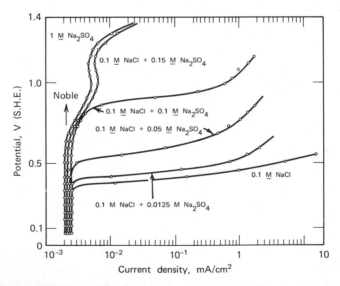

Fig. 8. Potentiostatic polarization curves for 18-8 stainless steel in $0.1M$ NaCl showing increasingly noble values of the critical potential with additions of Na_2SO_4, 25°C.

NaNO$_3$, NaClO$_4$ (Fig. 8). If sufficient extraneous salt is added to shift the critical potential to a value more noble than the prevailing corrosion potential, pitting does not occur at all on exposure of the stainless steel to the aqueous salt mixture. The added salts under these conditions become effective inhibitors. Addition of 3% NaNO$_3$ to 10% FeCl$_3$, for example, has prevented pitting of 18-8 stainless steel or any appreciable weight loss for a period of more than 25 years, whereas in the absence of NaNO$_3$ the alloy is observed to corrode seriously by pitting within a matter of hours.[33] The 3% NaNO$_3$ addition moves the critical pitting potential to a value more noble than the open-circuit cathode potential or the oxidation-reduction potential for $Fe^{++} \leftrightarrows Fe^{3+} + e^-$. Along the same lines, Leckie[34] polarized 18-8 stainless steel at a constant 0.1 V below the critical potential in 0.1N NaCl for 14 weeks without observable pitting.

Critical potentials of several metals in 0.1N NaCl are listed in Table 1, values of which were obtained from anodic polarization curves. Most of the data were obtained allowing 5 min or more at a given potential and observing whether the resultant current increases or decreases in

TABLE 1
Critical Pitting Potentials in 0.1N NaCl, 25°C

	Potential (V) S.H.E.	Reference
Al*	−0.37	35
Ni	0.28	36
Zr	0.46	37
18-8 Stainless steel†	0.26	36
30% Cr-Fe	0.62	36
12% Cr-Fe	0.20	36
Cr	>1.0	38
Ti	>1.0 (1N NaCl)	39
	≃1.0 (1N NaCl; 200°C)	

* $\phi_{critical}$, (V) s.h.e. $= -0.124 \log (Cl^-) - 0.504$ where (Cl^-) is activity of Cl^-.
† $\phi_{critical}$ (V) s.h.e. $= -0.088 \log (Cl^-) + 0.168$.

[33] H. Uhlig and J. Gilman, *Corrosion*, **19**, 261t (1963).
[34] H. Leckie, *J. Electrochem. Soc.*, **117**, 1152 (1970).
[35] H. Kaesche, *Z. Physik. Chem.*, N.F., **34**, 87 (1962); H. Böhni and H. Uhlig, *J. Electrochem. Soc.*, **116**, 906 (1969).
[36] J. Horvath and H. Uhlig, *J. Electrochem. Soc.*, **115**, 791 (1968).
[37] Y. Kolotyrkin, *Corrosion*, **19**, 261t (1963).
[38] N. Greene, C. Bishop, and M. Stern, *J. Electrochem. Soc.*, **108**, 836 (1961).
[39] C. Hall, Jr., and N. Hackerman, *J. Phys. Chem.*, **57**, 262 (1953); F. Posey and E. Bohlmann, *2nd Symp. Fresh Water from the Sea*, Athens, Greece, May, 1967.

time. The critical potential is then the most noble potential for which the current decreases or remains constant; it is usually confirmed by holding the potential at the critical value for 12 hr or more and observing absence of pits under a low-power microscope.

It is observed that an increase in the chromium content of stainless steels—and, to a lesser extent, an increase of nickel content—shifts the critical potential to more noble values (corresponding to increased resitance to pitting).[36, 37] The noble critical potentials for chromium and titanium (more noble than the oxygen electrode potential in air (0.8V) in accord with $2H_2O \leftrightarrows O_2(0.2 \text{ atm}) + 4H^+(10^{-7}) + 4e^-$ indicate that these metals are not expected to undergo pitting corrosion in aerated saline media at normal temperatures. At elevated temperatures and high Cl^- concentrations, however, the critical potentials become more active so that corrosion pitting of titanium, for example, is observed in concentrated hot $CaCl_2$ solutions despite its immunity to pitting in seawater.

The critical potential has been explained, from one point of view, as that value needed to build up an electrostatic field within the passive or oxide film sufficient to induce Cl^- penetration to the metal surface.[40] Other anions may also penetrate the oxide depending on size and charge, contaminating the oxide and thereby making it a better ionic conductor favoring oxide growth. Eventually either the oxide is undermined by condensation of migrating vacancies, or cations of the oxide undergo dissolution at the electrolyte interface; in both cases, pitting results. The induction period preceding pitting is related to the time required for the supposed penetration of Cl^- through the oxide film.

Alternatively, the critical potential is explained in terms of competitive adsorption of Cl^- with oxygen of the passive film (adsorption theory).[32, 37] The metal has typically greater affinity for oxygen than for Cl^-, but as the potential is made more noble, the concentration of Cl^- ions at the metal surface increases, eventually reaching a value that allows Cl^- to displace adsorbed oxygen. The observed induction period is the time required for successful competitive adsorption at favored sites of the metal surface. Adsorbed Cl^-, compared to adsorbed oxygen, as described earlier, results in lower anodic overvoltage for metal dissolution, which accounts for a higher rate of corrosion at any site where the exchange has taken place. Extraneous anions (e.g., NO_3 or SO_4^{--}) that do not break down the passive film or cause pitting, compete with Cl^- for sites on the passive surface, making it necessary to shift the

[40] T. Hoar, D. Mears, and G. Rothwell, *Corros. Sci.*, **5**, 279 (1965).

potential to a still more noble value in order to increase Cl^- concentration sufficient for successful exchange with adsorbed oxygen.

Below the critical potential, Cl^- cannot displace adsorbed oxygen so long as the passive film remains intact, and hence pitting does not occur. Should passivity break down because of factors other than those described, (e.g., reduced oxygen or depolarizer concentration at a crevice, or cathodic polarization of local shielded areas), pitting could then initiate independent of whether the overall prevailing potential is above or below the critical value. But under conditions of uniform passivity for all the metal surface, application of cathodic protection to avoid pitting corrosion need only shift the potential of the metal below the critical value. This is in contrast to the usual procedure of cathodic protection, which necessitates polarization of a metal to the much more active open-circuit anode potential.

The relation between minimum anion activity necessary to inhibit pitting of 18-8 stainless steel, aluminum, and probably many other passive metals in a solution of given Cl^- activity, follows the relation $\log (Cl^-) = k \log (\text{anion}) + \text{const.}$ The latter expression can be derived assuming that ions adsorb competitively in accord with the Freundlich adsorption isotherm.[33]

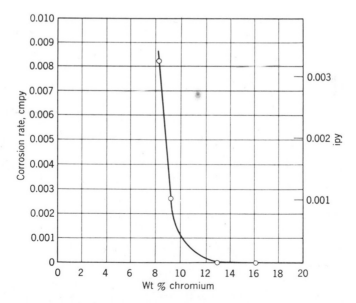

Fig. 9. Corrosion rates of chromium-iron alloys in intermittent water spray, room temperature [W. Whitman and E. Chappell, *Ind. Eng. Chem.*, **18,** 533 (1926)].

PASSIVITY IN ALLOYS

Several metals (e.g., chromium) are naturally passive when exposed to the atmosphere and they remain bright and tarnish-free for years, in contrast to iron or copper which corrode or tarnish in short time. It is found that the passive property of chromium is conferred on alloys of Cr-Fe provided that 12 wt.% Cr or more is present. Ironbase alloys containing a minimum of 12% Cr are known as the stainless steels. Typical corrosion, potential, and critical-current density behavior of Cr-Fe

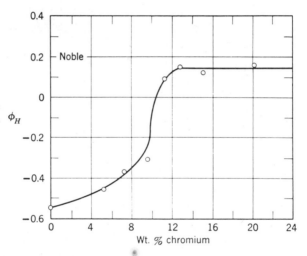

Fig. 10. Potentials of chromium-iron alloys in 4% NaCl [H. Uhlig, N. Carr, and P. Schneider, *Trans. Electrochem. Soc.*, **79**, 111 (1941)].

alloys are shown in Figs. 9 to 11. Note in Fig. 11 that $i_{critical}$ for passivation of Cr-Fe alloys at pH 7 reaches a minimum at about 12% Cr in the order of 2 $\mu A/cm^2$. This value is so low that corrosion currents in aerated aqueous media easily achieve or exceed this value, illustrating why >12% Cr-Fe alloys are self-passivating. In addition, the passive film becomes more stable with chromium content of the alloy as was discussed earlier.

Several other alloy systems exhibit critical compositions for passivity, as was first described by Tammann.[41] Examples of approximate critical

[41] G. Tammann, *Z. Anorg. Chem.*, **169**, 151 (1928); (with E. Sotter), *Ibid.*, **127**, 257 (1923).

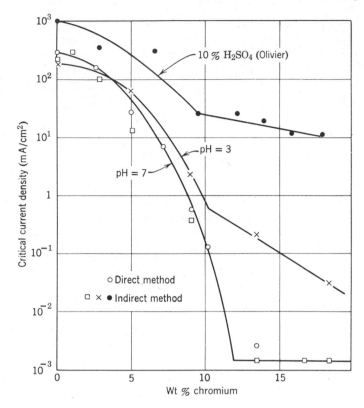

Fig. 11. Critical current densities for passivation of chromium-iron alloys in deaerated 3% Na_2SO_4 at pH 3 and 7, 25°C [P. King and H. Uhlig, *J. Phys. Chem.*, **63**, 2026 (1959). Data in 10% H_2SO_4, room temperature, by R. Olivier, *Int. Committee Electrochem. Therm. Kinet.*, 6th Meeting Poitiers, p. 314, Butterworths, London (1955)].*

compositions determined from plots of $i_{critical}$ versus alloy composition are 35% Ni-Cu, 15% Mo-Ni, 8% Cr-Co, and 14% Cr-Ni. Critical alloy compositions for passivity have also been observed in three or four component alloys (e.g. Fe-Cr-Ni-Mo,[42] Fe-Ni-Mo, and Cr-Ni-Fe[43]).

Critical compositions based on $i_{critical}$ values are not sensitive to the

---

* Critical current densities by the direct method are obtained from potentiostatic polarization curves or similar measurements. The indirect method makes use of the relation $i - i_{critical} = K/t$, where t is the time at current density i to reach the critical passive potential, and K is the number of coulombs expended in reaching the passive potential. Both methods lead to the same value of $i_{critical}$.

[42] H. Uhlig, *Trans. Electrochem. Soc.*, **85**, 307 (1944).
[43] H. Feller and H. Uhlig, *J. Electrochem. Soc.*, **107**, 864 (1960).

pH of the electrolyte in which the polarization data are measured (Fig. 11). On the other hand, critical compositions based on corrosion data usually vary with the environment to which the alloys are exposed. In 33% HNO_3, for example, the critical chromium concentration for passivity of Cr-Fe alloys is decreased to 7% Cr, whereas in $FeSO_4$ solution it is increased to 20% Cr; only in aqueous solutions of about pH 7 is the critical composition equal to 12%. This effect of environment is to be expected because whether an alloy becomes passive depends on whether the corrosion rate equals or exceeds the specific $i_{critical}$ for passivity in the given environment. As seen in Fig. 11, increasing chromium content reduces $i_{critical}$, but decreasing pH (and increasing temperature) have the opposite effect. The corrosion rate in turn depends on factors of metal and environment affecting both the rate of cathodic and anodic reactions.

It can be shown from first principles[44] that $i_{critical} = K(H^+)^\lambda$ where K and λ are constants whose values depend on the anion. The cathodic reaction rate in an aerated solution when controlled by oxygen reduction, on the other hand, is given by the diffusion current i_{diff} corresponding to the reaction $\frac{1}{2}O_2 + H_2O \rightarrow 2OH^- - 2e^-$. For an unstirred air-saturated solution, it can be shown that $i_{diff} = 0.039$ A/cm². Hence a critical pH exists for each alloy composition at which $i_{critical} = i_{diff}$ and above which, but not below, passivity is stable. The observed critical pH values, for example, of 18% Cr, 8% Ni and of 12% Cr stainless steels in $0.1M$ Na_2SO_4[45] are 1.4 and 5.0, respectively, in close agreement with the calculated values. With iron, for which $i_{critical}$ is much higher than for the stainless steels, the critical pH is approximately 10.

The structure of the passive film on alloys, as with passive films in general, has been described both by the oxide film theory and by the adsorption theory. It has been suggested that protective oxide films form above the critical alloy composition for passivity, but nonprotective oxide films form below the critical composition. The preferential oxidation of passive constituents (e.g., chromium) may form protective oxides (e.g., Cr_2O_3) above a specific alloy content but not below. No quantitative predictions have been offered based on this point of view.

By the adsorption theory, it is considered that, in the presence of water, oxygen chemisorbs on chromium-iron alloys above the critical composition corresponding to passivity, but immediately reacts below the critical composition to form a nonprotective oxide or other type film. Whether the alloy favors a chemisorbed or reaction product film depends on the

[44] H. H. Uhlig, *J. Electrochem. Soc.*, **108**, 327 (1961).
[45] H. Leckie, *Corrosion*, **24**, 70 (1968).

electron configuration of the alloy surface, in particular on *d*-electron interaction. The so-called *electron configuration theory* describes the specific alloying proportions corresponding to a favorable *d*-electron configuration accompanying chemisorption and passivity. It suggests the nature of electron interaction that determines which alloyed component dominates in conferring its chemical properties on those of the alloy; for example, why the properties of nickel dominate above those of copper at >30–40% Ni in the nickel-copper alloys.

Nickel-Copper Alloys

Nickel, containing 0.6 *d*-electron vacancy per atom (as measured magnetically), when alloyed with copper, a nontransition metal containing no *d*-electron vacancies, confers passivity on the alloy above approximately 30 to 40 at.% Ni. Initiation of passivity beginning at this composition is indicated by corrosion rates in sodium chloride solution (Figs. 12 and 13), corrosion pitting behavior in seawater (Fig. 13), and, more quantitatively, by measured values of $i_{critical}$ and $i_{passive}$ (Fig.

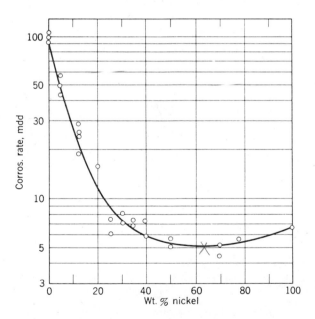

Fig. 12. Corrosion rates of copper-nickel alloys in aerated 3% NaCl, 80°C, 48-hr tests.

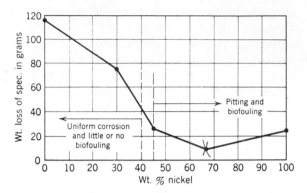

Fig. 13. Behavior of copper-nickel alloys in seawater (F. LaQue, *J. Amer. Soc. Nav. Engrs.*, **53,** 29 (1941)].

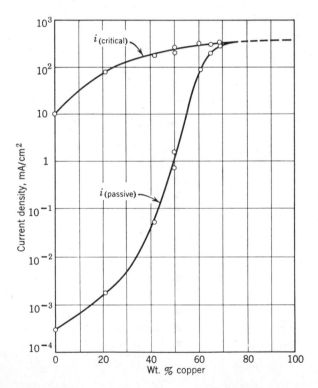

Fig. 14. Values of critical and passive current densities obtained from potentiostatic anodic polarization curves for copper-nickel alloys in 1N H_2SO_4, 25°C (Osterwald and Uhlig).

14),[46-48] or by decay (Flade) potentials (Fig. 15)[49] in $1N$ H_2SO_4.

Corrosion pitting in seawater is observed largely above 40% Ni because pit growth is favored by passive-active cells (see p. 75) and such cells can operate only when the alloy is passive in the high nickel composition range. Practically, this distinction is observed in the specification of materials for seawater condenser tubes in which pitting attack must

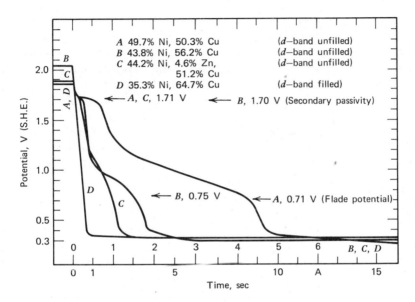

Fig. 15. Potential decay curves for alloys of nickel-copper and nickel-copper-zinc in $1N$ H_2SO_4, 25°C (two time scales). Pure copper behaves like alloy *D*. (Mansfeld and Uhlig[49]).

be rigorously avoided. The cupro-nickel alloys are used (10–30% Ni) but not Monel (70% Ni-Cu).

Similarly, marine fouling organisms are much less successful in establishing themselves on the surface of nonpassive nickel-copper compositions because the latter corrode uniformly at rates that release enough

[46] H. Uhlig, *Z. Elektrochem.*, **62**, 700 (1958).
[47] J. Osterwald and H. Uhlig, *J. Electrochem. Soc.*, **108**, 515 (1961).
[48] F. Mansfeld and H. Uhlig, *J. Electrochem. Soc.*, **117**, 427 (1970).
[49] F. Mansfeld and H. Uhlig, *Corros. Sci.*, **9**, 377 (1969).

Cu^{++} to poison fouling organisms.* But on passive nickel-copper compositions, for which the overall corrosion rate is much less, fouling organisms in general can gain a foothold and flourish (Fig. 13).

Potentiostatic anodic polarization behavior of nickel-copper alloys in $1N$ H_2SO_4 (Fig. 14) establishes that the passive current density largely disappears above 60% Cu and vanishes completely at about 70% Cu. Polarization curves of alloys containing >70% Cu or <30% Ni resemble those of pure copper, hence such alloys are no longer passive. Potential decay curves (Fig. 15) confirm that a passive film is formed on anodically passivated alloys containing >40% Ni, but not otherwise. In other words, alloys containing copper above the critical composition lose their transition-metal characteristics; that is, they no longer contain d-electron vacancies. In this connection, the magnetic saturation moment, which is also a function of d-electron vacancies in the alloy, becomes zero at >60% Cu. This observation is interpreted as a filling of vacancies in the d band of electron energy levels of nickel by electrons donated by copper. Physicists have made the assumption that if copper and nickel atoms are considered to be alike, except that copper contains one more electron per atom than nickel, then the 0.6 d-electron vacancy per nickel atom is expected, as observed, to be just filled by electrons from copper at 60 at. % Cu.[50, 51]

One can also start with the alternative assumption that the two atoms maintain their individuality and that the vacancies per atom of nickel are a function of alloy concentration.[52] In the gaseous state, nickel has the configuration $3d^84s^2$ corresponding to two d-electron vacancies, or to the equivalent of two uncoupled d-electrons in the third shell of the atom. [The maximum number of d-electrons that can be accommodated is 10, corresponding to copper: $(3d^{10}4s)$]. In the process of condensing to a solid and forming the metallic bond, the uncoupled electrons of any single nickel atom tend to couple with uncoupled electrons of neighboring atoms. This results in a smaller number of electron vacancies in the solid compared to the gas, accounting for the measured 0.6 vacancy or 0.6 uncoupled electron per nickel atom. If we assume

---

* The minimum concentration of Cu^{++} required to poison marine organisms corresponds to a corrosion rate for copper of about 0.001 ipy or 5 mdd. [F. LaQue and W. Clapp, *Trans. Electrochem. Soc.*, **87**, 103 (1945)].

[50] N. Mott and H. Jones, *The Theory of the Properties of Metals and Alloys*, pp. 196–198, Oxford University Press, 1936.

[51] C. Kittel, *Introduction to Solid State Physics*, 3rd Ed., p. 581, Wiley, New York, 1968.

[52] H. H. Uhlig, presented at 3rd Int. Symp. Passivity, Cambridge, England, July, 1970. *Electrochim. Acta*, In press.

that the intercoupling of d-electrons increases with proximity of nickel atoms in the alloy and is a linear function of nickel concentration, then the vacancies per nickel atom can be set equal to $2 - (2 - 0.6)$ at. % Ni/100, corresponding to 2 vacancies for 0% Ni and 0.6 vacancy for 100% Ni. The alloy loses its transition-metal characteristics beginning at the composition for which the total number of d vacancies equals the total number of donor electrons (1 per copper atom) or: at. % Ni $(2 - 0.014$ at. % Ni$) = 1 \times$ at. % Cu. Setting at. % Cu = $(100 -$ at. % Ni) and solving, the critical composition is found to be 41 at. % Ni. This value corresponds closely to the observed value derived from magnetic saturation data.

This model was checked by alloying small amounts of other nontransition elements Y, or transition elements Z, with nickel-copper alloys and noting the specific compositions at which i_{critical} and i_{passive} merged, or Flade potentials disappeared. Nontransition-metal additions of valence >1 should shift the critical composition for passivity to higher percentages of nickel, whereas transition-metal additions should have the opposite effect. For example, one zinc atom of valence 2 or one aluminum atom of valence 3 should be equivalent in the solid solution alloy to two or three copper atoms, respectively. This has been confirmed experimentally.[53, 54] The relevant equations become

$$\text{at. % Ni } (2 - 0.014 \text{ at. % Ni}) = 1 \times \text{ at. % Cu} + n \text{ at. % Y}$$

$$\text{at. % Ni } (2 - 0.014 \text{ at. % Ni}) + v \text{ at. % Z} = 1 \times \text{ at. % Cu}$$

where n is the number of electrons donated per atom of Y and v is the number of vacancies introduced per atom of Z.

By plotting $1 \times$ at. % Cu $-$ at. % Ni $(2 - 0.014$ at. % Ni$)$ with n at. % Y, a straight line is predicted of unit negative slope for nontransition-element additions. If plotted instead with v at. % Z for transition-metal additions, a unit positive slope should result. A plot for the alloying additions so far studied are shown by data summarized in Fig. 16.[52] In order for the line to pass through the origin with unit slope, it was necessary to assume approximately 80% instead of 100% donation of valence electrons. This means that valence electrons from copper and from other nontransition elements are presumably donated in large part to nickel, but not entirely. Assuming 0.8 electron donor per copper atom in binary nickel-copper alloys, the critical nickel composition below which the d band is filled becomes 35 at. % instead

[53] N. Stolica and H. Uhlig, *J. Electrochem. Soc.*, **110**, 1215 (1963).
[54] F. Mansfeld and H. Uhlig, *J. Electrochem. Soc.*, **115**, 900 (1968).

of 41 at. % as calculated earlier.* This value is consistent with the composition at which i_{passive} and i_{critical} intersect in Fig. 14. Whether 80% donation applies only to the electron interaction of surface metal atoms on which passive films form is not yet clarified.

Values of n for germanium, aluminum, and zinc shown in Fig. 16 are 4, 3, and 2, respectively, in accord with their normal valence. For gal-

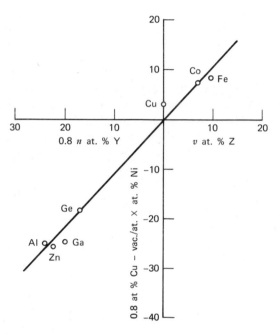

Fig. 16. Plot of excess electrons or d-electron vacancies in nickel-copper alloys at their critical compositions versus electron vacancies or electrons donated by alloying additions unit slope.

lium, a valence of 2 is in better accord with the other data, reflecting perhaps, the known tendency of gallium to form chemical compounds having a valence lower than 3, but other explanations are also possible. For iron and cobalt, the number of vacancies per atom is equated to $v_g - (v_g - v_s)$ at. % Z/100 where v_g and v_s are d-electron vacancies per atom in the gas and solid respectively. Values of v_g for iron and cobalt are 4 and 3, and for v_s they are 2.2 and 1.7, respectively.

*The calculated number of d vacancies/Ni atom at this composition equals $2 - 0.014 \times 35 = 1.51$ which is close to the value 1.6 assumed earlier, e.g., references 46, 53, 54.

There are no observed phase changes or major discontinuities in the thermodynamic properties of nickel-copper alloys at 60 to 70% Cu, whereas chemisorption on any metal is known to be favored by an unfilled *d*-band configuration.[55] The good agreement, therefore, between observed and predicted critical alloy compositions supports not only an effect of electron configuration on passivity but also an adsorbed structure of the passive film.

Other Alloys

Because present-day theory of the metallic state does not treat the situation, the electron configuration of alloys made up of two or more transition metals with relation to their passive behavior is not as well understood as for the copper-nickel system. Nevertheless, useful simplifying assumptions can be made. For example, the most passive component of an alloy is assumed to be the acceptor element, which tends to share electrons donated by the less passive components.

Accordingly, for stainless steels, the *d*-electron vacancies of chromium are assumed to fill with electrons from alloyed iron.[46] At the critical composition at which vacancies of Cr are apparently filled, which occurs for alloys containing less than 12% Cr, the corrosion behavior of the alloy is like that of iron. Above 12% Cr, the *d*-electron vacancies of chromium are unfilled, and the alloy behaves more like chromium.

The critical compositions for passivity in the Cr-Ni and Cr-Co alloys equal to 14% Cr and 8% Cr, respectively, can also be related to the contribution of electrons from nickel or cobalt to the unfilled *d* band of chromium.[46, 56] In the ternary Cr-Ni-Fe solid solution system, electrons are donated to chromium mostly by nickel above 50% Ni but by iron at lower nickel compositions.[57] Similarly molybdenum alloys retain in large part the useful corrosion resistance of molybdenum (e.g., to chlorides) so long as the *d* band of energy levels for molybdenum remains unfilled. In type 316 stainless steel (18% Cr, 10% Ni, 2–3% Mo), for example, the weight ratio of Mo/Ni is best maintained at or above 15/85 corresponding to the observed critical ratio for passivity in the binary molybdenum-nickel alloys equal to 15 wt. % Mo.[46, 58] At this ratio or above, passive properties imparted by molybdenum appear to be optimum.

[55] D. Hayward and B. Trapnell, *Chemisorption,* p. 8, Butterworth's, Washington, D. C., 1964.

[56] A. Bond and H. Uhlig, *J. Electrochem. Soc.,* **107,** 488 (1960).

[57] H. Feller and H. Uhlig, *J. Electrochem. Soc.,* **107,** 864 (1960).

[58] H. Uhlig, P. Bond, and H. Feller, *J. Electrochem. Soc.,* **110,** 650 (1963).

EFFECT OF CATHODIC POLARIZATION AND CATALYSIS

When chromium, stainless steels, and passive iron are cathodically polarized, passivity is destroyed. This occurs by reduction of the passive oxide or adsorbed oxygen film, according to whichever viewpoint of passivity is adopted. In addition, according to the adsorption theory, hydrogen ions discharged on transition metals tend to enter the metal as hydrogen atoms. Hydrogen so dissolved is partly dissociated into protons and electrons, the latter being available to fill vacancies in the d band of the metal. A transition metal containing sufficient hydrogen, therefore, is no longer able to chemisorb oxygen or become passive because the d band is filled.

Often catalytic properties also depend on ability of a metal or alloy to chemisorb certain components of its environment. It is not surprising, therefore, that transition metals are typically good catalysts and that the principles of electron configuration in alloys favoring catalytic activity are similar to those favoring passivity. For example, when palladium, which contains 0.6 d-electron vacancy per atom in the metallic state, is charged with hydrogen cathodically, it loses its efficiency as a catalyst for the ortho-para hydrogen conversion.[59] This behavior is explained by filling of the d band by electrons of dissolved hydrogen, as a result of which the metal no longer chemisorbs hydrogen. Correspondingly, the catalytic efficiencies of palladium-gold alloys resemble

[59] A. Couper and D. Eley, *Discussions Faraday Soc.*, **8**, 172 (1950).

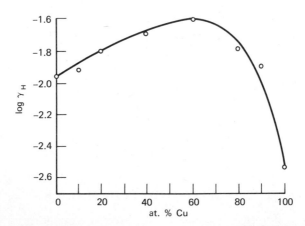

Fig. 17. Catalytic efficiency of nickel-copper alloys for $2H \rightarrow H_2$ as a function of alloy composition (efficiency increases with values of γ_H); (Hardy and Linnett).

palladium until the critical concentration of 60 at. % Au is reached, at which composition and above the alloys become poor catalysts. Gold, a nontransition metal, supplies electrons to the unfilled *d* band of palladium, and magnetic measurements confirm that the *d* band is just filled at the critical bold concentration.

The catalytic effect of copper-nickel alloys as a function of composition for the reaction $2H \to H_2$ is shown in Figure 17.[60] Above 60 at. % Cu, the filled *d* band is less favorable to hydrogen adsorption and hence favorable collisions of gaseous H with adsorbed H are less probable, and the reaction rate falls off. The similarity to passive behavior of copper-nickel alloys, which also decreases above 60 at. % Cu, can be noted. The parallel conditions affecting passivity and catalytic activity lend further support to the viewpoint that the passive films on transition metals and their alloys are chemisorbed.

GENERAL REFERENCES

Papers presented at 1st. and 2nd. Int. Symp. on Passivity: *Z. Elektrochem.*, **62**, 619–827 (1958) ; *J. Electrochem. Soc.*, **110**, 596–708 (1963).
Corrosion Handbook, edited by H. H. Uhlig, pp. 20–23, Wiley, New York, 1948.
N. Tomashov and G. Chernova, *Passivity and Protection of Metals Against Corrosion,* Plenum Press, New York, 1967.

---

[60] W. Hardy and J. Linnett, *Trans. Faraday Soc.*, **66**, 447 (1970).

chapter 6

Iron and steel

The electrochemical theory of corrosion as described earlier relates corrosion to a network of short-circuited galvanic cells on the metal surface. Metal ions go into solution at anodic areas in amount chemically equivalent to the reaction at cathodic areas. At anodic areas the following reaction takes place:

$$Fe \rightarrow Fe^{++} + 2e^- \tag{1}$$

This reaction is rapid in most media, as shown by lack of pronounced polarization when iron is made anode employing an external current. When iron corrodes, the rate is usually controlled by the cathodic reaction, which in general is much slower (cathodic control). In deaerated solutions, the cathodic reaction is

$$H^+ \rightarrow \tfrac{1}{2}H_2 - e^- \tag{2}$$

This reaction proceeds rapidly in acids,* but only slowly in alkaline or neutral aqueous media. The corrosion rate of iron in deaerated water at room temperature, for example, is less than 1 mdd. The rate of hydrogen evolution at a specific pH depends on presence or absence of low hydrogen overvoltage impurities in the metal. For pure iron, the metal surface itself may provide sites for H_2 evolution, hence high purity iron continues to corrode in acids but at a measurably lower rate than does commercial iron.

---

* With lowering of pH, the equilibrium potential of the hydrogen electrode becomes more noble, hence $\phi_c - \phi_A$ becomes larger and the resulting corrosion current increases.

The cathodic reaction can be accelerated by dissolved oxygen in accord with the following reaction, a process called *depolarization:*

$$2H^+ + \tfrac{1}{2}O_2 \rightarrow H_2O - 2e^- \tag{3}$$

Dissolved oxygen reacts with hydrogen atoms adsorbed at random on the iron surface, independent of the presence or absence of impurities in the metal. The oxidation reaction proceeds as rapidly as oxygen reaches the metal surface.

Adding (1) and (3), making use of the reaction $H_2O \rightarrow H^+ + OH^-$

$$Fe + H_2O + \tfrac{1}{2}O_2 \rightarrow Fe(OH)_2 \tag{4}$$

Hydrous ferrous oxide ($FeO \cdot nH_2O$) or ferrous hydroxide ($Fe(OH)_2$) composes the diffusion barrier layer next to the iron surface through which O_2 must diffuse. The pH of saturated $Fe(OH)_2$ is about 9.5, so that the surface of iron corroding in aerated pure water is always alkaline. The color of $Fe(OH)_2$, although white when the substance is pure, is normally green to greenish black because of incipient oxidation by air. At the outer surface of the oxide film, access to dissolved oxygen converts ferrous oxide to hydrous ferric oxide or ferric hydroxide, in accord with

$$Fe(OH)_2 + \tfrac{1}{2}H_2O + \tfrac{1}{4}O_2 \rightarrow Fe(OH)_3 \tag{5}$$

Hydrous ferric oxide is orange to red brown in color and comprises most of ordinary rust. It exists as nonmagnetic αFe_2O_3 (hematite) or as magnetic γFe_2O_3, the α form having the greater negative free energy of formation (greater stability). Saturated $Fe(OH)_3$ is nearly neutral in pH. A magnetic hydrous ferrous ferrite, $Fe_3O_4 \cdot nH_2O$, often forms a black intermediate layer between hydrous Fe_2O_3 and FeO. Hence rust films normally consist of three layers of iron oxides in different states of oxidation.

AQUEOUS ENVIRONMENTS

Effect of Dissolved Oxygen

AIR-SATURATED WATER

At ordinary temperatures in neutral or near-neutral water, dissolved oxygen is necessary for appreciable corrosion of iron. In air-saturated water, the initial corrosion rate may reach a value of about 100 mdd. This rate diminishes over a period of days as the iron oxide (rust) film is formed and acts as a barrier to oxygen diffusion. The steady-state corrosion rate may be 10 to 25 mdd, tending to be higher the greater the relative motion of water with respect to iron. Since the diffusion rate at steady state is proportional to oxygen concentration, it follows from (3) that the corrosion rate of iron is also proportional to oxygen concentration. Typical data are shown in Fig. 1. In the

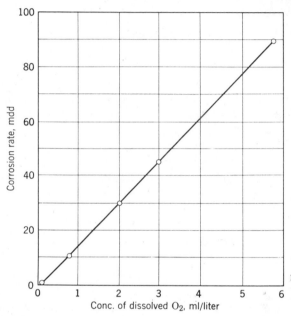

Fig. 1. Effect of oxygen concentration on corrosion of mild steel in slowly moving water containing 165 ppm $CaCl_2$, 48-hr test, 25°C (Uhlig, Triadis, and Stern).

absence of dissolved oxygen, the corrosion rate at room temperature is negligible both for pure iron and for steel.

HIGHER PARTIAL PRESSURES OF OXYGEN

Although increase in oxygen concentration at first accelerates corrosion of iron, it is found that beyond a critical concentration the corrosion rate drops again to a low value.[1] Deviations from the linear relation between corrosion rate and oxygen concentration occur sooner in distilled water than when Cl^- ions are present as in Fig. 1. In distilled water,

[1] H. Uhlig, D. Triadis, and M. Stern, *J. Electrochem. Soc.*, **102**, 59 (1955); see also, E. Groesbeck and L. Waldron, *Proc. Amer. Soc. Testing Mat.*, **31**, pt. II, 279–291 (1931).

the critical concentration of oxygen above which corrosion decreases again, is about 12 ml O_2/liter (Fig. 2). This value increases with dissolved salts and with temperature, and decreases with increase in velocity and pH. At pH of about 10, the critical oxygen concentration reaches the value for air-saturated water (6 ml O_2/liter) and is still less for more alkaline solutions.

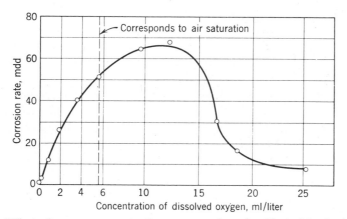

Fig. 2. Effect of oxygen concentration on corrosion of mild steel in slowly moving distilled water, 48-hr test, 25°C (Uhlig, Triadis, and Stern).

The decrease of corrosion rate is caused by passivation of iron by oxygen as shown by potentials of iron in air-saturated H_2O of −0.4 to −0.5 V (s.h.e.) and 0.1 to 0.4 V in oxygen-saturated H_2O (28 ml O_2/liter). Apparently at higher partial pressures, more oxygen reaches the metal surface than can be reduced by the corrosion reaction—the excess therefore is available to form the passive film.* On the basis of the oxide film theory of passivity, the excess oxygen supposedly oxidizes the ferrous oxide film to one having greater protective value as a diffusion barrier. According to the adsorption theory, the excess oxygen is available to chemisorb on the iron surface forming an adsorbed passive film.

Because passivity accompanies higher oxygen pressures, passive-active cells are established in the event passivity breaks down locally (e.g., at crevices). Such breakdown is accompanied by severe pitting, particu-

* Or looked at another way, increasing the cathodic reaction rate by higher oxygen concentration increases polarization of anodic areas until the critical current density for passivity is reached (see Fig. 1, p. 62). H. Uhlig, *J. Electrochem. Soc.*, **108,** 327 (1961).

larly at higher temperatures, in the presence of halide ions, or at a critical pressure of oxygen where passivity is on the verge of either forming or breaking down. This behavior limits the practical use of high partial oxygen pressures as a means for reducing corrosion of steel. In appreciable concentrations of chlorides (e.g., seawater) passivity of iron is not established at all, and in such media increased oxygen pressure results only in an increased corrosion rate.

ACTION OF ANAEROBIC BACTERIA

Although in deaerated water iron does not corrode appreciably, the corrosion rate in some natural environments is found to be abnormally high. These high rates have been traced to the presence of sulfate-reducing bacteria (*Sporovibrio desulfuricans*). Their relation to an observed accelerated corrosion rate in soils low in dissolved oxygen was first observed in Holland by von Wolzogen Kühr.[2] The bacteria are curved, measuring about 1×4 μ, and are found widespread in many waters and soils. They thrive only under conditions of poor or no aeration, and in the pH range of about 5.5 to 8.5. Certain varieties multiply in fresh waters and in soils containing sulfates, others in brackish waters and sea water, and still others are stated to exist in deep soils at temperatures as high as 60 to 80°C (140 to 175°F).

Sulfate-reducing bacteria easily reduce inorganic sulfates to sulfides in presence of hydrogen or organic matter, and are aided in this process by the presence of an iron surface. The aid which iron provides in this reduction is probably to supply hydrogen which is normally adsorbed on the metal surface, and which the bacteria make use of in the reduction of SO_4. For each equivalent of hydrogen atoms they consume, one equivalent of Fe^{++} enters solution to form rust and FeS. The bacteria, therefore, probably act essentially as depolarizers.

The reaction sequence can be outlined as follows:

Anode: $4Fe \rightarrow 4Fe^{++} + 8e^-$

Cathode: $8H_2O \rightarrow 8H_{ads.\ on\ Fe} + 8OH^- - 8e^-$

$8H_{ads} + Na_2SO_4 \xrightarrow{\text{bacteria}} 4H_2O + Na_2S$

$Na_2S + 2H_2CO_3 \rightarrow 2NaHCO_3 + H_2S$

Summary: $4Fe + 2H_2O + Na_2SO_4 + 2H_2CO_3 \rightarrow 3Fe(OH)_2 + FeS$
$+ 2NaHCO_3$

Note that ferrous hydroxide and ferrous sulfide are formed in the relation of 3 moles to 1 mole. Analysis of a given rust in which sulfate-reducing bacteria were active shows this approximate ratio of oxide

[2] C. von Wolzogen Kühr, *Water and Gas*, 7, 26, 277 (1923).

Bacteria → depolarizers

to sulfide. In the event corrosion was caused instead by dissolved hydrogen sulfide or soluble sulfides, it is not likely that a similar ratio of oxide to sulfide would be found, and furthermore the corrosion rate would be lower. Qualitatively, the action of sulfate-reducing bacteria as cause of corrosion in a water initially free of sulfides can be detected by adding a few drops of hydrochloric acid to the rust and noting the smell of hydrogen sulfide.

Severe damage by sulfate-reducing bacteria has been observed particularly in oil-well casing, buried pipelines, water-cooled rolling mills, or in pipe from deep water wells. A well water in the Middle West caused failure of 2-in. diameter galvanized water pipe within 2 years by action of sulfate-reducing bacteria, whereas municipal water employing similar wells, but which was chlorinated beforehand, was much less corrosive.

Chlorination is one way to reduce the damage caused by these bacteria, but it is apparently not always effective according to some reports. Aeration of water reduces activity of the bacteria since they are unable to thrive in presence of dissolved O_2. Addition of certain bacteriocides can also be beneficial. Streptomycin has no effect, but it is reported that tannates, potassium tellurite, cetyl pyridinium bromide, *o*-nitrophenol, or inorganic selenates are effective.[3, 4]

Effect of Temperature

When corrosion is controlled by diffusion of oxygen, the corrosion rate at a given oxygen concentration approximately doubles for every 30°C (55°F) rise in temperature.[5] In an open vessel, allowing dissolved oxygen to escape, the rate increases with temperature to about 80°C (175°F) and then falls to a very low value at the boiling point (Fig. 3). The falling off of corrosion above 80°C is related to a marked decrease of oxygen solubility in water as the temperature is raised, and this effect eventually overshadows the accelerating effect of temperature alone. In a closed system, on the other hand, oxygen cannot escape and the corrosion rate continues to increase with temperature until all the oxygen is consumed.

When corrosion is attended by hydrogen evolution, the rate increase is more than double for every 30°C rise in temperature. The rate for iron corroding in hydrochloric acid, for example, approximately doubles for every 10° rise in temperature.

[3] K. Butlin, W. Vernon and L. Whiskin, *Symposium on Corrosion of Buried Metals,* p. 29, Iron and Steel Inst., London, 1951.

[4] T. Farrer, L. Bick and F. Wormwell, *J. Appl. Chem.,* 3, pt. 2, 80 (1953).

[5] G. Skaperdas and H. Uhlig, *Ind. Eng. Chem.,* 34, 748 (1942).

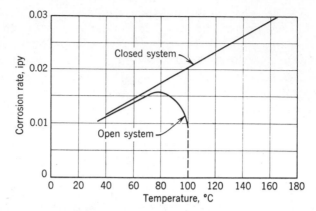

Fig. 3. Effect of temperature on corrosion of iron in water containing dissolved oxygen (*Corrosion, Causes and Prevention*, F. Speller, p. 168, McGraw-Hill, New York, 1951; with permission).

Effect of pH

The effect of pH of an aerated pure (or soft*) water on corrosion of iron at room temperature is shown in Fig. 4. Sodium hydroxide and hydrochloric acid were used to adjust pH in the alkaline and acid ranges.[6]

Within the range of about pH 4 to 10, the corrosion rate is independent of pH, and depends only on how rapidly oxygen diffuses to the metal surface. The major diffusion barrier of hydrous ferrous oxide is continuously renewed by the corrosion process. Regardless of the observed pH of water within this range, the surface of iron is always in contact with an alkaline solution of saturated hydrous ferrous oxide, the observed pH of which is about 9.5.

Within the acid region (pH < 4), the ferrous oxide film is dissolved, the surface pH falls, and iron is more or less in direct contact with the aqueous environment. The increased rate of reaction is then the sum of both an appreciable rate of hydrogen evolution and oxygen depolarization.

Above pH 10, increase in alkalinity of the environment raises the pH of the iron surface. The corrosion rate correspondingly decreases because iron becomes increasingly passive in presence of alkalies and dissolved oxygen, as explained under "Higher Partial Pressures of Oxy-

---

* As contrasted with a hard water, for which the effect of pH may be influenced by a protective $CaCO_3$ film. See p. 115.

[6] W. Whitman, R. Russell, and V. Altieri, *Ind. Eng. Chem.*, **16**, 665 (1924).

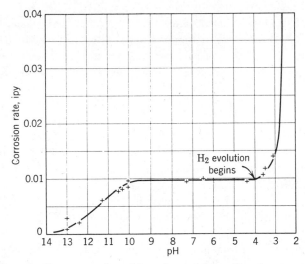

Fig. 4. Effect of pH on corrosion of iron in aerated soft water, room temperature (Whitman, Russell, and Altieri).

gen." Confirming the occurrence of passivity by Def. 1, Chapter 5, the potential of iron in water of pH < 10 changes from an active value of -0.4 to -0.5 V, to a noble value of 0.1 V in $1N$ NaOH, with an accompanying decrease in the corrosion rate. If the alkalinity is markedly increased, for example, to $16N$ NaOH (43%), passivity is disrupted and the potential achieves the very active value of -0.9 V. The corrosion rate correspondingly increases slightly to 0.0001 to 0.004 ipy (0.5–20 mdd), but that is still a relatively low rate. In this region, iron corrodes with formation of soluble sodium ferrite ($NaFeO_2$). In absence of dissolved oxygen, the reaction proceeds with hydrogen evolution forming sodium hypoferrite, Na_2FeO_2.[7] The fact that Fe^{++} is complexed by OH^- in strong alkalies to form FeO_2^- with accompanying reduction in activity of Fe^{++} accounts for the observed active potential of iron. Although the rate of formation of FeO_2^- in concentrated alkalies at room temperature is low, caused by marked polarization of probably both anodic and cathodic areas, the rate becomes excessively high at boiler temperatures.

As noted previously, in the region of pH 4 to 10 the corrosion rate depends only on the rate of diffusion of oxygen to the available cathodic surface. The extent of the cathodic surface is apparently not impor-

[7] R. Scholder, *Angew. Chem.*, **49**, 255 (1936).

tant.   This was made clear by an experiment of Whitman and Russell[8] who exposed to Cambridge tap water, steel specimens that were copper plated over three-quarters of the surface.   The total weight loss of these specimens compared to control specimens not plated was found to be the same.   All oxygen reaching the copper surface, acting as cathode, was reduced in accord with the reaction

$$H_2O + \tfrac{1}{2}O_2 \rightarrow 2OH^- - 2e^-$$

and an equivalent amount of Fe^{++} was formed at the iron surface acting as anode.*   In addition, all oxygen reaching cathodic areas of the iron surface itself produced an equivalent amount of Fe^{++}.   The total amount of iron corroding, therefore, was the same regardless of whether copper was plated over part of the specimens.   However, penetration of iron in the case of the plated specimens was four times that of the bare specimens.

It is obvious, therefore, that so long as oxygen diffusion is controlling, which is the case within pH 4 to 10, any small variation in composition of a steel and its heat treatment, or whether it is cold worked or annealed, has no bearing on corrosion properties.   Oxygen concentration, temperature, and velocity of the water alone determine the reaction rate.   These facts are important because almost all natural waters fall within the pH range 4 to 10.   This means that whether a high or low carbon steel, or similarly a low-alloy steel (e.g., 1–2% Ni, Mn, Mo, etc.), wrought iron, cast iron, or cold-rolled mild steel is exposed to fresh water or seawater, all the observed corrosion rates in a given environment are essentially the same.   Many laboratory and service data obtained with a variety of irons and steels support the validity of this conclusion.   A few typical data are summarized in Table 1.   These observations answer the once vociferous argument that wrought iron, for example, is supposedly more corrosion resistant than steel.

In the acid range (pH $<$ 4) oxygen diffusion is no longer controlling, and the corrosion reaction is established in part by the rate of hydrogen evolution.   The latter in turn depends on the hydrogen overvoltage of various impurities or phases present in specific steels or irons.   The rate becomes sufficiently high in this pH range to make anodic polarization a possible contributing factor (mixed control).   Because cementite (Fe_3C) is a phase of low hydrogen overvoltage, a low-carbon steel corrodes in acids at a lower rate than does a high-carbon steel.   Similarly,

---

* This equivalence is valid in waters of low conductivity only if anode and cathode areas are in close proximity.   In seawater, anodes and cathodes may be several feet apart.

[8] W. Whitman and R. Russell, *Ind. Eng. Chem.*, **16**, 276 (1924).

TABLE 1
Corrosion Rates of Various Steels When Oxygen Diffusion is Controlling

% Carbon	Treatment	Environ- ment	Temperature of Test	Corros- ion Rate, ipy
	Effect of Heat Treatment			
0.39	Cold drawn, annealed 500°C (930°F)	Distilled water	65°C (150°F)	0.0036
0.39	Normalized 20 min. at 900°C (1650°F)	Distilled water	65°C (150°F)	0.0034
0.39	Quenched 850°C (1560°F). Various specimens tempered 300°C (570°F) to 800°C (1470°F)	Distilled water	65°C (150°F)	0.0033
	Effect of Carbon			
0.05	Not stated	3% NaCl	Room	0.0014
0.11	Not stated	3% NaCl	Room	0.0015
0.32	Not stated	3% NaCl	Room	0.0016
	Effect of Alloying			
0.13	Not stated	Sea water		0.004
0.10, 0.34% Cu	Not stated	Sea water		0.005
0.06, 2.2% Ni	Not stated	Sea water		0.005
Wrought Iron	Not stated	Sea water		0.005

From *Corrosion, Causes and Prevention,* F. N. Speller, McGraw-Hill, New York, 1951; and *Corrosion Handbook,* edited by H. H. Uhlig, Wiley, New York, 1948.

heat treatment affecting the presence and size of cementite particles can appreciably alter the corrosion rate. Furthermore, cold-rolled steel corrodes more rapidly in acids than does an annealed or stressed-relieved steel because cold working produces finely dispersed low-overvoltage areas originating largely from interstitial nitrogen and carbon.

Since iron is not commonly used in strongly acid environments, the factors governing its corrosion in media of low pH are less important than those in natural waters and in soils. Nevertheless, there are certain applications where such factors must be considered, e.g., in steam-return lines containing carbonic acid, or in food cans where fruit and vegetable acids corrode the container with accompanying hydrogen evolution.

Less is known about the effect of impurities and metallurgical factors on the corrosion rate in the very alkaline region (pH $\approx$ 14) where corrosion is again accompanied by hydrogen evolution. In the passive region, approximately pH 10 to 13, the effect on passivity by impurities in their usual concentrations, or the effect of metallurgical factors, are not expected to be pronounced. In general any condition producing a large cathode-to-anode ratio facilitates achievement of passivity and increases the stability of the passive state once it is achieved.

CORROSION OF IRON IN ACIDS

We have seen that in strong acids (e.g., hydrochloric, sulfuric) the diffusion barrier oxide film on the surface of iron is dissolved below pH 4. In weaker acids (e.g., acetic or carbonic) dissolution of the oxide occurs at a higher pH, hence the corrosion rate of iron increases accompanied by hydrogen evolution at pH 5 or 6. This difference is explained[6] by the higher total acidity or neutralizing capacity of a partially dissociated acid compared with a totally dissociated acid at a given pH. In other words, at a given pH there is more available H^+ to react with and dissolve the barrier oxide film using a weak acid compared to a strong acid.

The increased corrosion rate of iron as pH decreases is not caused alone by increased hydrogen evolution; in fact, greater accessibility of oxygen to the metal surface on dissolution of the surface oxide favors oxygen depolarization which is often the more important reason. The sensitivity of the corrosion rate of iron or steel in nonoxidizing acids to dissolved oxygen concentration is shown by data of Table 2. In 6% acetic acid at room temperature the ratio of corrosion rate with oxygen present, to corrosion rate with oxygen absent, is 87. In oxidizing acids (e.g., nitric acid), which act as depolarizers and for which the

TABLE 2[9]
Effect of Dissolved Oxygen on Corrosion of Mild Steel in Acids

Corrosion Rate, ipy

Acid	Under O_2	Under H_2	Ratio
6% acetic	0.55	0.006	87
6% H_2SO_4	0.36	0.03	12
4% HCl	0.48	0.031	16
0.04% HCl	0.39	0.0055	71
1.2% HNO_2	1.82	1.57	1.2

[9] W. Whitman and R. Russell, *Ind. Eng. Chem.*, **17**, 348 (1925).

corrosion rate is therefore independent of dissolved oxygen, the ratio is almost unity. In general the ratios are larger the more dilute the acid. In more concentrated acids the rate of hydrogen evolution is so pronounced that oxygen has difficulty in reaching the metal surface. Hence depolarization in more concentrated acids contributes less to the overall corrosion rate than in dilute acids in which diffusion of oxygen is impeded to a lesser extent.

It is interesting that traces of oxygen in dilute H_2SO_4, or substantial amounts in more concentrated acid in which the corrosion rate is higher, inhibit the corrosion reaction. For zone-refined iron, the average rate in aerated $1.0N$ H_2SO_4, 25°C, was found to be 415 mdd, whereas in hydrogen-saturated acid the rate was 680 mdd.[10] Similar effects are shown by pure 9.2% Co-Fe alloy in $1.0N$ H_2SO_4 both aerated and deaerated, for which the corrosion rates are high and the diffusion of oxygen is impeded by rapid hydrogen evolution (Fig. 5).[10] It appears from poten-

[10] A. P. Bond, Sc.D. thesis, Dept. of Metallurgy, M.I.T., 1958.

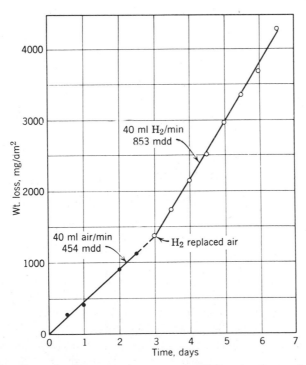

Fig. 5. Corrosion of 9.2% Co-Fe alloy in $1N$ H_2SO_4, showing an inhibiting effect of dissolved oxygen, 25°C (Bond).

tial and polarization measurements that oxygen in small concentrations at the metal surface increases cathodic polarization, thereby decreasing corrosion; in higher concentrations it acts mainly as a depolarizer, increasing the rate. The detailed mechanism of inhibition requires further study.

The important depolarizing action of dissolved oxygen suggests that the velocity of an acid should markedly affect the corrosion rate. This effect is observed, particularly with dilute acids for reasons previously stated (Figs. 6, 7). In addition, the inhibiting effect of dissolved oxygen

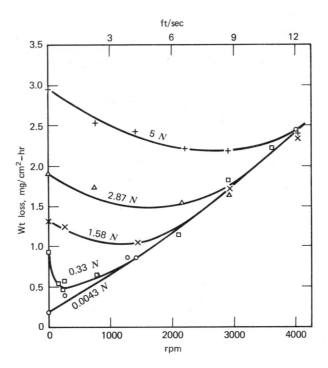

Fig. 6. Effect of velocity on corrosion of mild steel (0.12 % C) in sulfuric acid under air (23 ± 2°C, 45-min test; rotating spec, 0.7 in. diam, 2.19 in. long).[11]

is observed within a critical velocity range, the critical velocity becoming higher the more rapid the inherent reaction rate of steel with the acid. Relative motion of acid with respect to metal sweeps away hydrogen bubbles and reduces the thickness of the stagnant liquid layer at the metal surface, allowing more oxygen to reach the metal surface. Accordingly, the corrosion rate of steel in presence of air in $0.0043N$ H_2SO_4

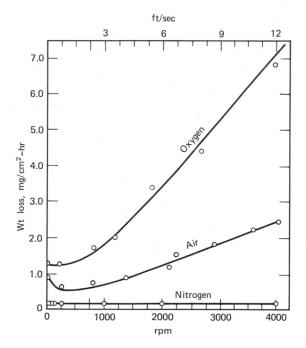

Fig. 7. Effect of velocity on corrosion of mild steel (0.12% C) in 0.33N sulfuric acid under air and oxygen.[11]   (Same conditions as for Fig. 6); 0.009% C steel under nitrogen.[12]

at a velocity of 12 ft/sec, is the same as that of 5N acid at the same velocity.   At rest, the ratio of corrosion rates is about 12.[11]   In absence of dissolved oxygen, only hydrogen evolution occurs at cathodic areas, and an effect of velocity is no longer observed (Fig. 7)[12].   This is expected because hydrogen overvoltage (activation polarization) is insensitive to velocity of the electrolyte.   The small decrease in reaction rate at low velocities reported in early measurements[11] was apparently caused by oxygen contamination of the nitrogen or hydrogen atmospheres used to deaerate the acids, or by the introduction of oxygen by the specimen itself.   At first the velocity increase in such instances brought sufficient oxygen or Fe^{+3} to the metal surface to inhibit corrosion, and later at higher velocities enough oxygen or Fe^{+3} to depolarize cathodic areas and slightly increase corrosion again.

For aerated acid, the minimum rate occurs at higher velocities the

[11] W. Whitman, R. Russell, C. Welling, and J. Cochrane, *Ind. Eng. Chem.*, **15**, 672 (1923).

[12] Z. Foroulis and H. Uhlig, *J. Electrochem. Soc.*, **111**, 13 (1964).

more concentrated the acid because rate of hydrogen evolution is more pronounced thereby impeding oxygen diffusion to the metal surface. Similarly, the minimum shifts to higher velocities the higher the carbon content of the steel because the corrosion rate and accompanying hydrogen evolution increase with carbon content.

Effect of Galvanic Coupling

The experiment of Whitman and Russell proving that weight loss of iron coupled to copper is the same as if all the surface had been iron, also showed that the actual penetration of iron increases when iron is coupled to a more noble metal. This experiment, therefore, provides information concerning the effect of coupling on the corrosion rate of the less noble component of a couple. For the situation where diffusion of a depolarizer is controlling, the general relation between penetration p (proportional to corrosion rate) of a metal having area A_a coupled to a more noble metal of area A_c, where p_0 is the normal penetration of the metal uncoupled, is given by

$$p = p_0 \left(1 + \frac{A_c}{A_a}\right) \tag{6}$$

If the ratio of areas A_c/A_a is large, the increased corrosion caused by coupling can be considerable. Conductivity of the electrolyte and geometry of the system enter the problem because only that part of the cathode area is effective for which resistance between anode and cathode is not a controlling factor. In soft tap water, the critical distance between copper and iron may be 0.5 cm; in seawater it may be several decimeters. The critical distance is greater the larger the potential difference between anode and cathode. All more noble metals accelerate corrosion similarly, except when a surface film (e.g., on lead) acts as a barrier to diffusion of oxygen or when the metal is a poor catalyst for reduction of oxygen.

In the case of coupled metals exposed to deaerated solutions for which corrosion is accompanied by hydrogen evolution, increased area of the more noble metal also increases corrosion of the less noble metal. Figure 8 shows polarization curves for an anode which polarizes little in comparison to a cathode at which hydrogen is evolved (cathodic control). Slope 1 represents polarization of a noble metal area having high hydrogen overvoltage. Slopes 2 and 3 represent metals with lower hydrogen overvoltage. The corresponding galvanic currents are given by projecting the intersection of anode-cathode polarization curves to the log I axis. Note that in general any metal on which hydrogen discharges

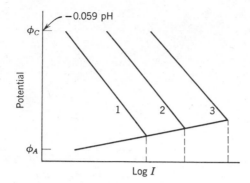

Fig. 8. Effect of hydrogen overvoltage of cathode on galvanic corrosion in deaerated nonoxidizing acids.

acts as a hydrogen electrode with an equilibrium potential at 1 atm hydrogen pressure of -0.059 pH volt. When a corroding metal is coupled to a more noble metal of variable area, the situation is shown in Fig. 9 where log current density is plotted instead of log total current. If the anode of area A_a is coupled to more noble metal of area A_c, then the galvanic current density at the anode produced by coupling can be shown to be

$$\log i_{\text{galv}} = \frac{-\phi_{\text{corros}} - 0.059 \text{ pH}}{\beta} + \log \frac{A_c}{A_a} i_0 \qquad (7)$$

where ϕ_{corros} is the corrosion potential (S.H.E.) of the couple (measured at a large distance compared to dimensions of the couple), and β and i_0 are the Tafel constants for hydrogen ion discharge on the noble metal. The increased penetration at the anode is given by $p - p_0 = i_{\text{galv}}/k$. If we assume that the anode is polarized very little as a result of coupling, we can set $(-\phi_{\text{corros}} - 0.059 \text{ pH})/\beta = \log K$ and then approximately:

$$p - p_0 = \frac{K}{k} \frac{A_c}{A_a} i_0 \qquad (8)$$

From this equation it appears that, other than the proportional effect of area ratio A_c/A_a, the corrosion rate of the less noble metal of a couple increases with increase in exchange current density i_0 for the more noble metal and with decrease in β (corresponding to low hydrogen overvoltage). For iron as anode in the acid range, $p - p_0$ would be almost the same for any value of pH because the potential ϕ_A changes with pH almost the same and in the same direction as does the hydrogen electrode.[13]

[13] H. H. Uhlig, *Proc. Nat. Acad. Sci. (U. S.)*, **40**, 276 (1954).

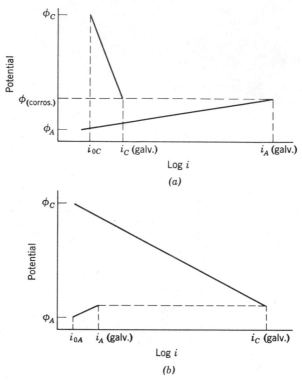

Fig. 9. Effect of anode-cathode area ratio on corrosion of galvanic couples in deaerated nonoxidizing acids. (a) Large cathode coupled to small anode. (b) Large anode coupled to small cathode.

Effect of Velocity on Corrosion in Natural Waters, Cavitation-Erosion

In natural fresh waters, the pH is usually too high for hydrogen evolution to play an important role, and relative motion of the water at first increases the corrosion rate by bringing more oxygen to the surface. At sufficiently high velocities, enough oxygen may reach the surface to cause partial passivity. If this happens, the rate decreases again after the initial increase (Fig. 10). Should the velocity increase still further, the mechanical erosion of passive or corrosion-product films again increases the rate. The maximum in rate preceding passivity comes at a velocity that varies with smoothness of the metal surface and with impurities in the water. In presence of high concentration of Cl⁻, as in seawater, passivity is not established at any velocity and the corrosion rate increases without an observed decrease at some inter-

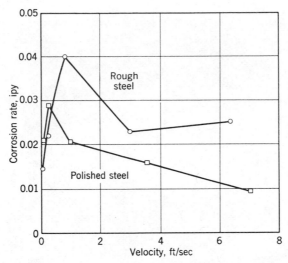

Fig. 10. Effect of velocity on corrosion of mild steel tubes containing Cambridge water, 21°C, 48-hr tests [Russell, Chappell, and White, *Ind. Eng. Chem.,* **19,** 65 (1927)].

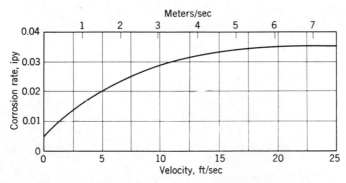

Fig. 11. Effect of velocity on corrosion of steel in seawater (F. LaQue, in *Corrosion Handbook,* p. 391).

mediate velocity (Fig. 11). The same behavior is expected at elevated temperatures that preclude the possibility of passivity by dissolved oxygen.

CAVITATION-EROSION

If conditions of velocity are such that repetitive low (below atmospheric) and high-pressure areas are developed, bubbles form and collapse at the metal-liquid interface. This phenomenon is called cavita-

Fig. 12. Cavitation-erosion damage to cylinder liner of diesel engine (Speller and LaQue).

tion. The damage to a metal caused by cavitation is called cavitation-erosion, or cavitation damage. The damage can be reproduced in the laboratory using metal probes undergoing forced high-frequency mechanical oscillations normal to the metal surface. The surface becomes deeply pitted and appears to be spongy (Fig. 12). Damage from this cause may be purely mechanical, as is experienced with glass or plastics or when damage to a metal occurs in organic liquids. The damage, however, may involve both chemical and mechanical factors, particularly if protective films are destroyed, allowing corrosion to proceed at a higher rate. But even in such instances, the mechanical factor of damage is apparent from the distorted layer of metals grains at the immediate surface, as if the metal had been severely cold worked (Fig. 3, p. 599, *Corrosion Handbook*). On the other hand, the part played by chemical factors is apparent, for example, from increased metal loss in laboratory tests employing seawater compared to fresh water. It has been suggested that localized corrosion fatigue may account for higher metal loss in the presence of corrosive environments.

Cavitation-erosion occurs typically on rotors of pumps, on the trailing faces of propellers and of water turbine blades, and on the water-cooled side of diesel engine cylinders.[14] Damage can be reduced by operating rotary pumps at the highest possible head of pressure in order to avoid bubble formation. For turbine blades, aeration of water serves to cushion the damage caused by collapse of bubbles. Also, Neoprene, rubber, or similar elastomer coatings on metals are reasonably resistant to damage from this cause. Since 18-8 stainless steel is one of the relatively resistant alloys (Fig. 13), it is used as a facing for water turbine blades. To reduce damage of diesel engine cylinder liners, addition of 2000 ppm sodium chromate to the cooling water has proved effective.[15]

Effect of Dissolved Salts

The effect of sodium chloride concentration on corrosion of iron in air-saturated water at room temperature is shown in Fig. 14. The corrosion rate first increases with salt concentration, then decreases, the value falling below that for distilled water when saturation is reached (26% NaCl).

Since oxygen depolarization controls the rate throughout the sodium

[14] F. Speller and F. LaQue, *Corrosion*, **6**, 209 (1950).
[15] W. Leith and A. Thompson, *Trans. Amer. Soc. Mech. Eng., J. Basic Engrs.*, p. 795, (December, 1960).

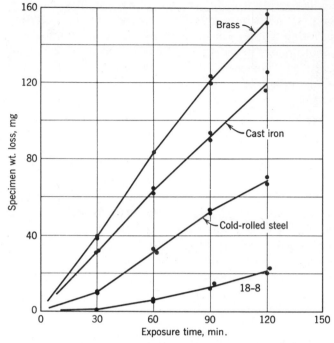

Fig. 13. Resistance of metals to cavitation-erosion in laboratory tests, Cambridge water, room temperature [Schumb, Peters, and Milligan, *Metals and Alloys*, p. 3 (May, 1937)].

chloride concentration range, it is of some interest to understand why the rate first increases, reaching a maximum at about 3% NaCl (sea-water concentration) and then decreases. Oxygen solubility in water decreases continuously with sodium chloride concentration, explaining the falling off of corrosion rate at the higher sodium chloride concentrations. The initial rise appears to be related to a change in the protective nature of the diffusion barrier rust film that forms on corroding iron. In distilled water having low conductivity, anodes and cathodes must be located relatively near each other. Consequently, OH$^-$ ions forming at cathodes in accordance with modified (3),

$$\tfrac{1}{2}O_2 + H_2O \rightarrow 2OH^- - 2e^- \tag{9}$$

are always in the proximity of Fe^{++} ions forming at nearby anodes, resulting in a film of Fe(OH)$_2$ adjacent to and adherent to the metal surface. This provides an effective diffusion-barrier film.

In sodium chloride solutions, on the other hand, the conductivity is

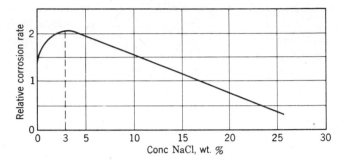

Fig. 14. Effect of sodium chloride concentration on corrosion of iron in aerated solutions, room temperature (composite of data, several investigations).

greater, hence additional anodes and cathodes can operate much further removed one from the other. At such cathodes, NaOH does not react immediately with $FeCl_2$ formed at anodes; instead, these substances diffuse into the solution and react to form $Fe(OH)_2$ away from the metal surface. Obviously, any $Fe(OH)_2$ so formed does not provide a protective barrier layer on the metal surface. Hence iron corrodes more rapidly in dilute sodium chloride because more dissolved oxygen can reach cathodic areas. Above 3% NaCl, the continuing decreased solubility of oxygen becomes more important than any change in the diffusion barrier layer, hence corrosion decreases.

Alkali metal salts (e.g., KCl, LiCl, Na_2SO_4, KI, NaBr, etc.) affect the corrosion rate of iron and steel in approximately the same manner as sodium chloride. Chlorides appear to be slightly more corrosive in the order Li, Na, and K.[16] Alkaline earth salts (e.g., $CaCl_2$, $SrCl_2$) are slightly less corrosive than alkali metal salts. Nitrates appear to be slightly less corrosive than chlorides or sulfates at low concentrations (0.1–0.25 N) but not necessarily at higher concentrations.[17] The small differences for all these solutions may arise, for example, from their specific effect on the $Fe(OH)_2$ diffussion-barrier layer, or perhaps from the different adsorptive properties of the ions at a metal surface giving rise to differing anode-cathode area ratios or differing overvoltage characteristics for oxygen reduction.

Acid salts, which are salts that hydrolyze to form acid solutions, cause corrosion with combined hydrogen evolution and oxygen depolarization at a rate paralleling that of the corresponding acids at the same pH

[16] C. Borgmann, *Ind. Eng. Chem.,* **29,** 814 (1937).
[17] O. Bauer, O. Kröhnke, and G. Masing, *Die Korrosion metallischer Werkstoffe,* vol. 1, p. 240, S. Hirzel, Leipzig, 1936.

value. Examples of such salts are $AlCl_3$, $NiSO_4$, $MnCl_2$, and $FeCl_2$. *Ammonium salts* (e.g., NH_4Cl) are also acid, but produce a higher corrosion rate than corresponds to their pH[17] (at NH_4Cl concentrations > $0.05N$). Increased corrosivity is accounted for by the ability of NH_4^+ to complex iron ions, thereby reducing activity of Fe^{++} and increasing tendency of iron to corrode. Ammonium nitrate in high concentration is more corrosive (as much as eight times more) than the chloride or sulfate, in part because of the depolarizing ability of NO_3^-.

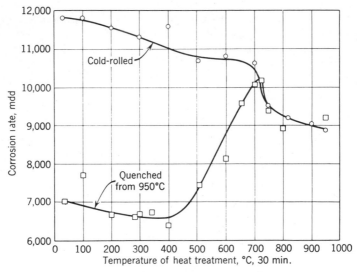

Fig. 15. Corrosion of mild steel in aqueous solution of 44.4% NH_4NO_3, 5.9% NH_3, room temperature, as a function of metal heat treatment (Schick and Uhlig).

In presence of excess NH_3, common to some synthetic fertilizer solutions, the corrosion rate in ammonium nitrate at room temperature may reach the very high value of 2 ipy (5 cmpy)[18-21] (Fig. 15). The complex[21] formed in this case has the structure $[Fe(NH_3)_6](NO_3)_2$. Since coupling mild steel to an equal area of platinum has no effect on the

[18] E. Weitz and H. Muller, *Ber. Deut. Chem. Ges.,* **58,** 363 (1925).

[19] I. Libinson, I. Kukushkin, and Z. Morozova, *Zh. Khim. Prom.,* **12,** (6), 590 (1935).

[20] N. Hackerman, H. Hurd, and E. Snavely, *Corrosion* 14, 203t (1958); J. Goodrich and N. Hackerman, *J. Electrochem. Soc.,* **108,** 1092 (1961); D. Jones and N. Hackerman, *Corros. Sci.,* 8, 565 (1968).

[21] G. Schick and H. Uhlig, *J. Electrochem. Soc.,* **111,** 1211 (1964).

rate, the reaction is apparently anodically controlled. Metallurgical structure affects the rate, a cold-worked mild steel reacting much more rapidly than one quenched from elevated temperature. This indicates that the reaction is not diffusion controlled, but depends instead on rate of metal ion formation at the anode and perhaps also to some extent on rate of depolarization at the cathode.

Alkaline salts, which hydrolyze to form solutions of pH > 10 act as corrosion inhibitors. They passivate iron in the presence of dissolved oxygen in the same manner as NaOH (Fig. 4, p. 99). Examples of such salts are trisodium phosphate (Na_3PO_4), sodium tetraborate ($Na_2B_2O_7$), sodium silicate (Na_2SiO_3), and sodium carbonate (Na_2CO_3). In addition to favoring passivation of iron by dissolved oxygen, they may form corrosion-product layers of ferrous or ferric phosphates in the case of Na_3PO_4, or analogous compounds in the case of Na_2SiO_3, such layers acting as more efficient diffusion barriers than hydrous FeO. They may, on this account, also inhibit corrosion below pH 10 and may provide better inhibition above pH 10 than NaOH or Na_2CO_3.

Oxidizing salts are either good depolarizers and therefore corrosive, or they are passivators and efficient inhibitors. Examples of the first class are $FeCl_3$, $CuCl_2$, $HgCl_2$, and sodium hypochlorite. They represent the most difficult class of chemicals to handle in metal equipment. Examples of the second class are Na_2CrO_4, $NaNO_2$, $KMnO_4$, and K_2FeO_4. The differences in properties accounting for an oxidizing salt being either a depolarizer or a passivator are discussed under *Inhibitors and Passivators,* Chapter 16.

NATURAL WATER SALTS

Natural fresh waters contain dissolved calcium and magnesium salts in varying concentrations, depending on the source and location of the water. If the concentration of such salts is high, the water is called hard; otherwise it is soft. It was recognized for many years before the causes were clearly understood that a soft water was more corrosive than a hard water. For example, a galvanized-iron hot-water tank was observed to last 10 to 20 years before failing by pitting in Chicago Great Lakes water (34 ppm Ca^{++}, 157 ppm dissolved solids), whereas in Boston water (5 ppm Ca^{++}, 43 ppm dissolved solids) a similar tank lasted only one to two years.

The mechanism of protection afforded by a hard water is the natural deposition of a thin diffusion-barrier film, composed largely of calcium carbonate ($CaCO_3$), on the metal surface. This film retards diffusion of dissolved oxygen to cathodic areas, supplementing the natural corrosion barrier of $Fe(OH)_2$ mentioned earlier (p. 98). In soft waters no such protective film of $CaCO_3$ can form. But hardness alone is not the only factor that determines whether a protective film is possible. Ability of $CaCO_3$ to precipitate on the metal surface also depends on total acidity or alkalinity, pH, and concentration of dissolved solids in the water.

Langelier in 1936,[22] following preliminary investigations of Tillmans[23] and Baylis[24] along similar lines, divided natural fresh waters into two groups—those oversaturated with $CaCO_3$ and those undersaturated. Since only near- or oversaturated waters tended to form a protective film of $CaCO_3$ on iron, an estimate of the corrosivity of a water was established through analytical criteria for under- or oversaturation. Using certain simplifications, Langelier showed that the value of pH, called pH_s, at which a water is in equilibrium with solid $CaCO_3$ (neither tends to dissolve nor precipitate), can be calculated from the relation*

$$pH_s = (pK_2' - pK_s') + p_{Ca^{++}} + p_{alk} \qquad (10)$$

where K_2' is the ionization constant $(H^+)(CO_3^{--})/(HCO_3^-)$, K_s' is the solubility product of calcium carbonate $[(Ca^{++})(CO_3^{--})]$, the concentration of calcium ions (Ca^{++}) is in moles/1000 grams H_2O, and alk (alkalinity) represents the equivalents per liter of titratable base to the methyl orange end point (often reported as ppm $CaCO_3$) according to the relation

$$(alk.) + (H^+) \rightarrow 2CO_3^{--} + HCO_3^- + OH^-$$

The letter p refers to the negative logarithm of all these quantities. The Saturation Index is then defined as the difference between the measured pH of a water and the equilibrium pH_s for $CaCO_3$, or

$$\text{Saturation Index} = pH_{measured} - pH_s$$

Charts (see Appendix) and nomographs have been prepared to obtain values of pH_s for waters varying widely in composition and at various temperatures.[25, 26]

An estimate of over- or undersaturation can also be obtained in the laboratory by measuring the pH of a water before and after exposure to pure $CaCO_3$ powder for a time adequate to achieve equilibrium. An increase in pH corresponds to undersaturation. Because the composition

---

* For derivation of this and a more exact equation, see Appendix, p. 394.

[22] W. F. Langelier, *J. Amer. Water Works Assoc.*, **28**, 1500 (1936).

[23] J. Tillmans, *Die chemische Untersuchung von Wasser und Abwasser*, 2nd ed., W. K. Halle, 1932.

[24] J. Baylis, *J. Amer. Water Works Assoc.*, **27**, 220 (1935).

[25] C. P. Hoover, *J. Amer. Water Works Assoc.*, **30**, 1802 (1938).

[26] S. Powell, H. Bacon and J. Lill, *Ind. Eng. Chem.*, **37**, 842 (1945).

of the water is altered by any $CaCO_3$ taken up or precipitated during the test, the final pH is not necessarily the same as the calculated pH_s. Alternatively, the water after saturation with $CaCO_3$ can be titrated with acid, the increase in required milliliters of acid compared to that required by the original water being a measure of undersaturation (marble test).[27]

It follows that any fresh water can be placed in one of the following categories:

Saturation Index		Characteristic of Water
Positive	Oversaturated with respect to $CaCO_3$	Protective film of $CaCO_3$ forms
Zero	In equilibrium	
Negative	Undersaturated with respect to $CaCO_3$	Corrosive

Chicago water, for example, has an Index of 0.2, whereas the value for Boston water is —3.0.

A soft water with negative Saturation Index can be treated with lime $(Ca(OH)_2)$, or soda ash (Na_2CO_3), or both, depending on prevailing conditions, in order to raise the Saturation Index to a positive value, thereby making the water less corrosive. A Saturation Index of about +0.5 is considered by some engineers to be satisfactory—higher than this may cause excessive deposition of $CaCO_3$ (scaling), particularly at elevated temperatures.

At above-room temperatures, the Saturation Index may become more positive, and, in any case, the rate of $CaCO_3$ deposition is higher, should there be any tendency toward deposition. Hence to provide corrosion protection at all temperatures without excessive scaling requires adjustment of water composition to a Saturation Index that is at least constant over the entire operating temperature range. Powell et al. showed[26, 28] that for any specific alkalinity there is a corresponding pH value at which the decrease of measured pH with temperature is almost exactly compensated by decrease in the factor $(pK_2' - pK_s')$. Under these conditions, the Saturation Index is nearly constant with temperature, and scaling tends to be the same in hot or cold water (Table 3). The amount of possible scaling, of course, depends on the value of the Saturation Index. The alkalinity and the pH of waters can be adjusted to bring

[27] F. Speller, *Corrosion, Causes and Prevention*, p. 634, McGraw-Hill, New York, 1951.
[28] S. Powell, H. Bacon, and J. Lill, *J. Amer. Water Works Assoc.*, **38**, 808 (1946).

TABLE 3
Alkalinity-pH Limits for Uniform Scale Deposition at Various Temperatures[28]

Alkalinity (ppm, as $CaCO_3$)	pH (as measured, room temp.)
50	8.10–8.65
100	8.60–9.20
150	8.90–9.50
200	8.90–9.70

them into the proper composition range using $Ca(OH)_2$, Na_2CO_3, NaOH, H_2SO_4, or CO_2.

Limitations of Saturation Index. If a natural water contains colloidal silica or organic matter (e.g., algae), $CaCO_3$ may precipitate on the colloidal or organic particles instead of on the metal surface. If such is the case, the corrosion rate will remain high even though the Saturation Index is positive. Another limitation is that for waters high in dissolved salts such as NaCl, or for waters at elevated temperatures, the $CaCO_3$ film may lose its protective character at local areas, resulting in corrosion pitting. Furthermore, if complexing ions are added to a chemically treated water, such as polyphosphates which retard precipitation of $CaCO_3$, the Saturation Index may no longer apply as an index of corrosivity.

Other than these exceptions, the Saturation Index is a useful qualitative guide to the relative corrosivity of a fresh water in contact with metals whose corrosion rate depends on diffusion of dissolved oxygen to the surface, such as iron, copper, brass, and lead. The Index does not apply to corrosivity of a water in contact with passive metals that corrode less the *higher* the surface concentration of oxygen, such as aluminum or the stainless steels.

METALLURGICAL FACTORS

Varieties of Iron or Steel

As discussed under effect of pH, the corrosion rate of iron or steel in natural waters is controlled by diffusion of oxygen to the metal surface. Hence, whether a steel is manufactured by the Bessemer, oxygen furnace, or open-hearth process, and whether it is a wrought iron, or a cast iron, makes little or no difference to the corrosion rate in natural

waters, including seawater. The same statement applies to corrosion in
a variety of soils, because factors determining the corrosion rate underground are similar to those of total submersion. In general, therefore,
the least expensive steel or iron of a given cross-sectional thickness having the required mechanical properties is the one that should be specified
for those environments.

On the other hand, in the acid range of pH (approximately <4)
and probably also the extreme alkaline range (>13.5) where impurities
play a role in the hydrogen evolution reaction, differences in manufacture
affect the corrosion rate. A relatively pure iron corrodes in acids at
a much lower rate than an iron or steel high in residual elements such
as carbon, nitrogen, sulfur, and phosphorus.

The high nitrogen content of Bessemer steel makes it more sensitive
than open-hearth steels to stress-corrosion cracking in hot caustic or
nitrate solutions. For this reason, open-hearth steel is usually specified
for boilers.

Cast iron in natural waters or in soils corrodes initially at the expected
normal rate, but may provide much longer service life than steel. Other
than the factor of considerable metal thickness common to cast structures, this advantage occurs because cast iron, composed of a mixture
of ferrite phase (almost pure iron) and graphite flakes, forms corrosion
products which in some soils or waters cement together the residual
graphite flakes. The resulting structure (e.g., water pipe), although corroded completely, may have sufficient remaining strength, despite low
ductility, to continue functioning under the required operating pressures
and stresses. This type of corrosion is called *graphitic corrosion*. It
occurs only with gray cast iron (or "ductile" cast iron containing sphe-
roidal graphite) but not with white cast iron (cementite + ferrite).
Graphitic corrosion can be reproduced in the laboratory over a period
of weeks or months by exposing gray cast iron to a very dilute sulfuric
acid which is renewed periodically.

Effects of Composition

As has been discussed, composition of an iron or steel within the usual
commercial limits of carbon and low-alloy steels, has no practical effect
on the corrosion rate in natural waters or soils. (See Table 1, p. 101;
also Table 1, p. 179; *Corrosion Handbook*, Fig. 3, p. 452.) Only when
a steel is alloyed in the proportions of a stainless steel ($>12\%$ Cr)
or a high-silicon iron or high-nickel iron alloy for which oxygen diffusion
no longer controls the rate, is corrosion appreciably reduced. For atmospheric exposures, the situation is changed because the addition of

certain elements in small amounts (e.g., 0.1–1% Cr, Cu, or Ni) has a marked effect on the protective quality of naturally-formed rust films (see "Atmospheric Corrosion," Chapter 8).

Although *carbon* content of a steel has no effect on the corrosion rate in fresh waters, a slight increase in rate (maximum 20%) has been observed in seawater as the carbon content is raised from 0.1 to 0.8%.[29] The cause of this increase is probably related to greater importance of the hydrogen evolution reaction in chloride solutions (with complexing of Fe^{++} by Cl^-) supplementary to oxygen depolarization as the cathodic surface of cementite (Fe_3C) increases.

In acids, the corrosion rate increases with both carbon and nitrogen content of a steel. The extent of the increase depends largely on prior heat treatment (see "Effect of Heat Treatment," p. 124) and is greater for cold-worked steels (Fig. 3, p. 129).

Alloyed *sulfur* and *phosphorus* markedly increase rate of attack in acids. These elements form compounds that apparently have low hydrogen overvoltage; in addition, they tend to decrease anodic polarization so that the corrosion rate of iron is stimulated by these elements at

[29] C. Chappell, *J. Iron Steel Inst.*, **85,** 270 (1912).

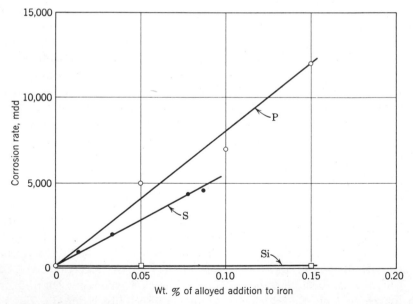

Fig. 16. Effect of alloyed phosphorus, sulfur, and silicon in iron on corrosion in deaerated 0.1*N* HCl, annealed specimens, 25°C (Foroulis and Uhlig).

TABLE 4
Corrosion Rates of Iron Alloys in Deaerated Citric Acid
and in 4% NaCl + HCl, 25° C (Stern)

	0.1M Citric Acid pH = 2.06	4% NaCl + HCl pH = 1
Pure Iron (0.005% C)	29 mdd	30 mdd
+0.02% P	165	1000
+0.015% S	706	2830
+0.11% Cu	41	390
+0.10% Cu + 0.03% P	376	606
+0.08% Cu + 0.02% S	32	186

both anodic and cathodic sites. Rates in deaerated citric acid are given in Table 4.[30] In strong acids, the effect of these elements is still more pronounced[31] (Fig. 16; Table 4).

Arsenic is present in some steels in small amounts. In quantities up to 0.1% it increases the corrosion rate in acids (less than sulfur and phosphorus); in larger amounts (0.2%) it decreases the rate.[31]

Manganese, in amounts normally present, effectively decreases acid corrosion of steel containing small amounts of sulfur. Inclusions of MnS have low electrical conductivity compared to FeS; in addition, manganese reduces the solid solubility of sulfur in iron, thereby probably restoring the anodic polarization of iron which is lowered by presence of sulfur.[32] *Silicon* only slightly increases the rate in dilute hydrochloric acid (Fig. 16).

Copper alloyed with pure iron to the extent of a few tenths of one per cent moderately increases the corrosion rate in acids. However, in the presence of phosphorus or sulfur, which are normal components of commercial steels, copper counteracts the accelerating effect of these elements. Copper-bearing steels, therefore, usually corrode in nonoxidizing acids at lower rates than do copper-free steels.[33, 34] The data of Table 4 for several relatively pure iron alloys show that 0.1% Cu reduces corrosion in 4% NaCl + HCl when 0.03% P or 0.02% S are present in the iron, but not for the phosphorus alloy in citric acid. These par-

[30]M. Stern, *J. Electrochem. Soc.* **102**, 663 (1955).

[31] Z. Foroulis and H. Uhlig, *J. Electrochem. Soc.*, **112**, 1177 (1965).

[32] G. Wranglen, *Corros. Sci.*, **9**, 585 (1969).

[33] P. Bardenheuer and G. Thanheiser, *Mitt. Kaiser-Wilhelm Inst. Eisenforschung Düsseldorf*, **14**, 1 (1932).

[34] E. Williams and M. Komp, *Corrosion*, **21**, 9 (1965).

ticular relations apply only to the specific compositions and experimental conditions reported; they are not necessarily general. Steels containing a few tenths of one per cent of copper are more resistant to the atmosphere but show no advantage over copper-free steels in natural waters or buried in the soil where oxygen diffusion controls the rate.

An amount up to 5% *chromium* (0.08% C) was reported to decrease weight losses in seawater at the Panama Canal[35] at the end of one year. A sharp increase in rates was observed between two to four years; after 16 years the chromium steels lost 22 to 45% more weight than did 0.24% C steel. Depth of pits was less for the chromium steels after one year, but comparable to pit depth in carbon steel after 16 years. Hence, for long exposures to seawater, low chromium steels apparently offer no advantage over carbon steel. Low-alloy chromium steels ($< 5\%$ Cr) have improved resistance to corrosion fatigue in oil-well brines free of hydrogen sulfide.

An amount up to 5% *nickel* (0.1% C) did not appreciably change weight losses of steels exposed to seawater in tests at the Panama Canal[35] extending up to 16 years. Depth of pits, although less for one year's exposure, was greater for long exposure times for the nickel steels compared to 0.24% C steel (77% deeper for 5% Ni steel after eight years). Low-nickel steels ($<5\%$ Ni) have improved resistance to corrosion fatigue in oil-well brines containing hydrogen sulfide.[36] Nickel also decreases the corrosion rate of steels exposed to alkalies, the effect increasing with nickel content.[37]

GALVANIC EFFECTS THROUGH COUPLING OF DIFFERENT STEELS

Unimportant though low-alloy components may be in determining overall corrosion rates in waters or in soils, composition is nevertheless of considerable importance to the galvanic relations and consequent corrosion of steels coupled to each other. For example, because of increased anodic polarization, a low-nickel, low-chromium steel is cathodic to mild steel in most natural environments. The reason is obvious from relations shown in Fig. 17. Both mild steel and a low-alloy steel, not coupled, corrode at about the same rate, i_{corros}, established by the limiting rate of oxygen reduction. When coupled, however, the potentials of both steels, which are originally different, become equal to $\phi_{coupled}$. The corrosion rate of mild steel, estimated from the extension of its anodic polarization curve, is now increased to i_2, and that of the low-alloy

[35] C. Southwell and A. Alexander, *Materials Prot.*, **9**, 14 (1970).

[36] B. Wescott, in *Corrosion Handbook*, pp. 578–590.

[37] *Corrosion Resistance of Metals and Alloys*, edited by F. L. LaQue, and H. R. Copson, p. 334, Reinhold, New York, 1963.

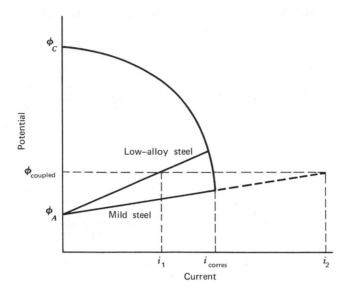

Fig. 17. Polarization diagram showing effect of coupling of a low-alloy steel to mild steel on subsequent corrosion rates.

steel is decreased to i_1. Note that the precise value of $\phi_{coupled}$ depends on the relative areas of the two steels. Accordingly, steel bolts and nuts used to couple underground mild steel pipes, or a weld rod used for steel plates on the hull of a ship, should always be of a low-nickel, low-chromium steel or similar composition which is cathodic to the major area of the structure (small cathode, large anode). Should the reverse polarity occur, serious corrosion damage would be caused quickly either to the bolts or to the critical area of weld metal.[38]

Cast iron is initially anodic to low-alloy steels and not far different in potential from mild steel. As cast iron corrodes, however, especially if graphitic corrosion takes place, exposed graphite on the surface shifts the potential in the noble direction. After some time, therefore, depending on the environment, cast iron may achieve a potential cathodic to both low-alloy steels and mild steel. This behavior is important in designing valves, for example. The trim of valve seats must maintain dimensional accuracy and be free of pits; consequently, the trim must always be chosen cathodic to the valve body making up the major internal area of the valve. For this reason, valve bodies of steel are

[38] E. Uusitalo in *Proc. 2nd Int. Congr. Metallic Corrosion*, p. 812, Nat. Assoc. Corros. Engrs., Houston, Texas, 1966.

often preferred to cast iron for aqueous media of high electrical conductivity.

Effect of Heat Treatment

The effect on corrosion rate in dilute sulfuric acid resulting from heat treatment of a carbon steel is typified by data of Heyn and Bauer[39] (Fig. 18). A carbon steel quenched from high temperatures has a structure called *martensite*. This is single phase, with carbon in interstitial

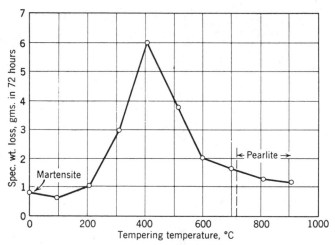

Fig. 18. Effect of heat treatment on corrosion of 0.95% C steel in 1% H_2SO_4 Polished specimens 2.5 × 2.5 × 0.6 cm, tempering time probably 2 hr (Heyn and Bauer).

positions of the body-centered tetragonal lattice of iron atoms. Random distribution of carbon atoms accompanied by electronic interaction of carbon atoms with neighboring iron atoms limits their effectiveness as cathodes of local-action cells, consequently in dilute acid the corrosion rate of martensite is relatively low. Interstitial carbon reacts in large part with acid to form a complex hydrocarbon gas mixture (accounting for the odor of pickled steel) and some residual amorphous carbon, which is observed as a black smut on the steel surface (Fig. 19). When martensite is heated at low temperatures and then air cooled (called temper-

[39] E. Heyn and O. Bauer, *J. Iron Steel Inst.*, **79**, 109 (1909).

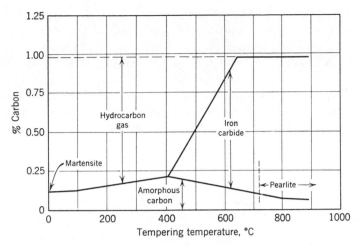

Fig. 19. Effect of heat treatment of 0.95 % C steel on relative distribution of carbon in corrosion products as gas, carbon, and carbides. Corrosion in 10% H_2SO_4 (Heyn and Bauer).

ing), decomposition to ϵ iron carbide of unknown composition takes place. This two-phase structure sets up galvanic cells which accelerate the corrosion reaction. Some finely divided cementite (Fe_3C) also appears by decomposition of the ϵ phase, the amount of which for a two-hr heat treatment of 0.95% C steel reaches a maximum at about 400°C (300°C for 0.07% C steel). After tempering at this temperature, cementite acting as cathode offers maximum peripheral surface adjoining ferrite, and galvanic action is at a maximum. Beyond this temperature, cementite coalesces to larger particle size, and the corrosion rate decreases. The particles of cementite are now large enough to resist complete dissolution in acid and can be detected in the residue of corrosion products. At the same time, there is a corresponding decrease in hydrocarbon gas formation. On slowly cooling a carbon steel from the austenite region above 723°C (face-centered cubic lattice), cementite in part assumes a lamellar shape, forming a structure called *pearlite*. This structure again corrodes at a comparatively low rate because of the relatively massive form of cementite formed by decomposition of austenite compared with smaller size cementite particles resulting from decomposition of martensite. The importance of both the amount of cementite acting as cathode and its state of subdivision bear out the assumption of the electrochemical mechanism of corrosion. The rate is cathodically controlled and hence depends on hydrogen overvoltage and interfacial area of cathodic sites.

In practice, an effect of heat treatment on corrosion is seldom observed because oxygen diffusion controls the rate in the usual environments. However, in handling of acid oil-well brines, marked localized corrosion is sometimes found near welds or "up-set" ends of steel oil-well tubing. This observed increase in corrosion encircling a limited inner region of the tube is called ring-worm corrosion. It is caused by heat treatment incidental to joining and fabrication techniques, and it can be minimized by a final heat treatment of the tubing or by addition of suitable inhibitors to the brine.[40]

GENERAL REFERENCES

CORROSION OF IRON IN AQUEOUS MEDIA

Corrosion Handbook, edited by H. H. Uhlig, pp. 125–143, Wiley, New York, 1948.

ACTION OF BACTERIA

The Corrosion and Oxidation of Metals, U. R. Evans, pp. 275–278, 1960; 1st suppl. vol., pp. 121–123, Ed. Arnold, London, 1968.

CAVITATION-EROSION

S. Kerr, in *Corrosion Handbook,* edited by H. H. Uhlig, pp. 597–601; 993–997, Wiley, New York, 1948.
H. Preiser and B. Tytell, *Corrosion,* **17,** 535t (1961).

METALLURGICAL FACTORS

Corrosion, Causes and Prevention, F. Speller, pp. 58–140, McGraw-Hill, New York, 1951.
Corrosion Handbook, edited by H. H. Uhlig, pp. 139–141, Wiley, New York, 1948.
H. Uhlig, "Effect of Microstructure on Corrosion" in *Relation of Properties to Microstructure,* pp. 189–208, American Society for Metals, Cleveland, 1954.
H. H. Uhlig, *Corrosion,* **19,** 231t (1963).

---

[40] M. Holmberg, *Corrosion,* **2,** 278 (1946); R. Manuel, *Ibid.,* **3,** 415 (1947).

chapter 7

Effect of stress

COLD WORKING

In accord with previous discussions, a cold-worked commercial steel is found to corrode in natural waters at the same rate as an annealed steel.[1] In acids, however, cold working increases the corrosion rate several fold (Fig. 1).[2] Many authors have traditionally ascribed the effect to residual stress within the metal, which serves to increase the corrosion tendency. But the residual energy produced by cold working, as measured in a calorimeter (usually <7 cal/g), is less than sufficient to account for an appreciable change in free energy, and hence this intuitive concept is probably wrong.[3] The observed increase of rate is apparently caused instead by segregation of carbon or nitrogen atoms at imperfection sites produced by plastic deformation (Fig. 2) rather than by the presence of the imperfections themselves (Fig. 3). Such sites exhibit lower hydrogen overvoltage than does either cementite or iron[2] and probably constitute the most important factor. Entering to a lesser extent are (1) the increased surface area of cementite lamellae fractured by the cold-work process and (2) the preferred orientation of ferrite grains, the latter either increasing or decreasing corrosion depending on the particular crystal faces that lie parallel to the metal surface.

Subsequent heat treatment of cold-worked steel at 100°C induces additional diffusion of interstitial carbon atoms to imperfections in the metal lattice, thereby increasing the cathodic area of low hydrogen over-

[1] *Corrosion Handbook*, p. 138.

[2] Z. Foroulis and H. Uhlig, *J. Electrochem. Soc.*, **111**, 522 (1964).

[3] H. H. Uhlig, in *Physical Metallurgy of Stress Corrosion Fracture*, pp. 1–17, edited by T. Rhodin, Interscience, New York, 1959. See Prob. 2, Ch. 7, p. 82.

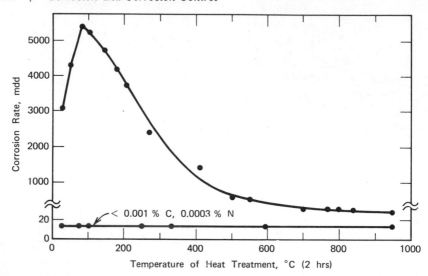

Fig. 1. Effect of heat treatment of cold-worked 0.076% C steel (85% red. thick.) and zone-refined iron (50% red. thick.) on corrosion in deaerated 0.1N HCl, 25°C, (Foroulis and Uhlig).

voltage and also accelerating the corrosion rate. Heat treatment at higher temperatures reduces the density of the imperfection sites, accompanied by precipitation of carbides or nitrides of increasing particle size the higher the temperature of heat treatment. Hence heat treatment above 100°C results in cathodic sites of higher hydrogen overvoltage and of decreasing peripheral area, leading to a corresponding reduction of corrosion. A severely cold-worked, zone-refined (pure) iron, on the other hand, lacking sufficient carbon or nitrogen, corrodes in dilute, de-

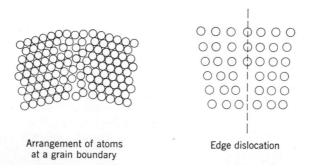

Arrangement of atoms
at a grain boundary

Edge dislocation

Fig. 2. Schematic illustration of imperfection sites in a metal. (From A. Cottrell, *Theoretical Structural Metallurgy*, Ed. Arnold, 1955.) It is at the enlarged distances between atoms that some impurity or alloyed atoms tend to concentrate.

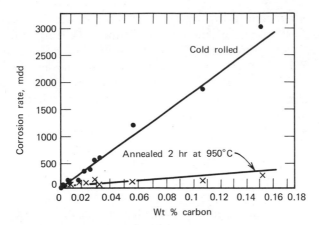

Fig. 3. Effect of carbon in steels cold-rolled 50%, or subsequently annealed, on corrosion in deaerated 0.1*N* HCl, 25°C.[2]

aerated hydrochloric acid at no higher rate than when annealed (Figs. 1, 3).[2]

When commercial nickel is severely cold worked, unlike mild steel it does not corrode appreciably more rapidly in acids,[4] which suggests that similar formation of low overvoltage cathodic sites derived from segregated, interstitial impurities does not occur.

STRESS CORROSION CRACKING OF IRON AND STEEL

Whenever mild steel is stressed in tension to levels near or beyond the elastic limit, and exposed simultaneously to hot concentrated alkaline or hot concentrated nitrate solutions,* it suffers cracking along inter-granular paths. This is called stress corrosion cracking (s.c.c.). The mechanism is quite different from that just described under "Cold Working." The required tensional stress may be applied or may be residual in the metal; a compressive stress is not damaging. Time to failure is a matter of minutes under severe conditions of stress and environment, or of years when conditions are less severe.

Stress corrosion cracking was first encountered in a practical way in riveted steam boilers. Stresses at rivets always exceed the elastic limit, and boiler waters are normally treated with alkalies to minimize corrosion. Crevices between rivets and boiler plate allow boiler water to concentrate, until the concentration of alkali suffices to induce s.c.c.,

---

* Boiling 60% Ca(NO₃)₂ + 3% NH₄NO₃ is one solution used in accelerated tests.

[4] H. Copson, in *Corrosion Handbook,* p. 577.

sometimes accompanied by explosion of the boiler. Because alkalies were recognized as one of the causes, failures of this kind were first called *caustic embrittlement*. With the advent of welded boilers and with improved boiler water treatment, s.c.c. has become less common. Its occurrence has not been eliminated entirely, however, because stresses, for example, may be set up at welded sections of boilers or in tanks used for storing concentrated alkalies.

The presence of silicates in an alkaline boiler water above 225°C accelerates s.c.c.; at lower temperatures silicates may act as inhibitors.[5] Laboratory tests in boiling 50% NaOH have shown that at atmospheric pressures, 0.2 to 1.0% PbO, $KMnO_4$, Na_2CrO_4, or $NaNO_3$ are also accelerators.[5] In this connection, it is significant that at the higher temperatures and at pressures common to boiler conditions, nitrates added in amounts equivalent to 20 to 40% of NaOH alkalinity act as inhibitors of s.c.c.[5] and are used for this purpose in practice. On the other hand, addition of 2% NaOH to boiling nitrate solutions may inhibit cracking.[6] This is one of several illustrations that can be cited of the necessity to thoroughly understand the fundamentals of a laboratory test before extrapolating results to large-scale applications.

In addition to trouble experienced by the chemical or nuclear chemical[7] industries in handling nitrate solutions when hot, stressed steel may also fail in contact with nitrates at room temperature. Such failure occurred, for example, in 0.7% C steel cables of the Portsmouth, Ohio, bridge after 12 years in service.[8] The cables cracked at their base where rain water, presumably containing trace amounts of ammonium nitrate from the atmosphere, had accumulated and concentrated. Subsequent tests conducted at the National Bureau of Standards showed that specimens of the cable stressed in tension failed within $3\frac{1}{2}$ to $9\frac{1}{2}$ months when immersed in $0.01N$ NH_4NO_3, or $NaNO_3$ at room temperature. No cracking occurred after similar tests in distilled water, or in $0.01N$ solutions of NaCl, $(NH_4)_2SO_4$, $NaNO_2$, or NaOH. There was evidence in this case that the steel cable was unusually susceptible to s.c.c.

Stress corrosion cracking of steel also occurs when the metal is in contact with anhydrous liquid ammonia at room temperature, for example at cold-formed heads and at welds of steel tanks used to contain

[5] W. Schroeder and A. Berk, "Intercrystalline Cracking of Boiler Steel and its Prevention," Bur. of Mines Bulletin 443, U. S. Govt. Printing Office (1941).

[6] R. Parkins, in *Stress Corrosion Cracking and Embrittlement,* edited by W. Robertson, p. 154, Wiley, New York, 1956.

[7] M. Holzworth et al., *Materials Prot.,* **7,** 36 (1968).

[8] R. Pollard, in *Symposium on Stress Corrosion Cracking of Metals,* p. 437, ASTM-AIME, Philadelphia, 1944.

the liquefied gas. Cracks appear to be intergranular, but they can also be transgranular. Cracking is avoided by stress-relief heat treatment of the steel, by avoiding air contamination of the NH_3, or by addition of about 0.2% H_2O which acts as an inhibitor.[9] Intergranular cracking of stressed steel has also been reported in contact with $SbCl_3$ + HCl + $AlCl_3$ in a hydrocarbon solvent.[10]

Transgranular s.c.c. was reported at room temperature on exposure of 0.1 to 0.2% C steel to an aqueous mixture of carbon dioxide and carbon monoxide at 100 psi.[11] Cathodic polarization in the latter solution prevented damage. Explosions have been reported involving cracks in steel tanks containing compressed illuminating gas. The failures at stresses below the elastic limit were transgranular and were traced to small amounts of HCN contained in the gas.[12] The trouble was overcome by removing traces of HCN and moisture from the gas. Whether CO and CO_2 also entered as a cause was not determined.

Based on tests in boiling nitrate solutions,[13] it was found that severely cold-worked mild steel (0.06% C, 0.001% N) is resistant to s.c.c. (Fig. 4). Along these lines, it is recognized in practice that cold-drawn steel wire is more resistant to s.c.c. than is oil-tempered wire having equal mechanical properties. Heat treatment of cold-rolled mild steel at 600°C (1110°F) for one-half hr, at 445°C (830°F) for 48 hours, or at lower temperatures for correspondingly longer times, induces susceptibility again. Plastically deformed steel, therefore, stress relieved in the range of 400 to 650°C (750–1200°F) and subsequently stressed is *more* rather than less susceptible as a result of heat treatment. Mild steel quenched from 900 to 950°C (1650–1740°F) is susceptible, but can be made resistant by tempering at 250°C (480°F) for one half hr (Fig. 4), or at 200°C (390°F) for 48 hr, even if the steel exposed to nitrates is highly stressed after heat treatment. This resistant state, however, is temporary; on further heating (in the unstressed state) at 445°C (830°F) for 70 hr, or at 550°C (1020°F) for 3 hr, and for correspondingly shorter times at higher temperatures, the steel becomes susceptible again and remains susceptible.

The overall evidence suggests that grain boundaries become suitable crack paths only when carbon (or nitrogen) atoms (not Fe_3C) segregate at grain-boundary regions. Pure iron is immune to stress-corrosion

[9] A. Loginow and E. Phelps, *Corrosion,* **18**, 299t (1962).

[10] R. Treseder and A. Wachter, *Corrosion,* **5**, 383 (1949).

[11] M. Kowaka and S. Nagata, *Corrosion,* **24**, 427 (1968); A. Brown, J. Harrison, and R. Wilkins, *Corros. Sci.* **10**, 547 (1970).

[12] H. Buchholtz and R. Pusch, *Stahl u. Eisen,* **62**, 21 (1942).

[13] H. H. Uhlig and J. Sava, *Trans. Amer. Soc. Metals,* **56**, 361 (1963).

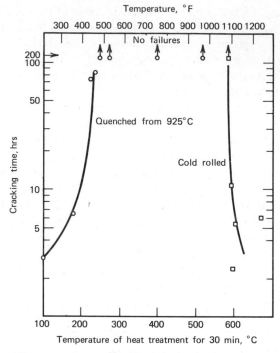

Fig. 4. Effect of heat treatment of mild steel after quenching or cold rolling (70% reduction of thickness) on resistance to stress corrosion cracking in boiling nitrate solution.

cracking. Iron ($>0.002\%$ C)[14] or mild steel (0.06% C) quenched from about 925°C contains a sufficient concentration of carbon atoms along the grain boundaries to induce susceptibility. Low-temperature heating, (e.g., 250°C for one-half hr) randomly nucleates iron carbides, which rob the grain boundaries of carbon, accounting for increased resistance or immunity. Longer heating, or at higher temperatures (e.g., 70 hr at 445°C) allows slow-moving lattice imperfections within the grain to migrate to the grain boundaries, transporting carbon atoms with them, and the steel again becomes susceptible. Cold working, on the other hand, produces immunity by destroying continuous grain-boundary paths, and, more important, by generating imperfections that have high affinity for carbon and deplete any continuous paths along which carbon atoms tend to segregate. Reasons for the production of crack-sensitive paths by carbon atoms localized at grain boundaries are discussed later on.

Remedial measures to avoid s.c.c. of steel in nitrate solutions, and probably also in alkalies for which the mechanism of cracking appears

[14] M. Long and H. Uhlig, *J. Electrochem. Soc.*, **112**, 964 (1965).

to be similar, include the following:

1. *Severe Cold Working.* Cold rolling to >50% reduction of thickness is found to impart relative immunity to a stressed mild steel in boiling nitrate solution. The resistant state is predicted to persist at low temperatures, e.g., 100 to 200°C (200–400°F) for thousands of hours.

2. *Heat Treatment.* Mild steel slowly cooled from 900 to 950°C (1650°–1740°F), or quenched from this temperature range and then tempered at about 250°C (480°F) for one-half hr, or at higher temperatures for shorter times, is resistant. This resistance is temporary and short-lived (200 hr) at temperatures in the order of 400°C (750°F), but it is predicted to reach the order of thousands of hours at 300°C (570°F) or below.[13]

3. *Surface Peening or Shot Blasting.* Compressive stresses are produced at the surface of the metal which are effective in avoiding damage so long as the compressive layers are continuous, remain intact, and are not dissolved by general corrosion.

4. *Cathodic Protection.* Schroeder and Berk[5] found that cathodic polarization of steel stressed in hot sodium hydroxide–sodium silicate solution greatly delayed or prevented cracking. Parkins[15] found similar protection in hot nitrate solution. Bohnenkamp[16] reported maximum susceptibility of various carbon steels (0.003–0.11% C) in 33% NaOH boiling at 120°C at —0.66 to —0.75 V (s.h.e.) with orders of magnitude longer life at potentials 0.1 V either noble or active to this range. Anodic as well as cathodic protection was found to be effective.

5. *Special Alloys.* Steels containing small alloying additions of aluminum, titanium, or niobium plus tantalum,[17] which react preferentially with carbon and nitrogen, exhibit improved resistance to s.c.c. but are not immune.

6. *Use of Inhibitors.* Sodium nitrate has already been mentioned as a practical inhibitor for steel exposed to boiler waters. Crude quebracho extract and waste sulfite liquor are also used. Buffer ions such as PO_4^{-3} are useful, because they avoid high concentrations of OH^- in concentrated boiler water.

In accelerated tests, the addition of 3% NaCl or 2% sodium acetate to boiling 60% $Ca(NO_3)_2$ + 3% NH_4NO_3, at 108°C, was found to inhibit cracking (>200 hr)[18] of mild steel.

[15] R. Parkins, *J. Iron Steel Inst.*, **172**, 149 (1952).

[16] K. Bohnenkamp, *Proc. Fundamental Aspects of Stress Corrosion Cracking*, p. 374., Nat. Assoc. Corros. Engrs., Houston, Texas, 1969.

[17] E. Baerlecken and W. Hirsch, *Stahl u. Eisen*, **73**, 785 (1953).

[18] V. Agarwala, M. S. thesis, Dept. of Metallurgy and Materials Science, M.I.T., 1966.

MECHANISM OF STRESS CORROSION CRACKING
OF STEEL AND OTHER METALS

The distinguishing characteristic of s.c.c. is the requirement of a tensile stress acting conjointly with a specific environment. Typical specific environments for several metals are listed in Table 1. It will be noted

TABLE 1
Some Specific Environments Causing Stress-Corrosion Cracking

Metal	Environment
1. Mild steel	NO_3^-, OH^-
High-strength steels*	H_2O
2. Stainless steels, austenitic	Cl^-, OH^-, Br^-
3. α Brass	NH_3, Amines
4. Titanium alloys	
8% Al, 1% Mo, 1% V	Cl^-, Br^-, I^-
5. Aluminum alloys	H_2O, NaCl solutions
6. <40 at.% Au-Cu	$FeCl_3$, Aqua regia

* >180,000 psi yield strength or approximately >Rockwell C 40 hardness value.

that the damaging anions have no clear-cut relation to general corrosion rates of the metals they affect. Chloride ions but not nitrate ions cause s.c.c. of 18-8 austenitic stainless steels (containing 74% Fe), but the opposite situation applies to mild steel. The necessity for a tensile stress is illustrated by the sensitivity of 70% Cu, 30% Zn brass to slow intergranular corrosion (not requiring stress) in a variety of electrolytes (e.g., dilute H_2SO_4, $Fe_2(SO_4)_3$ or $BiCl_3$ solutions[19]) but the much more rapid s.c.c., which is usually also intergranular, requires the presence of NH_3 or amines. Furthermore, an improperly heat-treated 18-8 stainless steel (e.g., sensitized at 650°C for 1 hr) will fail intergranularly in a wide variety of electrolytes regardless of whether it is stressed; but the same heat-treated steel stressed in tension and placed in a boiling solution of magnesium chloride fails *transgranularly* by s.c.c. despite well-defined corrosion paths along grain boundaries.[20]

There are five characteristics of s.c.c. that must be explained by any acceptable working theory.

[19] Wilson Lynes, *Corrosion*, **21**, 125 (1965).
[20] M. Scheil, in *Symp. Stress-Corrosion Cracking of Metals,* p. 433, ASTM-AIME, Philadelphia, Pa., 1945.

1. Pronounced specificity of damaging chemical environments.

2. General resistance or immunity of all pure metals. Pure alloys, on the other hand, (e.g., of Cu-Zn, Cu-Au, Mg-Al) are susceptible.

3. Successful use of cathodic polarization to avoid the initiation or continuation of s.c.c.

4. Inhibiting effect of various extraneous anions added to damaging environments. For example, 2% $NaNO_3$ or 1% sodium acetate or 3.5% NaI added to magnesium chloride solution boiling at 130°C (33 g/100 ml) inhibits s.c.c. of 18-8 stainless steel (>200 hours).[21] Similarly, Cl^- or acetates inhibit s.c.c. of mild steel in boiling nitrates,[18] and SO_4^{--} or NO_3^- can be used to inhibit a titanium alloy (8% Al, 1% Mo, 1% V) that otherwise cracks in 3.5% NaCl solution at room temperature.[22]

5. An appreciable effect of metallurgical structure. For example, ferritic stainless steels (body-centered cubic) are much more resistant to Cl^- than are austenitic stainless steels (face-centered cubic). Also β and γ brass (>40% Zn) may crack in water alone, but α brass requires NH_3 or an amine. Any metal of large grain size, whether failure is inter- or transgranular, is more susceptible to s.c.c. than is the same metal of small grain size.

Although several theories differing in detail have been proposed, the general mechanism of s.c.c. is covered by two major theories, namely, "electrochemical" and "stress-sorption cracking."

Electrochemical Theory

Dix in 1940[23] proposed that galvanic cells are set up between metal and anodic paths established by heterogeneous phases (e.g., $CuAl_2$ precipitated from a 4% Cu-Al alloy) along grain boundaries or along slip planes. When the alloy, stressed in tension, is exposed to corrosive environments, the ensuing localized electrochemical dissolution of metal opens up a crack; in addition, the applied stress effectively ruptures brittle oxide films at the tip of the crack, thereby exposing fresh anodic material to the corrosive medium. Supporting this mechanism was a measured potential of metal at grain boundaries that was negative or active to the potential of grains. Furthermore, cathodic polarization effectively prevented s.c.c.

This viewpoint was later extended to metals that do not form intermetallic precipitates, but for which phase changes or segregation of alloy-

[21] H. Uhlig and E. Cook, Jr., *J. Electrochem. Soc.*, **116**, 173 (1969).

[22] T. Beck and M. Blackburn, *Amer. Inst. Aeronaut. Astronaut. J.*, **6**, 326 (1968).

[23] E. Dix, *Trans. Amer. Inst. Mining Met. Eng.*, **137**, 11 (1940).

ing elements or impurities supposedly occurs during the process of plastic deformation of metal at the tip of the crack, the resulting composition gradient setting up galvanic cells. Modifications of the electrochemical theory include the suggestion that cracks are formed mechanically and that electrochemical dissolution is needed only to remove periodic barriers to crack growth.[24] But brittle fracture of a ductile metal is unlikely no matter how sharp the notch. Furthermore, it was observed that removal of $FeCl_3$ solution from the crack formed in a stressed single-crystal Cu_3Au accompanied by further extension of the crystal resulted in immediate cessation of cracking followed by ductile behavior.[25] Also, a crack propagating in stressed 18-8 stainless steel immersed in boiling magnesium chloride solution stopped at the interface of 18-8 welded to nickel, the latter metal being immune to s.c.c. in chlorides.[26]

It became obvious, in any event, that the electrochemical mechanism could not readily account for the observed specificity of environment as shown by examples of Table 1. In principle, a large number of related electrolytes of comparable ionic conductivity should induce s.c.c.; but this is not the case. Furthermore, there are no obvious electrochemical reasons to account for the marked inhibition conferred by small additions of nonoxidizing ions like acetates, most metal compounds of which are soluble, to media used for accelerated tests. Other difficulties appeared, such as the transgranular cracking of an 18-8 stainless steel described earlier, despite well-defined intergranular paths for electrochemical dissolution. Furthermore, titanium and aluminum alloys undergo stress cracking in carbon tetrachloride which is not an electrolyte,[22, 27] and stressed high-strength carbon steels undergo cracking in pure hydrogen at 1 atm (but not if the hydrogen contains 0.7 vol % O_2).[28]

A critical experiment bearing on the doubtful validity of the electrochemical mechanism was performed by measuring the critical potentials for s.c.c. of 18-8 stainless steel exposed to magnesium chloride solution boiling at 130°C with and without inhibiting anion additions.[21] Anodic polarization induces shorter cracking times the more noble the controlled potential; cathodic polarization, on the other hand, extends the observed

[24] F. Keating, *Symp. Internal Stresses in Metals and Alloys,* p. 311, Inst. Metals, London, 1947.

[25] R. Bakish and W. Robertson, *Acta Met.,* **4,** 342 (1956).

[26] D. Van Rooyen, *J. Electrochem. Soc.,* **107,** 715 (1960).

[27] H. Paxton and T. Procter, *Proc. Fundamental Aspects of Stress-Corrosion Cracking,* p. 509, Nat. Assoc. Corros. Engrs., Houston, Texas, 1969.

[28] G. Hancock and H. Johnson, *Trans. Amer. Inst. Mining Met. Eng.,* **236,** 513 (1966).

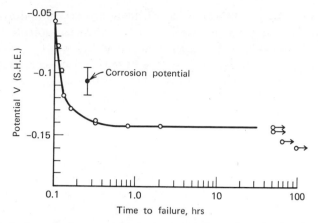

Fig. 5. Effect of applied potential on time to failure of stressed cold-rolled 18-8 stainless steel in magnesium chloride solution boiling at 130°C.[21]

cracking times. Below the critical value of -0.145 V (s.h.e.) the alloy becomes essentially immune (Fig. 5). Addition of various salts (e.g., sodium acetate) to the magnesium chloride solution is found to shift the critical potential to more noble values. When the amount of added salt shifts the critical potential to a value that is noble to the corrosion potential, s.c.c. no longer occurs (Fig. 6). If, therefore, the critical potential is interpreted as the open-circuit anode potential,

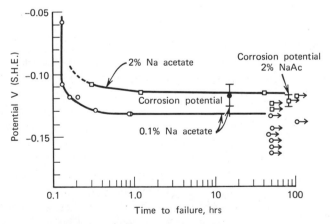

Fig. 6. Effect of applied potential on time to failure of stressed cold-rolled 18-8 stainless steel in magnesium chloride solution with sodium acetate additions, boiling at 130°C (2% sodium acetate addition is inhibiting).[21]

which is a characteristic of normal cathodic protection and a zero corrosion rate (see p. 58), it is impossible for the corrosion potential to assume any value that is active to the critical value. Hence the critical potential that is observed to be either active or noble to the corrosion potential must have another interpretation. This is discussed in the following section.

Stress-Sorption Cracking

In light of the accumulated evidence, it was proposed[3, 29] that s.c.c. in general proceeds not by electrochemical dissolution of metal but instead by weakening of the cohesive bonds between surface metal atoms through adsorption of damaging components of the environment. The name suggested for this mechanism is stress-sorption cracking. Since chemisorption is specific, damaging components are also specific. The surface energy of the metal is said to be reduced, increasing the tendency of the metal to form a crack under tensile stress. Hence the mechanism is related to the Griffith criterion[30] of crack formation in glass and similar brittle solids, for which the strain energy of the stressed solid must be greater than the total increased surface energy generated by an incipient crack. Adsorption of any kind which reduces surface energy should obviously favor crack formation, hence water adsorbed on glass understandably reduces the stress needed to cause fracture.

Langmuir[31] showed that only a monolayer of adsorbate is needed to markedly decrease the affinity of surface metal atoms for themselves or for their environment. Such adsorbed films prevent cold welding of metal surfaces placed in contact with each other. Stress corrosion cracking involves a similar decrease of surface affinities, with the distinction that only those specific adsorbates are effective which successfully reduce the attractive force of adjoining metal atoms for each other located at the extreme root of a notch subject to high tensile stress and experiencing some plastic deformation. Certain adsorbed ions, and in some instances water alone, do this more effectively for many metals than does oxygen. Oxygen adsorbs strongly, but presumably retains a high affinity for adjacent adsorbed oxygen, unlike damaging species, thereby preventing release of the neighboring bonds necessary to crack growth. It appears likely that damaging adsorption of the kind described occurs on mobile defect sites generated continuously at the crack tip. Hence the kinds of anions that adsorb and their subsequent effect

[29] E. Coleman, D. Weinstein, and W. Rostoker, *Acta Met.,* 9, 491 (1961).
[30] A. Griffith, *Phil. Trans.,* A220, 163 (1920).
[31] I. Langmuir, *J. Amer. Chem. Soc.,* 38, 2221 (1916); *Ibid.,* 40, 1361 (1918).

TABLE 2
Susceptibility* of Solid Metals to Embrittlement by Liquid Metals[35]

Liquid→	Li	Hg	Bi	Ga	Zn
Steel	C	NC	NC	NC	C
Copper alloys	C	C	C	—	—
Aluminum alloys	NC	C	NC	C	C
Titanium alloys	NC	C	NC	NC	NC

* C = cracking; NC = no cracking.

on metal properties differ from the effects predicted from adsorption measurements on the unstressed lattice.

The critical potential for s.c.c. is interpreted accordingly as that value above which damaging ions adsorb on appropriate defect sites and below which they desorb. Such a potential in principle can be either active or noble to the corrosion potential. Inhibiting anions, which of themselves do not initiate cracks, compete with damaging species for adsorption sites, thereby making it necessary to apply a more positive potential in order to reach a surface concentration of the damaging species adequate for adsorption and cracking. Whenever the inhibiting anion drives the critical potential above the corrosion potential, cracking no longer occurs; this is because the damaging ion is no longer able to absorb. The mechanism of competitive adsorption parallels that described earlier applying to the critical pitting potential, which is also shifted in the noble direction by the presence of extraneous anions (p. 76).

Stress-sorption cracking is the basic mechanism applying to stress cracking of plastics by specific organic solvents[32, 33] and to the cracking of solid metals by specific liquid metals (*liquid metal embrittlement*). It is also the mechanism proposed earlier by Petch and Stables[34] to account for stress cracking of steel induced by interstitial hydrogen (*hydrogen cracking*, p. 142).

The mechanism of liquid metal embrittlement is analogous to that of s.c.c. in that only certain combinations of liquid metals and stressed solid metals result in intergranular failure (Table 2). This specificity

[32] R. Mears, R. Brown, and E. Dix, Jr., in *Symp. Stress-Corrosion Cracking of Metals*, p. 323, ASTM-AIME, Philadelphia, Pa., 1945.

[33] F. Fischer, *Kunststoffe,* **55,** 453 (1965).

[34] N. Petch and P. Stables, *Nature,* **169,** 842 (1952).

[35] W. Rostoker, J. McCaughey, and H. Markus, *Embrittlement by Liquid Metals,* Reinhold, New York, 1960.

has important consequences, one being that mercury boilers can be and are constructed of carbon steel, but not of titanium or its alloys or of brass, because catastrophic intergranular failure would result. It is apparently possible for adsorbed mercury atoms to reduce the metal bond strength within grain-boundary regions of stressed titanium or brass adequate to cause failure, but this is not true of iron.

In summary, stress-sorption cracking may be said to account for most of the factors that characterize s.c.c. listed earlier, as well as the obviously nonelectrochemical cracking of glass by water; of plastics by organic solvents; or of metals by organic media, liquid metals or hydrogen. In this sense, to the extent that the model is correct, the basic mechanism of environmental fracture is the same for all materials.

Effect of Applied Stress, Composition, and Structure

A linear relation is reported between applied stress and log time to fracture by s.c.c. for austenitic and martensitic stainless steels, carbon

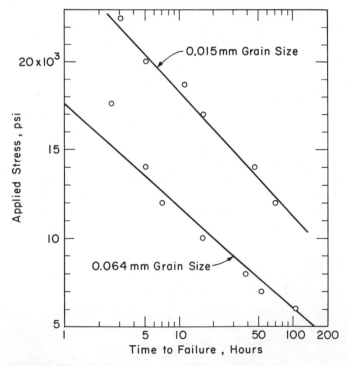

Fig. 7. Relation of applied stress to time for fracture of 66% Cu, 34% Zn brass exposed to ammonia (data of Morris[36] replotted).

steels, brass, and aluminum alloys. This relation and the usually greater resistance of small-grain-size alloys is shown for brass in Fig. 7.[36] For some metals the slope changes to one that is relatively shallow at low values of stress, indicating greater sensitivity of failure time to change of applied stress. It will be noted, however, that neither the general relation expressing fracture time versus stress nor the typical data of Fig. 7 support the concept of a threshold stress below which s.c.c. does not occur. A low surface tensile stress means only that time to fracture is relatively long.*

Possible explanations for the effects of applied stress, metal composition, and metal structure are still speculative. It was mentioned earlier that damaging adsorption appears to take place on mobile defect sites generated continuously at a notch or at the crack tip. Since creep of metals is also related to mobile defects in the metal lattice, a relation between creep behavior and time to fracture can be expected. The steady-state creep rate, R, follows the relation $\sigma = k \log R + $ const. where σ is the applied stress and k is a constant. The observed general relation for fracture time t by s.c.c. is $\sigma = -k \log t + $ const. These relations suggest that the fracture time is inversely proportional to the creep rate or $t = k'/R$.

Pure metals are not susceptible to s.c.c., perhaps because mobile imperfections move into and out of the surface areas at the root of a notch

---

* The actual stress at the base of a notch or crack is greater than the calculated average stress based on applied load. Hence the stress intensity factor, K_I, which is a measure of the elastic stress field at the tip of a crack, is sometimes plotted instead of the applied stress. For a previously fatigued sheet specimen with a single edge crack of length a

$$K_I = \frac{Pa^{1/2}}{BW}\left[1.99 - 0.41\left(\frac{a}{W}\right) + 18.70\left(\frac{a}{W}\right)^2 + \cdots \right]$$

where P is the applied uniform load, B is the specimen thickness, and W is the specimen width. See W. Brown, Jr., and J. Srawley, Spec. Tech. Report. No. 410, Amer. Soc. Testing Mater., Philadelphia, Pa., 1966.

A crack usually accelerates failure by acting as a stress raiser. In addition, anodic corrosion products accumulating at the base of a notch or crack tend to be acid and low in dissolved oxygen concentration and may further stimulate crack growth, particularly of stressed high-strength passive metals. For this reason, tests for s.c.c. of titanium alloys can avoid a long initiation time for breakdown of passivity by use of a so-called precracked specimen in which a short crack is formed initially by fatiguing the specimen.[37] It should be noted that although acid corrosion products accumulate at crevices because of operating concentration cells, it is not implied thereby that the mechanism of crack growth involves anodic dissolution of metal atoms at the tip of the crack.

[36] A. Morris, *Trans. Amer. Inst. Mining Met. Eng.*, **89**, 256 (1930).

[37] B. F. Brown, *Materials Res. and Std.*, **6** (3), 129 (1966).

too rapidly for adsorption to succeed. The presence of interstitial impurities—for example, carbon atoms along the grain boundaries of iron—slows down the movement of imperfections sufficiently (and perhaps also alters the chemical affinity of surface imperfections), favoring adsorption of damaging anions with consequent disruption of metallic bonds. Grain size becomes important because the greater number of imperfections available in large grains, on piling up along suitable barriers, favors increased adsorption at the barrier surface.

Body-centered cubic stainless steels are more resistant to s.c.c. by chlorides, presumably because the many operating slip planes of this lattice type favor tangled arrays of dislocations which make crack propagation difficult, whereas face-centered cubic stainless steels on which slip is largely constrained to (111) systems favor coplanar or parallel arrays of dislocations along which cracks move more easily.[38] The more important factor, however, may be that the corrosion potentials of ferritic stainless steels tend to be active to their critical potentials, whereas the reverse is true of austenitic stainless steels.[39] Addition of 2% Ni to a resistant ferritic 18% Cr stainless steel, for example, shifts the critical potential to a value that is active to the corrosion potential and hence such alloys fail within 2 hr in a solution of magnesium chloride boiling at 130°C. In austenitic stainless steels, on the other hand, increasing amounts of alloyed nickel shift the critical potential in the noble direction more than the corrosion potential; consequently, such alloys become more resistant to failure.[40] Above 45% Ni, the austenitic stainless steels are resistant regardless of applied potential or length of exposure to magnesium chloride solution. At this composition, metallurgical factors that suppress any tendency of the alloy to crack, such as tangled arrays of dislocations and a lower solubility of nitrogen in alloys of high nickel content, become important. Nitrogen, which tends to associate with dislocations more than carbon does, is one of the impurities that especially favors sensitivity of austenitic stainless steels to s.c.c. (see p. 318). As was previously suggested for carbon in iron, nitrogen may act to increase the life of dislocations moving into the metal surface on which adsorption of damaging anions occurs.

HYDROGEN CRACKING

Some metals, when stressed, crack on exposure to a variety of corrosive aqueous solutions with no evidence that the damaging solutions need be specific. For example, a stressed high-strength carbon steel or a martensitic stainless steel immersed in dilute sulfuric or hydrochloric

[38] P. Swann, *Corrosion*, **19**, 102t (1963).
[39] R. Newberg and H. Uhlig, to be published.
[40] H. Lee and H. Uhlig, *J. Electrochem. Soc.*, **117**, 18 (1970).

acid may crack within a few minutes. The failures have the outward appearance of s.c.c.; but if the alloy is cathodically polarized, cracking still occurs or happens in shorter time. This is contrary to the behavior of austenitic stainless steels in boiling magnesium chloride—such steels are cathodically protected under parallel circumstances. Also, if catalyst poisons that favor entrance of hydrogen into the metal lattice (e.g., sulfur or arsenic compounds) are added to the acids, cracking is intensified.[41] In practice, many stressed high-strength steels (e.g., carbon steels or 9% Ni steels) have failed from this source within days or weeks after exposure to oil-well brines[42] or to natural gas[43] containing hydrogen sulfide. Failure has also occurred within hours when martensitic 12% Cr-steel self-tapping screws (cathode) were applied in contact with an aluminum roof (anode) in a moist atmosphere. Cracking of steel springs, sometimes observed during pickling in sulfuric acid or after electroplating, is another example.

The evidence in all these cases is that cracking is caused by hydrogen entering the metal either through a corrosion reaction or by cathodic polarization[44] which liberates hydrogen at the metal surface. Steels containing interstitial hydrogen are not always damaged. They almost always lose ductility (hydrogen embrittlement), but only under conditions of sufficiently high applied or residual tensile stress does cracking take place. Failures of this kind are called hydrogen stress cracking or hydrogen cracking. The cracks tend to be mostly transgranular. In martensite they may follow former austenite grain boundaries.[44]

Carbon steels are especially susceptible to hydrogen cracking when heat treated to form martensite, but are less so if the structure is pearlitic. A carbon steel heat treated to form a spheroidized carbide structure is reported to be less susceptible than pearlite, bainite, or martensite.[43] Austenitic steels, e.g., 18-8 or 14% Mn steel[45] (face-centered cubic), in which hydrogen is more soluble than in ferrite and the diffusion rate is lower, are immune under most conditions of exposure.

Mechanism of Hydrogen Cracking

The mechanism of cracking has been explained by development of internal pressure[46] on the assumption that interstitial atomic hydrogen

[41] H. H. Uhlig, *Metal Progr.*, p. 486, April (1950).

[42] "Symposium on Sulfide Stress Corrosion," *Corrosion*, **8**, 326 (1952).

[43] P. Bastien, in *Physical Metallurgy of Stress Corrosion Fracture*, edited by T. Rhodin, p. 311, Interscience, New York, 1959.

[44] A. Schuetz and W. Robertson, *Corrosion*, **13**, 437t (1957).

[45] H. H. Uhlig, *Trans. Amer. Inst. Mining Met. Eng.*, **158**, 183 (1944).

[46] C. Zapffe and C. Sims, *Trans. Amer. Inst. Mining Met. Eng.*, **145**, 225 (1941). See also Ref. 43 and F. de Kazinczy, *J. Iron Steel Inst.*, **177**, 85 (1954); J. Morlet, H. Johnson, and A. Troiano, *Ibid.*, **189**, 37 (1958).

is released as molecular hydrogen at voids or other favored sites under extreme pressure. Such an effect certainly occurs, as is shown by the formation of visible blisters containing hydrogen when occasional ductile metals are cathodically polarized or exposed to certain corrosive media. Under similar conditions a less ductile metal would crack instead. Specimens of cathodically charged single-crystal 3% Si-Fe alloy showed microscopic rifts about 0.02 cm long parallel to (100) planes caused by release of hydrogen at favored sites of the crystal.[47]

Another proposed mechanism of hydrogen cracking is that hydrogen diffuses to and adsorbs at imperfections at the apex of the crack, reducing surface energy of metal atoms subject to a tensile force[34] (stress-sorption cracking).

An interesting characteristic of hydrogen cracking is a specific delay time for appearance of cracks after stress is applied. The delay time is only slightly dependent on stress; it decreases with increasing hydrogen concentration in the steel and with increase in hardness or tensile strength.[48] For small concentrations of hydrogen, fracture may occur some days after the stress is applied.

A critical minimum stress exists, below which delayed cracking will not take place in any time. The critical stress decreases with increase in hydrogen concentration. These effects are shown in Fig. 8 for SAE 4340 steel (0.4% C) charged with hydrogen by cathodic polarization in sulfuric acid, then cadmium plated to help retain hydrogen, and finally subjected to a static stress.[49] The hydrogen concentration was reduced systematically by baking.

Delay in fracture apparently results because of the time required for hydrogen to diffuse to specific areas near a crack nucleus until the concentration reaches a damaging level. These specific areas are presumably arrays of imperfection sites produced by plastic deformation of metal just ahead of the crack. Hydrogen atoms preferably occupy such sites because they are then in a lower energy state compared to their normal interstitial positions. The crack propagates discontinuously, because plastic deformation occurs first and then hydrogen diffuses to imperfection arrays produced by deformation, whereupon the crack propagates one step further. A sharp notch in a steel surface favors plastic deformation at its base, hence lowering the critical minimum stress and shortening the delay time. Below −110°C or at high deformation speeds, hydrogen embrittlement and cracking are minimized because diffusion of hydrogen is too slow.

[47] A. Tetelman and W. Robertson, *Trans. Met. Soc. (AIME)*, **224,** 775 (1962).

[48] A. Troiano, *Trans. Amer. Soc. Metals,* **52,** 54 (1960).

[49] H. Johnson, J. Morlet, and A. Troiano, *Trans. Amer. Inst. Mining Met. Eng.*, **212,** 528 (1958).

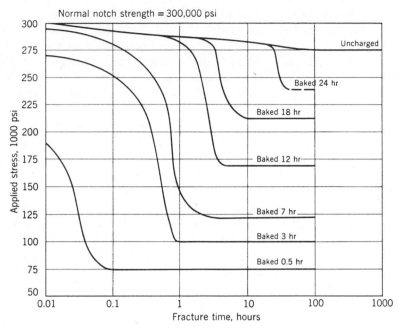

Fig. 8. Delayed fracture times and minimum stress for cracking of 0.4% C steel as a function of hydrogen content. Specimen initially charged cathodically, baked at 150°C for varying times to reduce hydrogen content (Johnson, Morlet, and Troiano).

Crack nuclei appear to be formed by separation at the interface of precipitated phases (e.g., Fe_3C, or intermetallic compounds such as those occurring in Maraging steels) from the plastically deformed matrix. Nuclei of this kind generated by Fe_3C in low-carbon 10% Ni-Fe alloys can be oriented by cold rolling so that resistance to hydrogen cracking is greatly increased in specimens stressed parallel to the rolling direction, but not in specimens stressed at right angles.[50] The usual necessity of a precipitate or second phase for hydrogen cracking to occur explains in part the resistance of pure iron, and also of austenitic steels in which carbon is more soluble than in the corresponding ferritic steels of the same approximate composition.

RADIATION DAMAGE

Metals exposed to intense radiation in the form of neutrons or other energetic particles undergo lattice changes resembling in many respects

[50] J. Marquez, I. Matsushima, and H. Uhlig, *Corrosion,* **26,** 215 (1970).

those produced by severe cold work. Lattice vacancies, interstitial atoms, and dislocations are produced, and these increase the diffusion rate of specific impurities or alloyed components. During radiation, a local temperature rise, called "temperature spike," may occur. There are two kinds of spikes: *thermal spikes* in which few or no atoms leave their lattice sites, and *displacement spikes* in which many atoms move into interstitial positions.

Few data are available on the consequences of radiation to the corrosion behavior of metals. Except for chemical products produced in the environment by radiation (e.g., HNO_3 or H_2O_2), which have a secondary effect on corrosion, or formation of localized displacement spikes during radiation, the effect of radiation may be expected to parallel that of cold work. That is, metals whose corrosion rate is controlled by oxygen diffusion should suffer no marked change in rate after irradiation. In acids, on the other hand, irradiated steel (but not pure iron) would presumably suffer a greater increase in rate than would irradiated nickel, which is less sensitive to cold working.

Austenitic stainless steels often become more sensitive to stress corrosion cracking after cold working; on this basis they might be expected to become more sensitive after irradiation. Davies at al.[51] found in fact that a stressed 17% Cr, 11% Ni, 2.5% Mo (type 316) stainless steel after irradiation with fast neutrons failed in much shorter time (1 hr) in boiling 42% $MgCl_2$ compared to unirradiated specimens (10 hr). Fracture time after irradiation, but not before, was independent of applied stress (5000–22,000 psi) suggesting that irradiation damage imposed severe residual stresses in the alloy to which the further contribution of an applied stress was negligible. The authors, however, preferred to explain their results in terms of a change in properties of a surface oxide film. A 20% Cr, 25% Ni, 1% Nb stainless steel did not fail in their tests either before or after irradiation.

The effect of irradiation on corrosion of some uranium alloys is found to be appreciable for reasons that are not yet entirely clear. For example, a 3%Cb-U alloy having moderate resistance to water at 260°C disintegrated within one hour after irradiation. Furthermore, the corrosion of a zirconium alloy (*Zircaloy*-2, see p. 370) at 250°C in dilute uranyl sulfate solution containing small amounts of H_2SO_4 and $CuSO_4$ was very much increased by reactor irradiation.[52] In a review of the

[51] M. Davies, D. Landsman, and W. Seddon, *Rept.*, AERE-R. 5014, Harwell, England, 1965.

[52] M. Simnad, in *The Effects of Radiation on Materials*, edited by J. Harwood et al., pp. 129–133, Reinhold, New York, 1958.

subject, Cox[53] stated that both fast neutron irradiation and presence of dissolved oxygen or an oxidizing electrolyte must be present simultaneously for any observed acceleration of corrosion to occur in high-temperature water. Accelerated corrosion of Zircaloys induced by irradiation is not observed above 400°C (750°F). The effects are explained in terms of changes in the physical properties of a protective oxide film.

CORROSION FATIGUE

A metal that progressively cracks on being stressed alternately or repeatedly is said to fail by fatigue. The greater the applied stress at each cycle, the shorter is the time to failure. A plot of stress versus number of cycles to failure, called the *S–N* curve, is shown in Fig. 9. A

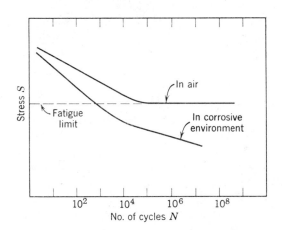

Fig. 9. *S–N* curve for steels subjected to cyclical stress.

number of cycles at the corresponding stress to the right of the upper solid line results in failure, but no failure occurs for an infinite number of cycles at or below the so-called *endurance limit* or *fatigue limit*. For steels, but not necessarily for other metals, a true endurance limit exists which is approximately half the tensile strength. The *fatigue strength** of any metal, on the other hand, is the stress below which failure does not occur within a stated number of cycles. Frequency of stress applica-

* This is called "endurance limit" by some authors.
[53] D. Cox, *J. Nuclear Mat.*, **28,** 1 (1968). See also, R. Asher et al., *Corros. Sci.*, **10,** 695 (1970).

tion is sometimes also stated because this factor influences the number of cycles to failure.

In a corrosive environment, failure at a given stress level usually occurs within fewer cycles, and a true fatigue limit is no longer observed (Fig. 9). In other words, failure occurs at any applied stress if the number of cycles is sufficiently large. Cracking of a metal resulting from the combined action of a corrosive environment and repeated or alternate stress is called *corrosion fatigue*. The damage is almost always greater than the sum of the damage by corrosion and fatigue acting separately.

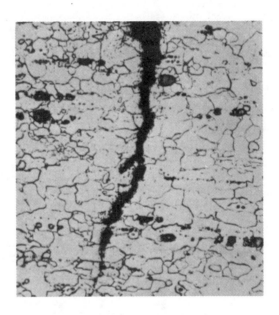

Fig. 10. Corrosion fatigue crack through mild steel sheet, resulting from fluttering of the sheet in a flue gas condensate. 250X.

Corrosion fatigue cracks are typically transgranular. They are often branched (Fig. 10; also Fig. 7, p. 1106, *Corrosion Handbook*) and several cracks are usually observed at the metal surface in the vicinity of the major crack accounting for failure. Fatigue cracks are similarly transgranular (exception: lead and tin), but rarely is there evidence of more than one major crack. In corrosion fatigue, corrosion pits may form at the metal surface at the base of which cracks initiate, but pitting is not a necessary precursor to failure.

Aqueous environments causing corrosion fatigue are numerous and are not specific, contrary to the situation for stress-corrosion cracking for which only certain ion-metal combinations result in damage. Steel suffers corrosion fatigue in fresh waters, seawater, combustion-product condensates, general chemical environments, etc., with the general rule that the higher the uniform corrosion rate, the shorter is the resultant fatigue life.

Corrosion fatigue is a common cause of unexpected cracking of vibrating metal structures designed to operate safely in air at stresses below the fatigue limit. For example, the shaft of a ship propeller slightly out of line will operate satisfactorily until a leak develops, allowing water to impinge on the shaft in the area of maximum alternating stress. Cracks may then develop within a matter of days, resulting in eventual parting and failure of the shaft. Steel oil-well sucker rods, used to pump oil from underground, have limited life because of corrosion fatigue resulting from exposure to oil-well brines. Despite use of high-strength medium-alloy steels and oversized rods, regular failures from this source are a loss to the oil industry in the order of millions of dollars annually. Wire cables commonly fail by corrosion fatigue. Pipes carrying steam or hot liquids of variable temperature may fail similarly because of periodic expansion and contraction (thermal cycling).

The usual fatigue test conducted in air on a structural metal is influenced by oxygen or moisture, and in part, therefore, always represents a measure of corrosion fatigue. In early tests, the endurance limit for copper in a partial vacuum was found to be increased 14% over that found in air. For mild steel, the increase was only 5%, but for 70-30 brass it was 26%.[54] In later tests,[55] fatigue life of OFHC copper at 10^{-5} mm-Hg pressure of air was found to be 20 times greater than at 1-atm air pressure. The main effect was attributed to oxygen; this had little effect on initiation of cracks, but had considerable effect on rate of crack propagation. Fatigue life of pure aluminum was also affected by air, but contrary to the situation for copper, water vapor in absence of air was equally effective. Gold, which neither chemisorbs oxygen nor oxidizes, had the same life whether fatigued in air or in a vacuum. High-strength steel ($>165,000$-psi yield strength) is found to have a much shorter fatigue life in moist air compared to dry air. On the other hand, the fatigue life of mild steel (68,000 psi yield strength) is not affected by moisture in air provided condensation does not occur.[56]

[54] H. Gough and D. Sopwith, *J. Inst. Metals,* **49,** 93 (1932).
[55] N. Wadsworth, in *Internal Stresses and Fatigue in Metals,* edited by G. Rassweiler and W. Grube, pp. 382–396, Elsevier, New York, 1959.
[56] H. Lee and H. Uhlig, to be published.

Fresh waters and particularly brackish waters have a greater effect on the corrosion fatigue of steels than of copper, the latter being a more corrosion-resistant metal. Stainless steels and nickel or nickel-base alloys are also better than carbon steels. In general, *resistance of a metal to corrosion fatigue is associated more nearly with its inherent corrosion resistance than with high mechanical strength.*

A few values of corrosion fatigue strength determined by McAdam[57] in fresh and brackish waters are listed in Table 3. These values, besides varying with environment, are found to vary with rate of stressing, with temperature, and with aeration, and hence they are useful only for qualitative comparison of one metal with another. Unlike the fatigue limit in air, they are not usually reliable for engineering design. Conclusions from data of Table 3 and similar data are:

1. There is no relation between corrosion fatigue strength and tensile strength.

2. Medium-alloy steels have only slightly higher corrosion fatigue strength than carbon steels.

3. Heat treatment does not improve corrosion fatigue strength of either carbon or medium-alloy steels; residual stresses are deleterious.

4. Corrosion-resistant steels, particularly steels containing chromium, have higher corrosion fatigue strength than other steels.

5. Corrosion fatigue strength of all steels is lower in salt water than in fresh water.

Remedial Measures

There are several means available for reducing corrosion fatigue. In the case of mild steel, thorough deaeration of a saline solution restores the normal fatigue limit in air (Fig. 11).[58] Cathodic protection to −0.49 V (S.H.E.) accomplishes the same result. Inhibitors are also effective; addition of 200 ppm $Na_2Cr_2O_7$ to tap water was found to reduce the corrosion fatigue of normalized 0.35% carbon steel wire to a level representing improvement over behavior of the steel in air.[59] Sacrificial coatings (e.g., zinc or cadmium electrodeposited on steel) are effective because they cathodically protect the base metal at defects in the coating. One of the very first of the observations and diagnoses of corrosion

---

[57] D. McAdam, Jr., *Proc. Amer. Soc. Testing Mat.,* **27,** II, 102 (1927). See also many other papers by the same author, who is largely responsible for the early systematic accumulation of corrosion fatigue data, and who coined the term "corrosion fatigue."

[58] D. Duquette and H. Uhlig, *Trans. Amer. Soc. Metals,* **61,** 449 (1968).

[59] F. Speller, L. McCorkle, and P. Mumma, *Proc. Amer. Soc. Testing Mat.,* **28,** pt. II, 159 (1928); *Ibid.,* **29,** pt. II, 238 (1929).

TABLE 3

Fatigue Limit and Corrosion Fatigue Strength of Various Metals (McAdam)

Metal	Fatigue Limit in Air (psi)	Corrosion Fatigue Strength (psi) (10^7–10^8 cycles at 1450 cycles/min)		Damage Ratio, Corros. Fatigue Strength/Fatigue Limit	
		Well Water*	Salt Water†	Well Water	Salt Water
0.11% C steel, annealed	25,000	16,000		0.64	
0.16% C steel, quenched, drawn‡	35,000	20,000	7,000	0.57	0.20
1.09% C steel, annealed	42,000	23,000		0.55	
3.5% Ni, 0.3% C steel, annealed	49,000	29,000		0.59	
0.9% Cr, 0.1% V, 0.5% C steel, annealed	42,000	22,000		0.52	
13.8% Cr, 0.1% C steel, quenched, drawn	50,000	35,000	18,000	0.70	0.36
17% Cr, 8% Ni, 0.2% C steel, hot-rolled	50,000	50,000	25,000	1.00	0.50
Nickel, 98.96%, annealed 760°C	33,000	23,500	21,500	0.71	0.65
Monel, 67.5% Ni, 29.5% Cu, annealed 760°C	36,500	26,000	28,000	0.71	0.77
Cupro-nickel, 21% Ni, 78% Cu, annealed 760°C	19,000	18,000	18,000	0.95	0.95
Copper, annealed 650°C	9,800	10,000	10,000	1.02	1.02
Aluminum, 99.4%, annealed	5,900		2,100		0.36
Aluminum, 98%, 1.2% Mn, hard	10,700	5,500	3,800	0.51	0.36
Duralumin, tempered	17,000	7,700	6,500	0.45	0.38
Brass, 60–40, annealed	21,000	18,000		0.86	

* 2 ppm $CaSO_4$, 200 ppm $CaCO_3$, 17 ppm $MgCl_2$, 140 ppm NaCl.
† Severn River water having about $\frac{1}{?}$ the salinity of seawater.
‡ "Drawn" means "tempered."

151

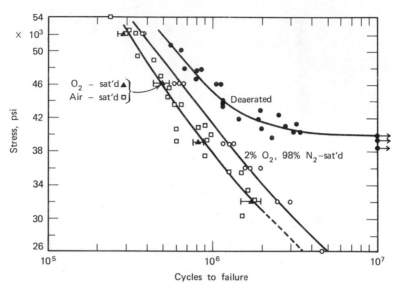

Fig. 11. Effect of dissolved oxygen concentration in 3% NaCl, 25°C, on fatigue behavior of 0.18% C steel (Duquette and Uhlig).

fatigue, made by B. Haigh in about 1916, involved premature failure of steel towing cables exposed to seawater, and galvanizing provided greatly increased life in this application.[60] Zinc coatings on aluminum are beneficial, but cadmium coatings are without effect.[61] Electrodeposits of tin, lead, copper, or silver on steel are also said to be effective without reducing normal fatigue properties; presumably they owe their effectiveness to exclusion of the environment.[62] Reports on the use of nickel or chromium coatings are conflicting. Organic coatings are useful if they contain inhibiting pigments (e.g., $ZnCrO_4$) in the prime coat. Shot peening the metal surface, or otherwise introducing compressive stresses, is beneficial.

Mechanism of Corrosion Fatigue

The mechanism of fatigue in air proceeds by localized slip within grains of the metal caused by stress, resulting in slip steps at the metal surface. Adsorption of air on the clean metal surface exposed at slip

[60] H. Gough, *J. Inst. Metals,* **49,** 17 (1932).

[61] I. Gerard and H. Sutton, *J. Inst. Metals,* **56,** 29 (1935).

[62] E. Gadd, in *Int. Conf. Fatigue of Metals,* p. 658, Institution Mech. Engr., London, 1956.

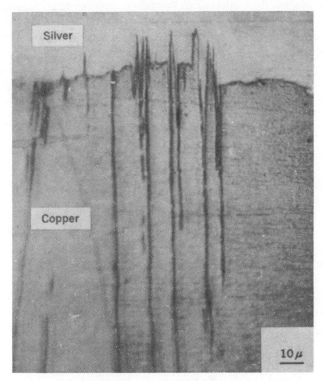

Fig. 12. Extrusions and intrusions in copper after 6×10^5 cycles in air. Specimen plated with silver after test and mounted at an angle to magnify surface protuberances by about 20X. Overall magnification about 500X. [W. Wood and H. Bendler, *Trans. Met. Soc., A IME,* **224,** 180 (1962)].

steps probably prevents rewelding on the reverse stress cycle. Continued slip produces displaced clusters of slip bands, which protrude above the metal surface (extrusions); corresponding incipient cracks (intrusions) form elsewhere (Fig. 12). Below the fatigue limit, but not above, work hardening accompanying each cycle of plastic deformation eventually impedes further slip, which in turn impedes the fatigue process. Hence a 0.19% C steel wire was found to be sensibly heated when cyclically stressed just beyond but not below the fatigue limit.[63]

A corrosive environment eliminates the fatigue limit (or shortens the fatigue life at stresses above the fatigue limit) probably by removing the barriers to plastic deformation such as dislocations piled up at the

[63] D. Whitwham and U. Evans, *J. Iron Steel Inst.,* **165,** 72 (1950).

metal surface at slip steps. The corrosion process may also conceivably induce plastic deformation by reducing surface energy, thereby favoring slip step formation, or by injecting dislocations along slip planes. In any event, when the metal undergoes corrosive attack, extrusions and intrusions accompanying the fatigue process resume their growth at stresses below the fatigue limit or they form at an accelerated rate above the fatigue limit. The lower the frequency of stress application, the more effectively does the corrosive environment decrease the number of cycles to failure (shorten fatigue life).[64]

Since pure metals are not immune to uniform corrosion, they are also not immune to corrosion fatigue. This behavior is contrary to their outstanding resistance to stress corrosion cracking, which operates by a different mechanism.

In order to affect fatigue life to a minimal extent, the uniform corrosion rate need not be high, but apparently it must exceed a critical value. The critical rate for mild steel stressed below the fatigue limit as measured by anodic current densities in deaerated 3% NaCl is 5 mdd ($2\mu A/cm^2$).[65] Below this rate, the approximate fatigue limit in air is observed. Above 5 mdd and below 75 mdd fatigue life is shorter the higher the corrosion rate, but within the range of about 75 to 750 mdd, fatigue life becomes constant. Hence, the fatigue life of mild steel in oxygen-saturated 3% NaCl solution is the same as that in an air-saturated solution despite a large difference in uniform corrosion rates (Fig. 11). In aerated 3% NaCl at pH 12, the normal fatigue limit is regained, despite observed pitting corrosion, presumably because the uniform corrosion rate now corresponds to that of passive steel, which corrodes at a value below the critical rate[65-67] (Fig. 13).

The critical corrosion rate of 5 mdd is equivalent to the removal of only 1×10^{-4} atom layer of iron for every stress cycle (1800 cycles/min). The relevant atoms presumably are those of the metal surface affecting plastic deformation at slip steps. Such atoms dissolve preferentially either because of kinetic factors operating during cyclic stressing (i.e., reduced anodic polarization) or because they are free of oxide films. They are probably not thermodynamically more active, because the potential required to regain the fatigue limit of mild steel in aerated 3% NaCl solution (cathodic protection) is about the same (−0.49 V, s.h.e.) as that required to attain constant life at stresses above the fatigue limit or to stop rusting of steel that is not stressed.[58]

[64] K. Endo and Y. Miyao, *Bull. Japan Soc. Mech. Engrs.*, **1**, 374 (1958).
[65] D. Duquette and H. Uhlig, *Trans. Amer. Soc. Metals*, **62**, 839 (1969).
[66] F. Radd, L. Crowder, and L. Wolfe, *Corrosion*, **16**, 415t (1960).
[67] V. Rollins et al., *Brit. Corros. J.*, **5**, 33 (1970).

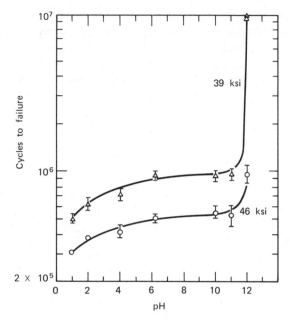

Fig. 13. Influence of pH of 3% NaCl, air-saturated, 25°C, on fatigue life of 0.18% C steel stressed above and below the fatigue limit (Duquette and Uhlig).

In other words, the open-circuit anode potential is independent of applied cyclic stress. The absence of an effect of cold work on the metal dissolution process is supported by the observation that previous cyclic stressing of steel in air (producing a certain amount of work hardening) has little or no effect on subsequent corrosion fatigue life in aerated $0.1N$ KCl solution.[63]

The main effect of the corrosive environment, therefore, is to reinitiate or induce plastic deformation of the cyclically stressed metal by whatever means, and this can succeed only when the corrosive environment acts conjointly with the applied stress. It has been proposed that previous formation of corrosion pits is a necessary condition to corrosion fatigue, but this appears to be unlikely. If pits are present, they can act under some conditions as notches or stress raisers favoring formation of fatigue cracks at their base. However, microscopic examination reveals that the intrusions formed initially during cyclic stressing establish crevices that may subsequently grow into pits because of differential aeration, passive-active cell formation, or acid corrosion-product accumulation within the crevice. In other words, a probable sequence of events is for the initial crack to develop into a pit.

Corrosion fatigue data (Table 3) show that differential aeration cells or passive-active cells, although important factors in accounting for observed pitting corrosion, cannot be primary factors in the fatigue mechanism. For example, copper forms differential aeration cells but is nevertheless more resistant to corrosion fatigue than are carbon steels. Stainless steels, nickel, and Monel, which are all passive and hence can form passive-active cells, are more resistant than are carbon steels, which do not form such cells. On the other hand, the shorter life of 0.16% C steel in aerated salt water compared to aerated well water (Table 3) is probably the result of enhanced differential aeration in the better conducting saline solution, resulting in a higher effective corrosion rate at the anodic crack tip. But the observed mechanism of crack growth is not one of electrochemical dissolution of metal atoms at the base of the crack. Instead, it involves as described previously, a process of plastic deformation which is accelerated by the corrosion process. The accelerating effect is not specific to Cl^- but occurs with any corrosive environment.

Copper is resistant to corrosion fatigue in fresh or salt waters, unlike many other metals (Table 3), because its normal corrosion rate at exposed slip steps or at incipient cracks lies below the critical value needed to accelerate plastic deformation. The critical corrosion rate for a fatigue life of 10^7 cycles is about 285 mdd (100 $\mu A/cm^2$) compared to 5 mdd for mild steel.[56] The ability of the corrosion process to accelerate plastic deformation explains why, in general, the property of corrosion resistance in a metal is more important than high tensile strength as a requirement for optimum resistance to corrosion fatigue.

FRETTING CORROSION

Fretting corrosion is another phenomenon that occurs because of mechanical stresses, and in the extreme it may lead to failure by fatigue or corrosion fatigue. It is defined as damage occurring at the interface of two contacting surfaces, one or both being metal, subject to slight relative slip. The slip is usually oscillatory as, for example, that caused by vibration. Continuous slip, as when one roll moves slightly faster than another roll in contact, leads to similar damage. Wear corrosion or friction oxidation are terms that have also been applied to this kind of damage.

Damage by fretting corrosion is characterized by discoloration of the metal surface and, in the case of oscillatory motion, by formation of pits. It is at such pits that fatigue cracks eventually nucleate. The rapid conversion of metal to metal oxide may in itself cause malfunctioning of machines, because dimensional accuracy is lost, or corrosion products cause clogging or seizing. The corrosion products are exuded from the faying surfaces, which in the case of steel are largely composed of αFe_2O_3 plus a small amount of iron powder.[68] In the case of nickel for continuous slip experiments, the products are NiO and a small amount of nickel; for copper they are Cu_2O with lesser amounts of CuO and copper.[69]

Fretting corrosion is frequently the cause of failure of suspension springs, bolt and rivet heads, king pins of auto steering mechanisms, jewel bearings, variable-pitch propellers, shrink fits, contacts of electrical relays, connecting rods, and many parts of vibrating machinery. It may cause discoloration of stacked metal sheets during shipment. One of the first examples of fretting corrosion was recognized when automobiles were shipped some years ago by railroad from Detroit to the West Coast. Because of vibration, the ball-bearing races of the wheels became badly pitted by fretting corrosion so that the automobiles were not operable. Damage was worse in winter than in summer, but could be avoided if the load on the wheels was relieved during shipment.

Laboratory experiments[68] have shown that fretting corrosion of steel versus steel requires oxygen but not moisture. Also, damage is less in moist air compared to dry air and is much less in a nitrogen atmosphere. Damage increases as temperature is lowered. The mechanism therefore is obviously not electrochemical. Increased load increases damage, accounting for the tendency of pits to develop at contacting surfaces because corrosion products, for example αFe_2O_3, occupy more volume (2.2 times as much in the case of iron) than the metal from which the oxide forms. Because the oxides are unable to escape during oscillatory slip, their accumulation increases the stress locally, thereby accelerating damage at specific areas of oxide formation. Fretting corrosion is also increased by increased slip, provided the surface is not lubricated. Increase in frequency for the same number of cycles decreases damage, but in nitrogen no frequency effect is observed. These effects are depicted in Fig. 14. Note that the initial rate of metal loss during the run-in period is greater than the steady-state value.

Mechanism of Fretting Corrosion

When two surfaces touch, contact occurs only at relatively few sites, called asperities, where the surface protrudes. Relative slip of the surfaces causes asperities to rub a clean track on the opposite surface which in the case of metal immediately becomes covered with adsorbed gas,

[68] I-Ming Feng and H. Uhlig, *J. Appl. Mech.*, **21**, 395 (1954).

[69] M. Fink and U. Hofmann, *Z. Anorg. Allg. Chem.*, **210**, 110 (1933) ; *Z. Metallk.*, **24**, 49 (1932).

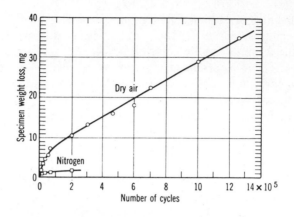

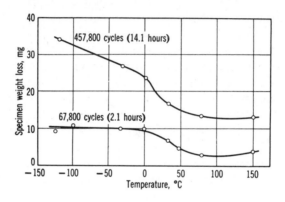

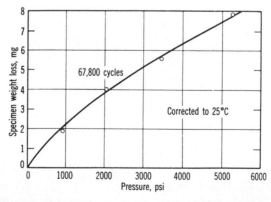

Fig. 14. Weight loss of mild steel versus mild steel by fretting corrosion in air (I-Ming Feng and H. Uhlig).

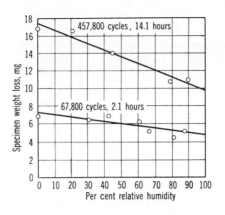

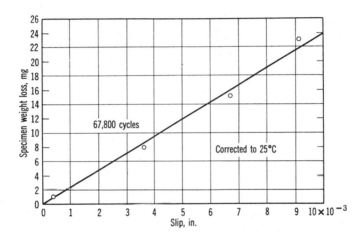

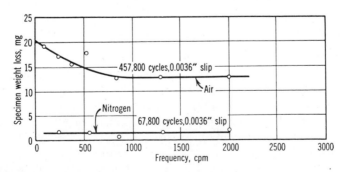

Fig. 14. (*Continued*)

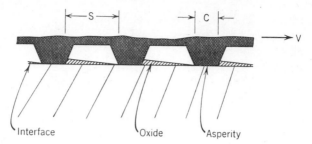

Fig. 15. Idealized model of fretting action at a metallic surface.

or it may oxidize superficially. The next asperity wipes off the oxide; or it may mechanically activate a reaction of adsorbed oxygen with metal to form oxide, which in turn is wiped off, forming another fresh metal track (Fig. 15). This is the chemical factor of fretting damage. In addition, asperities plow into the surface, causing a certain amount of wear by welding or shearing action through which metal particles are dislodged. This is the mechanical factor. Any metal particles eventually are converted partially into oxide by secondary fretting action of particles rubbing against themselves or against adjacent surfaces. Also the metal surface after an initial run-in period is fretted by oxide particles moving relative to the metal surface rather than by the mating opposite surface originally in contact (hence electrical resistance between the surfaces is at first low, then becomes high and remains so).

An equation for weight loss, W, of a metal surface undergoing fretting corrosion by oscillatory motion has been derived[70] (see Appendix, p. 404) on the basis of the model just described, which accounts reasonably satisfactorily for data of Fig. 14.

$$W = (k_0L^{1/2} - k_1L)\frac{C}{f} + k_2lLC$$

where L is the load, C is the number of cycles, f is the frequency, l is the slip, and k_0, k_1, and k_2 are constants. The first two terms of the right-hand side of the equation represent the chemical factor of fretting corrosion. These become smaller the higher the frequency, f, corresponding to less available time for chemical reaction (or adsorption) per cycle. The last term is the mechanical factor independent of frequency but proportionate to slip and load. It is found that either the mechanical or chemical factor may predominate in accounting for damage depending on specific experimental conditions. In nitrogen, the

[70] H. H. Uhlig, *J. Appl. Mech.*, **21**, 401 (1954).

mechanical factor alone is operable, the debris is metallic iron powder, and W is independent of frequency, f.

This mechanism is an extension of a similar one suggested by Fink.[71] Some investigators have proposed that only small-size metallic particles are dislodged (worn off) during the fretting process and that these are subsequently oxidized spontaneously in air.[72] But the effect of increasing frequency to decrease damage, the lesser damage in nitrogen even if the surface is initially covered with oxide,[68] and the lack of spontaneous oxidation of particles fretted in nitrogen and later exposed to air, for example, suggest that this mechanism does not apply.

Others have maintained that high temperatures produced by friction oxidize the metal, the oxide being subsequently rubbed off.[73] Although local high temperatures are undoubtedly produced by friction, the damage caused by fretting is not solely a high-temperature oxidation phenomenon. This is demonstrated by increased damage at below-room temperatures, by less damage at high frequencies for which surface temperatures are highest, by the fact that oxide produced in fretting corrosion of iron is αFe_2O_3 and not the high temperature form Fe_3O_4, and, finally, by steel being badly fretted in contact with polymethacrylate plastic, which melts at 80°C, and hence the surface could have reached temperatures of this order but not higher.[74]

The effect of moisture when adsorbed on the metal surface may be that of a lubricant. In addition, hydrated αFe_2O_3 is probably less abrasive than the anhydrous oxide. At low temperatures, damage is greater presumably because O_2 can adsorb more rapidly or more completely than at high temperatures. As in the case of corrosion fatigue, more fundamental research is needed to help clarify the general mechanism of fretting corrosion, which is now only partly understood.

Remedial Measures

1. *Combination of a Soft Metal with a Hard Metal.* At sufficiently high loads, soft metals serve to exclude air at the interface; furthermore, a soft metal may yield by shearing instead of sliding at the interface, thereby reducing damage. Tin-, silver-, lead-, indium-, and cadmium-coated metals in contact with steel have been recommended.

[71] M. Fink, *Trans. Amer. Soc. Steel Treating,* **18**, 1026 (1930); (with U. Hofmann), *Arch. Eisenhüttenw.,* **6**, 161 (1932).

[72] G. Tomlinson, *Proc. Roy. Soc. (London),* **A115**, 472 (1927); with P. Thorpe and H. Gough, *Proc. Inst. Mech. Engrs.,* **141**, 223 (1939); D. Godfrey, NACA Tech. Note 2039, February 1950.

[73] K. Dies, *Arch. Eisenhüttenw.,* **16**, 399 (1943).

[74] K. Wright, *Proc. Inst. Mech. Engrs. (B),* **1B**, (11), 556 (1952–1953).

Brass in contact with steel tends to be less damaging than steel versus steel. Combinations of stainless steels tend to be worst.

2. *Design of Contacting Surfaces to Avoid Slip Completely* (e.g., grit blasting, or otherwise roughening the surface). Intentional design to completely avoid slip is not always easy to accomplish because damage is presumably caused by relative movement approaching the order of atomic dimensions. Increased load is effective in this direction if it is high enough to prevent slip; otherwise damage is worse.

3. *Application of Cements* (e.g., rubber cement to the faying surfaces). Cements exclude air from the interface.

4. *Use of Lubricants.* Low-viscosity oils, particularly in combination with a phosphate-treated surface, can be helpful in reducing damage if the load is not too high. Low-viscosity oils diffuse more readily to the clean metal surface produced by oscillatory slip. Molybdenum sulfide is effective as a solid lubricant, particularly if baked onto the surface, but the beneficial effects tend to be temporary because the lubricant is eventually displaced by surface movement.

5. *Use of Elastomer Gaskets or Materials of Low Coefficient of Friction.* Rubber absorbs motion, thereby avoiding slip at the interface. Polytetrafluoroethylene (Teflon) has a low coefficient of friction and reduces damage. Because of their relatively poor strength, materials of this kind are expected to be effective only at moderate loads.

GENERAL REFERENCES

STRESS CORROSION CRACKING

H. Copson, in *Corrosion Handbook,* edited by H. H. Uhlig, pp. 569–578, Wiley, New York, 1948.

Symposium on Stress Corrosion Cracking of Metals, ASTM-AIME, Philadelphia, Pa., 1945.

Stress Corrosion Cracking and Embrittlement, edited by W. D. Robertson, Wiley, New York, 1956.

Physical Metallurgy of Stress Corrosion Fracture, edited by T. N. Rhodin, Interscience, New York, 1959.

H. L. Logan, *The Stress Corrosion of Metals,* Wiley, New York, 1966.

Stress Corrosion Testing, Spec. Tech. Publ. No. 425, Amer. Soc. Testing Mat., Philadelphia, Pa., 1967.

Proc. Fundamental Aspects of Corrosion Cracking, edited by R. Staehle et al., Nat. Assoc. Corros. Engrs., Houston, Texas, 1969.

FATIGUE AND CORROSION FATIGUE

B. Wescott, in *Corrosion Handbook,* edited by H. H. Uhlig, pp. 578–590, Wiley, New York, 1948.

Symposium on Basic Mechanisms of Fatigue, Tech. Publ. No. 237, Amer. Soc. Testing Mat., Philadelphia, Pa., 1959.

A. Gould, in "International Conference on Fatigue of Metals," pp. 341–347, Inst. Mech. Engrs., London, 1956.

H. Gough, *J. Inst. Metals,* **49,** 17 (1932).

P. Forrest, *Fatigue of Metals,* Pergamon Press, Oxford, 1962.

FRETTING CORROSION

J. Almen, in *Corrosion Handbook,* edited by H. H. Uhlig, pp. 590–597, Wiley, New York, 1948.

K. Wright, in *"Corrosion,"* edited by L. Schreir, pp. 8.87–8.94, Wiley, New York, 1963.

R. Waterhouse, *Proc. Inst. Mech. Engrs. (London),* **169,** 1157 (1955).

Symposium on Fretting Corrosion, Spec. Tech. Publ. No. 144, Am. Soc. Testing Mat., Philadelphia, Pa., 1952.

R. B. Waterhouse, *Fretting Corrosion,* Pergamon Press (in press)1971.

EFFECT OF RADIATION

The Effects of Radiation on Materials, edited by J. Harwood, H. H. Hausner, J. G. Morse, and W. G. Rauch, Reinhold, New York, 1958.

HYDROGEN CRACKING

M. Smialowski, *Hydrogen in Steel,* Pergamon Press, Oxford, 1962.

Hydrogen-Induced, Delayed, Brittle Failure of High Strength Steels, A. Elsea and E. Fletcher, DMIC Report 196, Battelle Mem. Inst., Columbus, Ohio, 1964.

chapter 8

Atmospheric corrosion of iron and other metals

In the absence of moisture, iron exposed to the atmosphere corrodes at a negligible rate. For example, steel parts abandoned in the desert remain bright and tarnish-free for long periods of time. Also, as discussed earlier, the corrosion process cannot proceed without an electrolyte; hence in climates below the freezing point of water or of aqueous condensates on the metal surface, rusting is negligible. Ice is a poor electrolytic conductor. Incidence of corrosion by the atmosphere depends, however, not only on the moisture content of air but also on the dust content and gaseous impurities which favor condensation of moisture on the metal surface.

TYPES OF ATMOSPHERES

Atmospheres vary considerably with respect to moisture, temperature, and contaminants, hence atmospheric corrosion rates vary markedly over all the world. Approaching the seacoast, air is laden with increasing amounts of sea salt, in particular NaCl. At industrial areas, appreciable amounts of SO_2, which converts to sulfuric acid, and lesser amounts of H_2S, NH_3, NO_2, and various suspended salts are encountered. A metal resisting one atmosphere may lack effective resistance elsewhere, and hence relative performance of metals changes with location. For example, galvanized iron performs well in rural atmospheres but is relatively less resistant to industrial atmospheres. On the other hand, lead performs in an industrial atmosphere at least as well as or better than elsewhere because a protective film of lead sulfate forms on the surface.

Recognition of marked differences in corrosivity has made it convenient to divide atmospheres into types. The major types are marine, industrial, tropical, arctic, urban, and rural. There are also subdivisions, such as wet and dry tropical with large differences in corrosivity. Also specimens exposed to a marine atmosphere corrode at greatly differing rates depending on proximity to the ocean. At Kure Beach, N. C., specimens of steel located 80 ft from the ocean where salt water spray is frequent, corroded about 12 times more rapidly than similar specimens located 800 ft from the ocean.[1]

CORROSION-PRODUCT FILMS

Rust films formed in the atmosphere tend to be protective; that is the corrosion rate decreases with time (Fig. 1). This is true to a lesser extent of pure iron for which the rate is relatively high, compared to the copper-bearing or low-alloy steels which are more resistant. Rust films on the latter steels tend to be compact and adherent, whereas on pure iron they are a powdery loose product. The corrosion rate

[1] F. L. LaQue, *Proc. Amer. Soc. Testing Mat.,* **51**, 495 (1951).

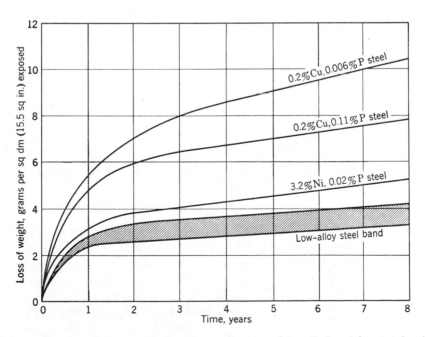

Fig. 1. Atmospheric corrosion of steels as a function of time (industrial atmosphere) (C. Larrabee, in *Corrosion Handbook,* p. 124).

TABLE 1
Average Atmospheric Corrosion Rates of Various Metals for 10- and 20-year Exposure Times [mils (in. $\times 10^{-3}$)/year]
(American Society for Testing and Materials)

	Atmosphere					
	New York City (urban-industrial)		La Jolla, Calif. (marine)		State College, Pa. (rural)	
	Years					
Metal	10	20	10	20	10	20
Aluminum	0.032	0.029	0.028	0.025	0.001	0.003
Copper	0.047	0.054	0.052	0.050	0.023	0.017
Lead	0.017	0.015	0.016	0.021	0.019	0.013
Tin	0.047	0.052	0.091	0.112	0.018	—
Nickel	0.128	0.144	0.004	0.006	0.006	0.009
65% Ni, 32% Cu, 2% Fe, 1% Mn (Monel)	0.053	0.062	0.007	0.006	0.005	0.007
Zinc (99.9%)	0.202	0.226	0.063	0.069	0.034	0.044
Zinc (99.0%)	0.193	0.218	0.069	0.068	0.042	0.043
0.2% C Steel* (0.02 P, 0.05 S, 0.05 Cu, 0.02 Ni, 0.02 Cr)	0.48					
Low-alloy steel* (0.1 C, 0.2 P, 0.04 S, 0.03 Ni, 1.1 Cr, 0.4 Cu)	0.09					

* Kearney, N. J. (near New York City); Values cited are one-half reduction of specimen thickness [C. P. Larrabee, *Corrosion*, **9,** 259 (1953)].

eventually reaches steady state and usually changes very little on further exposure. This is characteristic of other metals as well, as can be seen from data obtained by the American Society for Testing and Materials for various metals exposed 10 or 20 years to several atmospheres (Table 1).[2] Within the experimental error of such determinations, the rate for a 20-year period is about the same as for a 10-year period.

[2] *Symp. Atmospheric Corrosion of Non-Ferrous Metals,* Spec. Tech. Publ. No. 175, Amer. Soc. Testing Mat., Philadelphia, Pa., 1955. Also previous *Symp.,* 1946.

Metal surfaces located where they become wet or retain moisture but where rain cannot wash the surface may corrode more rapidly than specimens fully exposed. The reason for this is that sulfuric acid, for example, absorbed by rust will continue to accelerate corrosion, perhaps by means of the cycle

$$Fe \xrightarrow{H_2SO_4 + \frac{1}{2}O_2} FeSO_4 \xrightarrow{\frac{1}{2}O_2 + \frac{1}{2}H_2SO_4} \frac{1}{2}Fe_2(SO_4)_3 \xrightarrow{\frac{3}{2}H_2O} \frac{1}{2}Fe_2O_3 + \frac{3}{2}H_2SO_4 \quad (1)$$

Rust contaminated in this way catalyzes the formation of more rust.* Direct exposure of a metal to rain may therefore be beneficial, compared to a partially sheltered exposure. This advantage presumably would not extend to uncontaminated atmospheres.

The protective qualities of corrosion-product films are also evident from the weight losses of zinc specimens exposed for 28 days to an industrial atmosphere. Losses were found to depend largely on humidity and moisture conditions during the first five days of exposure and less on conditions arising later.[3] Similar results were found for steel, all specimens being exposed for 12 months, but the exposures were begun in different months of the year.[4] In winter the greater surface accumulation of combustion products, in particular sulfuric acid, produces a less protective initial corrosion product, which influences the subsequent corrosion rate. Schikorr[5, 6] confirmed that zinc exposed to Berlin or to Stuttgart atmospheres corrodes more rapidly in winter than in summer. For iron, he found that the corrosion rate was also higher in winter, but unlike zinc a decrease in rate occurred during cold spells, presumably either because $FeSO_4$ solution on the surface freezes or because oxidation of $FeSO_4$ is much retarded at low temperatures. Corresponding sulfate and sulfite solutions of Zn^{++} may freeze only at still lower temperatures (higher solubility) and, in any event, the corrosion process does not involve oxidation to higher valence ions as illustrated for iron by (1).

It has also been pointed out that, in evaluating aluminum alloys, the month in which atmospheric exposure tests are first begun is important to the final results.[7] Mayne,[8] along the same lines, found that

---

* G. Schikorr (private communication) points out that intermediate formation of ferric sulfate has not been demonstrated and that $FeSO_4$ may oxidize directly to Fe_2O_3.

[3] O. B. Ellis, *Proc. Amer. Soc. Testing Mat.*, **49**, 152 (1949).

[4] G. Schramm and E. Taylerson, *Symp. Outdoor Weathering of Metals and Metal Coatings*, Amer. Soc. Testing Mat., Philadelphia, Pa., 1934.

[5] G. Schikorr and I. Schikorr, *Z. Metallk.*, **35**, 175 (1943).

[6] G. Schikorr, *Schweiz. Arch. Angew. Wiss. Tech.*, **2**, 37 (1958).

[7] V. Carter and H. Campbell, p. 39; also H. Romans and H. Craig, Jr., p. 61 in *Metal Corrosion in the Atmosphere*, Spec. Tech. Publ. No. 435, Amer. Soc. Testing Mat., Philadelphia, Pa., 1968.

[8] J. Mayne, *J. Iron Steel Inst.*, **176**, 143 (1954).

paint applied to a rusted surface in December had a shorter life than the same paint applied in June. The reason proposed is that atmospheric pollutants from burning of fuels, which contaminate the surface during winter months, are washed away by rains during spring months.

FACTORS INFLUENCING CORROSIVITY OF THE ATMOSPHERE

It is important to remember that, in all except the most corrosive atmospheres, the average corrosion rates of metals are generally lower when exposed to air than when exposed to natural waters or to soils. This is illustrated by the data of Table 2 for steel, zinc, and copper

TABLE 2
Comparison of Atmospheric Corrosion Rates with Average Rates in Seawater and in Soils*

Environment	Corrosion Rate, mdd		
	Steel	Zinc	Copper
Rural atmosphere	–	0.17	0.14
Marine atmosphere	2.9	0.31	0.32
Industrial atmosphere	1.5	1.0	0.29
Sea water	25	10	8
Soil	5	3	0.7

* Atmospheric tests on 0.3% copper steel, $7\frac{1}{2}$-year exposure, from C. Larrabee, *Corrosion* **9**, 259 (1953). Atmospheric rates for zinc and copper, 10-year exposure, from *Symp. Atmospheric Exposure Tests on Non-Ferrous Metals*, ASTM, 1946. Seawater data from *Corrosion Handbook*. Soil data for steel are averaged for 44 soils, 12-year exposure; for zinc, 12 soils, 11-year exposure; for copper, 29 soils, 8-year exposure—from *Underground Corrosion*, M. Romanoff, Circ. 579, Nat. Bur. Std. (U. S.), 1957.

in three atmospheres compared to average rates in seawater and a variety of soils. In addition, atmospheric corrosion of passive metals (e.g., aluminum or stainless steels) tends to be more uniform and with less marked pitting than in waters or soils.

Specific factors influencing the corrosivity of atmospheres are dust content, gases in the atmosphere, and moisture (critical humidity).

Dust Content

The importance of atmospheric dust was emphasized by experiments of Vernon.[9] He exposed specimens of iron to an indoor atmosphere, some specimens being entirely enclosed by a cage of single-thickness muslin measuring several inches larger in size than the specimen. After several months, the unscreened specimens showed rust and appreciable gain in weight, whereas the muslin-screened specimens showed no rust whatsoever and had gained weight only slightly.

On a weight basis, dust is the primary contaminant of many atmospheres. The average city air contains perhaps 2 mg/m³, with higher values for an industrial atmosphere reaching 1000 mg/m³ or more.[10] It is estimated that more than 100 tons of dust per square mile may settle every month over an industrial city.* In contact with metallic surfaces, this dust influences the corrosion rate in an important way. Industrial atmospheres carry suspended particles of carbon and carbon compounds, metal oxides, H_2SO_4, $(NH_4)_2SO_4$, $NaCl$, and other salts. Marine atmospheres contain salt particles that may be carried many miles inland, depending on magnitude and direction of the prevailing winds. These substances, combined with moisture, initiate corrosion by forming galvanic or differential aeration cells; or because of their hygroscopic nature they form an electrolyte on the metal surface. Dustfree air, therefore, is less apt to cause corrosion than air heavily laden with dust, particularly if the dust consists of water-soluble particles or of particles on which H_2SO_4 is adsorbed.

Gases in the Atmosphere

The small amount of carbon dioxide normally present in air, contrary to the impressions of early investigators, neither initiates nor accelerates corrosion. Steel specimens rust in a carbon dioxide-free atmosphere as readily as in the normal atmosphere. Experiments by Vernon showed, in fact, that the normal carbon dioxide content of air actually decreases corrosion,[11] probably by favoring a more protective rust film.

A trace amount of hydrogen sulfide in contaminated atmospheres causes the observed tarnish of silver and may also cause tarnish of

---

* An estimated 475,000 tons of particulate emission per month is produced by electric utilities and industrial coal-fired power plants in the United States (*Chem. Eng. News,* p. 29, January 4, 1971).

[9] W. Vernon, *Trans. Faraday Soc.,* **23**, 113 (1927).

[10] A. Kutzelnigg, *Werks. Korros.* **8**, 492 (1957).

[11] W. Vernon, *Trans. Faraday Soc.,* **31**, 1668 (1935).

copper. The tarnish films are composed of Ag_2S, and a mixture of $Cu_2S + CuS + Cu_2O$, respectively.*

The most important corrosive constituent of industrial atmospheres is sulfur dioxide, which originates predominantly from the burning of coal, oil, and gasoline. In New York City it is estimated that 1.5 million tons of sulfur dioxide are produced every year from burning of coal and oil alone.[12] This is equivalent to burdening the atmosphere with an average of 6300 tons of H_2SO_4 every day.† Since fuel consumption is higher in winter, sulfur dioxide contamination is also higher (Fig. 2), consistent with the reference already made to the higher measured

Fig. 2. Variation of average sulfur dioxide content of New York City air with time of year (Greenburg and Jacobs).

corrosion rates of zinc and iron in winter compared to summer. It is also obvious from this cause that the average sulfur dioxide content of the air (and corresponding corrosivity) falls off with distance from the center of an industrial city, and that this effect is not as pronounced in the case of a residential city like Washington, D. C. (Table 3).

It is perhaps not surprising that the high sulfuric acid content of industrial and urban atmospheres should seriously shorten the life of metal structures. (See Tables 1 and 2.) The effect is most pronounced for metals that are not particularly resistant to sulfuric acid, such as zinc, cadmium, nickel, and iron. It is less pronounced for metals that

---

*Hence, a wrapping cloth impregnated with cadmium acetate prevents silver from tranishing by reacting with hydrogen sulfide to form insoluble cadmium sulfide $(Cd^{++} + H_2S \rightarrow CdS + 2H^+)$.

† This amount is now being reduced by limiting the allowable sulfur content of fuels burned within city limits.

[12] L. Greenburg and M. Jacobs, *Ind. Eng. Chem.*, **48**, 1517 (1956).

TABLE 3
Variation of SO$_2$ Content of Air with Distance from Center of City
(H. Meller, J. Alley, and J. Sherrick, quoted in Ref. 12)

		ppm					
City	Distance Miles	0–5	5–10	10–15	15–20	20–25	25–30
Detroit		0.023	0.012	0.006	0.004	0.004	0.005
Philadelphia-Camden		0.030	0.018	0.016	0.021	0.012	0.012
Pittsburgh		0.060	0.030	0.015	0.018	0.009	0.010
St. Louis		0.111	0.048	0.029	0.020	0.018	0.014
Washington		0.003	0.001	0.001	0.001	0.001	0.002

are more resistant to dilute sulfuric acid, such as lead, aluminum, and stainless steels. Copper, forming a protective basic copper sulfate film, is more resistant than nickel or 70% Ni-Cu alloy, on which the corresponding films are less protective. In the industrial atmosphere of Altoona, Pa., galvanized steel sheets (1.25 oz zinc/ft^2, 1.1 mil thick) began to rust after 2.4 years, whereas in the rural atmosphere of State College, Pa., rust appeared only after 14.6 years.[13]

Copper exposed to industrial atmospheres forms a protective green colored corrosion product called a *patina*, composed mostly of basic copper sulfate, $CuSO_4 \cdot 3Cu(OH)_2$. A copper-covered church steeple on the outskirts of a town may develop such a green patina on the side facing prevailing winds from the city but remain reddish-brown on the opposite side, where sulfuric acid is less readily available. Near the seacoast a similar patina forms, composed in part of basic copper chloride.

Nickel is quite resistant to marine atmospheres, but is sensitive to sulfuric acid of industrial atmospheres (Table 1), forming a surface tarnish composed of basic nickel sulfate. Corrosion in the industrial atmosphere of New York City is about 30 times higher than in the marine atmosphere of La Jolla, Calif., and about 20 times higher than in the rural atmosphere of State College, Pa. (Table 1).

Moisture (Critical Humidity)

From previous discussions, it is apparent that in an uncontaminated atmosphere at constant temperature, appreciable corrosion of a pure

---

[13] C. P. Larrabee, *Corrosion*, **9**, 259 (1953).

metal surface would not be expected at any value of relative humidity below 100%. Practically, however, because of normal temperature fluctuations (relative humidity increases on decrease of temperature) and because of hygroscopic impurities in the atmosphere or in the metal itself, the relative humidity must be reduced to values much lower than 100% in order to ensure that no water condenses on the surface. Vernon first discovered that a critical relative humidity exists below which corrosion is negligible.[14] Experimental values for the critical relative humidity are found to fall in general between 50 and 70% for steel, copper, nickel, and zinc. Typical corrosion behavior of iron as a function of relative humidity of the atmosphere is shown in Fig. 3.

REMEDIAL MEASURES

1. *Use of Organic, Inorganic, or Metallic Coatings.* Various coatings are discussed under separate chapters on mitigation of corrosion.

2. *Reduction of Relative Humidity.* Heating air, or better still reducing the moisture content, can serve to reduce relative humidity. Lowering the relative humidity to 50% suffices in many cases. If the presence of unusually hygroscopic dust or other surface impurities is suspected, the value should be reduced still further. This protective measure is effective except perhaps when corrosion is caused by acid vapors from nearby unseasoned wood or by certain volatile constituents of adjacent plastics, or paints.

3. *Use of Vapor-Phase Inhibitors and Slushing Compounds.* These are discussed in Chapter 16.

4. *Use of Alloys.* When alloyed with steel in small concentrations, copper, phosphorus, nickel, and chromium are particularly effective in reducing atmospheric corrosion. Copper additions are more effective in temperate climates than in tropical marine regions; chromium and nickel additions combined with copper and phosphorus are effective in both locations (Table 4). Corrosion rates of structural steels in tropical atmospheres (e.g., Panama) were found in general to be about two or more times higher than in temperate atmospheres (e.g., Kure Beach, N. C.), mainly because of the higher relative humidity and higher average temperatures.

The usefulness of low-alloy steels to effectively resist atmospheric corrosion through formation of protective rust films has resulted in the development of so-called *weathering steels.* These are employed for construction of buildings or bridges or for architectural trim without

[14] W. Vernon, *Trans. Faraday Soc.,* **23**, 113 (1927); *Ibid.,* **27**, 255 (1931); *Ibid.,* **31**, 1668 (1935); *Trans. Electrochem. Soc.,* **64**, 31 (1933).

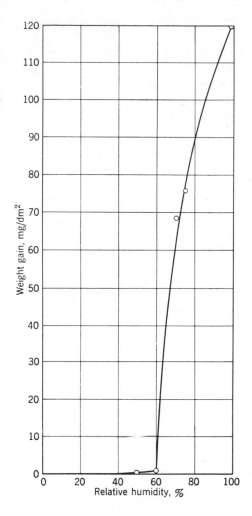

Fig. 3. Corrosion of iron in air containing 0.01% SO_2, 55 days' exposure, showing critical humidity (Vernon).

the necessity of painting, thereby saving appreciable amounts in maintenance costs over the life of the structure. A typical commercial composition (Corten) is as follows: 0.09% C, 0.4% Mn, 0.4% Cu, 0.8% Cr, 0.3% Ni, 0.09% P. Such steels do not have any advantage when buried in soil or totally immersed in water, because the corresponding rust films formed under continuously wet conditions are no more protective than those formed on carbon steels. (Their more noble corrosion potentials compared to carbon steels may make them useful in certain galvanic couples. See p. 122.) It is only by a process of alternate wetting and drying that the rust continues to be protective. Why alloying constituents of weathering steels should produce a protective rust is not yet known.

TABLE 4
Effect of Low-Alloy Components on Atmospheric Corrosion of Commercial Steel Sheet (Eight-Year Exposure)

Steel	Composition (%)				Loss of Thickness	
	C	P	Cu	Other	(mm)	(mils)
Industrial Atmosphere (Kearney, N. J.)[13]						
Carbon	0.2	0.02	0.03		0.20	8.0
Copper-bearing	0.2	0.02	0.3		0.11	4.4
Low-chromium	0.09	0.2	0.4	1 Cr	0.048	1.9
Low-nickel	0.2	0.1	0.7	1.5 Ni	0.051	2.0
Temperate Marine Atmosphere* (Kure Beach, N. C.)[13]						
Carbon	0.2	0.02	0.03		0.24	9.5
Copper-bearing	0.2	0.01	0.2		0.15	5.8
Low-chromium	0.1	0.14	0.4	1 Cr	0.069	2.7
Low-nickel	0.1	0.1	0.7	1.5 Ni	0.076	3.0
Tropical Marine Atmosphere (Panama Canal Zone)[15]						
Carbon	0.25	0.08	0.02		0.52	20.4
Copper-bearing	0.2	0.004	0.24		0.45	17.6
Low-chromium	0.07	0.008	0.1	3.2 Cr	0.23	9.1
Low-nickel	0.2	0.04	0.6	2.1 Ni	0.19	7.5

* 7.5-year exposure.

[15] C. Southwell, B. Forgeson, and A. Alexander, *Corrosion,* **14,** 435t (1958).

Stainless steels and aluminum resist tarnish in industrial, urban, and rural atmospheres, as is apparent from their satisfactory use over many years as architectural trim of buildings. Hastelloy C (54% Ni, 17% Mo, 5% Fe, 15% Cr, 4% W) is very resistant to tarnish in marine atmospheres, making it a useful alloy for reflectors on board ship.

GENERAL REFERENCES

Symposium on Atmospheric Corrosion of Non-Ferrous Metals, Spec. Tech. Publ. No. 175, Am Soc. Testing Mat., Philadelphia, Pa., 1955.

C. Larrabee, in *Corrosion Handbook,* edited by H. H. Uhlig, pp. 120–125, Wiley, New York, 1948. See also chapters on specific metals or alloys; coatings of zinc, nickel and chromium, cadmium, and lead.

H. Copson, "Design and Interpretation of Atmospheric Corrosion Tests," *Corrosion* **15,** 533t (1959).

Metal Corrosion in the Atmosphere, Spec. Tech. Publ. No. 435, Amer. Soc. Testing Mat., Philadelphia, Pa., 1968.

K. Chandler and M. Kilcullen, "Corrosion-Resistant Low-Alloy Steels," *Brit. Corros. J.,* **5,** 24 (1970).

chapter 9

Corrosion of iron and other metals in soils

Underground corrosion is of major importance because in the United States alone there are more than one million miles of buried oil, water, and gas pipelines. Corrosion of these structures necessitates regular maintenance and replacement costs of several hundred million dollars annually.

The corrosion behavior of iron and steel buried in the soil approximates, in some respects, their behavior on total immersion in water. Minor composition changes and structure of a steel, for example, are not important to corrosion resistance. Hence a copper-bearing steel, a low-alloy steel, a mild steel, and wrought iron are found to corrode at approximately the same rate in any given soil (see Fig. 3, p. 452, *Corrosion Handbook*). It is expected analogously that cold working or heat treatment would not affect the rate. Gray cast iron in soils, as well as in water, is subject to graphitic corrosion. Galvanic effects of coupling iron or steel of one composition to iron or steel of a different composition are important, as they are under conditions of total immersion (see p. 122).

In other respects, corrosion in soils resembles atmospheric corrosion in that observed rates, although usually higher than in the atmosphere, vary to a marked degree with the kind or type of soil. A metal may perform satisfactorily in some parts of the country but not elsewhere, because of specific differences in soil composition, pH, moisture content, etc. For example, a cast iron water pipe may last 50 years in New England soil, but only 20 years in the more corrosive soil of southern California.

FACTORS AFFECTING THE CORROSIVITY OF SOILS

Among the factors that govern corrosivity of a given soil are (1) porosity (aeration), (2) electrical conductivity, (3) dissolved salts, including depolarizers or inhibitors, (4) moisture, and (5) acidity or alkalinity. A porous soil may retain moisture over a longer period of time or may allow optimum aeration, and both factors tend to increase the initial corrosion rate. The situation is more complex, however, because corrosion products formed in an aerated soil may be more protective than those formed in an unaerated soil. In most soils, particularly if not well aerated, observed corrosion takes the form of deep pitting. Localized corrosion of this kind is obviously more damaging to a pipeline than a higher overall corrosion rate occurring more uniformly. Another factor to be considered is that in poorly aerated soils containing sulfates, sulfate-reducing bacteria may be found; these organisms often produce the highest corrosion rates normally experienced in any soil.

The effect of aeration has been summarized by Romanoff.[1]

In well-aerated soils the rate of pitting, although initially great, falls off rapidly with time because in the presence of an abundant supply of oxygen, oxidation and precipitation of iron as ferric hydroxide occur close to the metal surface, and the protective membrane formed in this manner tends to decrease the rate of pitting with time. On the other hand, in poorly aerated soils, the initial rate of pitting decreases slowly, if at all, with time. Under such conditions the products of corrosion, remaining in the deoxidized state, tend to diffuse outward into the soil, offering little or no protection to the corroding metal. The slope of the pit-depth–time curve may also be affected by the corrosiveness of the soil. Thus, even in a well-aerated soil an excessive concentration of soluble salt would prevent the precipitation of protective layers of corrosion products and the rate of corrosion would not be decreased with time.

In addition, it appears probable that aeration of soils may affect corrosion not only by the direct action of oxygen in forming protective films, but also indirectly through the influence of oxygen reacting with and decreasing concentration of the organic complexing agents or depolarizers naturally present in some soils which greatly stimulate local-action cells. In this regard, the beneficial effect of aeration extends to soils that harbor sulfate-reducing bacteria, because these bacteria become dormant in presence of dissolved oxygen.

Corrosion rates tend to increase with depth of burial near the soil surface, but not invariably. Bureau of Standards tests[1] on steel specimens exposed 6 to 12 years and buried 12 to 48 in. (30 to 120 cm)

---

[1] M. Romanoff, *Underground Corrosion,* Circ. 579, Nat. Bur. Std., (U.S.), 1957.

below the surface showed greater corrosion at greater depth for five soils but the reverse trend for two soils.

A soil containing organic acids derived from humus is relatively corrosive to steel, zinc, lead, and copper. The measured total acidity of such a soil appears to be a better index of its corrosivity than pH alone. High concentrations of sodium chloride and sodium sulfate in poorly drained soils such as are found in parts of southern California, make the soil very corrosive. In addition to increased activity of local-action cells, macrogalvanic cells or "long-line" currents established by oxygen concentration differences, by soils of differing composition, or by dissimilar surfaces on the metal, become more important when electrical conductivity is high. Anodes and cathodes may be thousands of feet or even miles apart. A poorly conducting soil, whether from lack of moisture or lack of dissolved salts or both, is in general less corrosive than a highly conducting soil. But conductivity alone is not a sufficient index of corrosivity; it has been demonstrated that the anodic or cathodic polarization characteristics of a soil are also a factor.[2]

Cinders constitute one of the most corrosive environments. For four- or five-year exposures, corrosion rates for steel and zinc in cinders was 5 times, for copper 8 times, and for lead 20 times higher than the average rates in 13 different soils.

BUREAU OF STANDARDS TESTS

The most extensive series of field tests on various metals and coatings in almost all types of soils were begun in 1910 by K. H. Logan of the National Bureau of Standards. These tests continued until 1955 and now constitute the most important source of information on soil corrosion available anywhere.[1] They showed similarity in corrosion rates in a given soil for various kinds of iron and steels; confirmation was obtained through five-year tests in England.[3] A few typical corrosion rates averaged for many soils are listed in Table 1. In addition, data are listed for two soils relatively corrosive to steel and for one that is relatively noncorrosive, showing how large are the variations in corrosion rate from one soil to another. For San Diego soil, the symbol > means that the thickness of test specimen was completely penetrated by pitting at the end of the exposure period.

Referring again to Table 1, we see that *copper* on the average corrodes at about one-sixth the rate of iron, but in tidal marsh, for example,

[2] N. Tomashov and Y. Mikhailovsky, *Corrosion,* **15,** 77t (1959).
[3] J. C. Hudson and G. Acock, in *Corrosion of Buried Metals,* Iron and Steel Inst. Spec. Report 45, London, 1952.

TABLE 1
Corrosion of Steels, Copper, Lead, and Zinc in Soils

National Bureau of Standards[1]

Maximum penetration in mils (1 mil = 0.001 inch) for total exposure period.

Average Corrosion Rates in mg/dm²/day

Soil	Open Hearth Iron 12 year exp.		Wrought Iron 12 year exp.		Bessemer Steel 12 year exp.		Copper 8 year exp.		Lead 12 year exp.		Zinc 11 year exp.	
	mdd	mils	mdd	mils	mdd	mils	mdd	mils	mdd	mils	mdd	mils
Average of several soils	4.5 (44 soils)	70	4.7 (44 soils)	59	4.5 (44 soils)	61	0.70 (29 soils)	<6	0.52 (21 soils)	>32	3.0 (12 soils)	>53
Tidal marsh, Elizabeth, N.J.	10.8	90	11.6	80	19.5	100	5.3	<6	0.2	13	1.9	36
Montezuma clay Adobe, San Diego, Calif.	13.7	>145	13.4	>132	14.3	>137	0.7	<6	0.6 (9.6 years)	10	...	...
Merrimac gravelly sandy loam, Norwood, Mass.	0.9	28	1.0	23	1.0	21	0.2 (13.2 years)	<6	0.13	19	...	...

ORGANIC GASES

the rate is comparatively higher than in most other soils, being one-half that of iron. The rate for copper is normal in otherwise corrosive San Diego soil. Pitting is not pronounced, the maximum depth reaching less than 0.006 in. (0.015 cm).

Lead also corrodes less on the average than does steel. In poorly aerated soils or soils high in organic acids, the corrosion rate may be much higher (four to six times) than the average. Pitting in some of these soils penetrated the test specimen thickness, accounting for an average maximum penetration greater than the average given in Table 1. *Zinc* also pitted in some soils to an extent greater than the specimen thickness. In five-year tests carried out in Great Britain, commercially pure *aluminum* was severely pitted in four soils (4 to >63 mils, 0.1 to >1.6 mm), but was virtually unattacked in a fifth soil.[4]

Increase in chromium content of steel decreases observed weight loss in a variety of soils; but above 6% Cr, depth of pitting increases. In 14-year tests, 12% Cr and 18% Cr steels were severely pitted. The 18% Cr, 8% Ni, type 304 stainless steel was not pitted or was only slightly pitted (<6 mils), nor was weight loss appreciable in 10 out of 13 soils. However, in three soils at least one specimen was perforated (16–32 mils thick) by pitting. Type 316 stainless steel did not pit in any of the 15 soils to which the alloy was exposed for 14 years. It is expected, however, that pitting would also occur with this alloy in longer time tests, since pitting occurs in seawater within about two and one-half years.

Zinc coatings are surprisingly effective in reducing weight loss and pitting rates of steel exposed to soils. In 10-year tests in 45 soils, coatings of 2.8 oz/ft^2 based on one side of the specimen (5 mils or 0.013 cm thick) protected steel against pitting with the exception of one soil (Merced silt loam, Buttonwillow, Calif.) in which some penetration of the base steel could be measured. In later tests extending up to 13 years, a coating of 3.1 oz/ft^2 effectively reduced (but did not prevent) corrosion, even in cinders in which the zinc coating was destroyed within the first two years. There is some evidence, perhaps not yet conclusive, that a major source of protection results from the alloy layer formed between zinc and the steel surface in the hot-dip galvanizing process. (See p. 234, Zn Coatings.)

Pitting Characteristics

It should be noted that since the dimensions of all field test specimens were in the order of inches up to about 1 ft, the reported pitting rates

[4] P. Gilbert and F. Porter, in *Corrosion of Buried Metals,* Iron and Steel Inst. Spec. Report 45, London, 1952.

represent minimum rather than maximum values. Actual depth of pits in a given time is found to increase slightly with size of test specimen, probably because cathodic area per pit increases, thus accounting for higher current densities at the pits. In addition to this factor, long-line currents or macrocells, if present, increase pit depth over the values obtained on small specimens where such cells do not operate.

The rate at which pits grow in the soil under a given set of conditions tends to decrease with time and follows a power law equation $P = kt^n$, where P = the depth of the deepest pit in time t, and k and n are constants. It is reported[5] that values of n for steels range from about 0.1 for a well-aerated soil, to 0.9 for a poorly aerated soil. The smaller the value of n, the greater is the tendency for the pitting rate to fall off with time. As n approaches unity, the pitting rate approaches a constant value, or penetration is proportional to time.

Pits tend to develop more on the bottom side of a pipeline than on the top side. This difference is sometimes sufficiently great to make it worthwhile rotating a pipeline 180° after a given period of exposure in order to increase pipe life. Pitting on the bottom side results from constant contact with the soil, whereas the top side, because of the pipe settling, tends to become detached, producing an air space between the pipe and the soil.

REMEDIAL MEASURES

1. Organic and Inorganic Coatings

Paint coatings of thickness normal to atmospheric protection fail within months when exposed to the soil. On the other hand, thickly applied coatings of coal tar, with reinforcing pigments or inorganic fibers to reduce cold flow, are found to be practical. They provide effective protection at reasonable cost, and are the coatings mostly used in practice. Bureau of Standards tests showed that soft rubber coatings 0.25 in. (0.6 cm) thick afforded excellent protection to steel in 11-year exposure tests.

Portland cement coatings applied over steel have been found in practice to be effectively protective for long years of exposure.[6] Bureau of Standards tests showed that vitreous enamel coatings, when free of pores, are also protective. Both these coatings are brittle and tend, therefore, to be readily damaged mechanically.

---

[5] Ref. 1, p. 39.
[6] F. N. Speller, *Corrosion, Causes and Prevention,* pp. 596–599, McGraw-Hill, New York, 1951.

2. Metallic Coatings

The effectiveness of zinc coatings has already been mentioned. Such coatings deteriorate more rapidly when galvanically coupled to large areas of bare iron, steel, or copper, in which case insulating couplings should prove useful.

3. Alteration of Soil

A soil high in organic acids can be made less corrosive by surrounding the metal structure with limestone chips. A layer of chalk ($CaCO_3$) surrounding buried pipes has been used in some soil formations liable to produce microbiological corrosion.[7]

4. Cathodic Protection

In general, all modern buried pipelines and tanks away from congested areas are protected by a combination of cathodic protection and reinforced coal-tar coating. This combination protects steel against corrosion in all soils, both effectively and economically, for as many years as cathodic protection is adequately maintained.

GENERAL REFERENCES

Kirk, H. Logan, *Underground Corrosion*, Nat. Bur. Std. (U. S.) Circ. C450 (1945).

M. Romanoff, *Underground Corrosion*, Nat. Bur. Std. (U. S.) Circ. 579 (1957).

Corrosion of Buried Metals, Spec. Report No. 45, Iron and Steel Inst., London, 1952.

K. Logan, in *Corrosion Handbook*, edited by H. H. Uhlig, pp. 446–466, Wiley, New York, 1948.

---

[7] U. R. Evans, *The Corrosion and Oxidation of Metals*, p. 272, Ed. Arnold, London, 1960.

chapter 10

Oxidation and tarnish

When a metal is exposed at room or elevated temperatures to an oxidizing gas (e.g., oxygen, sulfur or halogens) corrosion may occur in absence of a liquid electrolyte. This is sometimes called "dry" corrosion in contrast to "wet" corrosion which occurs when a metal is exposed to water or to the soil. In dry corrosion, a solid reaction-product film or scale (a scale is a thick film) forms on the metal surface through which the metal, the environment, or both must diffuse in order for the reaction to continue. It is found that usually ions rather than atoms migrate through solid oxides, sulfides, or halides; hence the reaction product can be considered an electrolyte. For copper oxidizing in O_2, or silver tarnishing in a contaminated atmosphere, Cu_2O and Ag_2S, respectively, are the solid electrolytes. The migrating ions are not hydrated and they diffuse simultaneously with electrons, but along different paths.

INITIAL STAGES

A clean reactive metal surface exposed to oxygen follows the sequence (1) adsorption of oxygen, (2) formation of oxide nuclei, (3) growth of a continuous oxide film. Because the free energy of adsorption of atomic oxygen exceeds the free energy of dissociation of oxygen, the first-formed adsorbed film consists of atomic oxygen. Low-energy electron diffraction data show that some metal atoms enter the approximate plane of adsorbed oxygen to form a relatively stable two-dimensional structure of mixed O ions (negative) and metal ions (positive). As was discussed earlier with relation to the passive film (p. 74), this initial adsorbed partial monolayer is thermodynamically more stable than is the metal oxide. For nickel, for example, the adsorbed film resists decomposition up to the melting point of nickel[1], whereas NiO decomposes accompanied by oxygen dissolving in the metal.* Continued exposure to low-pressure oxygen is followed by adsorption of O_2 molecules on metal atoms exposed through the first adsorbed layer. Since the second layer of oxygen is bonded less energetically than is the first layer, it is adsorbed without dissociation to its atoms. The resultant structure is usually more stable on transition than on nontransition metals.[2] Any additional layers of adsorbed oxygen are still less strongly bonded and the outer layers at elevated temperatures eventually become mobile, the corresponding diffraction pattern being that of an amorphous structure. It is likely that metal ions enter the multilayer adsorbed film in nonstoichiometric amounts and that such ions are also relatively mobile. For example, rate of surface diffusion of silver and copper atoms is observed to be higher in presence of adsorbed oxygen than in its absence.[3]

Because the free energy of adsorption per mole of oxygen decreases with amount of oxygen adsorbed, (the O–substrate bond becomes weaker) multilayer adsorbed oxygen on metal M eventually favors transformation to a crystalline stoichiometric oxide. In other words, ΔG for $O \cdot M_{ads} + nO_2 \rightarrow (nO_2) \cdot O \cdot M_{ads}$ becomes less negative per mole O_2 as n increases, whereas $\triangle G$ for $2nM + (nO_2) \cdot O \cdot M_{ads} \rightarrow (2n + 1)MO$ to form the oxide becomes correspondingly more negative. Oxide formation, therefore, preferably nucleates at surface sites where multilayer adsorption is favored, such as at surface vacancies, ledges, or other imperfections. When conditions are favorable, oxide nuclei form relatively suddenly at specific sites of the surface aided by rapid surface diffusion of M and O ions (Fig. 1).[4] Because the amount of adsorbed oxygen on any likely site increases with oxygen pressure, more sites are brought into play and the density of nuclei increases. Similarly, since multilayer adsorption decreases at elevated temperatures, the density of nuclei de-

---

* J. Moreau and J. Bénard (*Compt. Rend. Acad. Sci. (Paris)*, **242**, 1724 (1956) showed previously through observations of the metal surface in H_2O-H_2 mixtures at elevated temperatures that oxygen adsorbed on an 18% Cr stainless steel is thermodynamically more stable than is the metal oxide. For analogous data on iron, see A. Pignocco and G. Pellissier, *J. Electrochem. Soc.*, **112**, 1118 (1965); E. Hondros, *Acta Met.*, **16**, 1377 (1968).

[1] A. MacRae, *Surface Sci.*, **1**, 319 (1964).

[2] H. H. Uhlig, *Corros. Sci.*, **7**, 325 (1967).

[3] G. Rhead, *Acta Met.*, **11**, 1035 (1963); *Ibid.*, **13**, 223 (1965); *Trans. Faraday Soc.*, **61**, 797 (1965).

[4] L. Brockway and A. Rose, in *Fundamentals of Gas Surface Interactions*, edited by H. Saltsburg et al., p. 147, Academic Press, New York, 1967.

Fig. 1. Nuclei of Cu_2O formed on copper surface at 10^{-1} mm Hg oxygen pressure, 525°C, 20 sec. (17,600X). Black lines are bands of imperfections (stacking faults) in the copper lattice (Brockway and Rowe).

creases with increase of temperature.[5, 6] Nuclei grow rapidly to a certain height in the order of tens of angstroms,* then more rapidly in a lateral rather than in a vertical direction. On copper exposed to 1-mm Hg oxygen pressure at 550°C a continuous film of Cu_2O forms within 6 sec; at 1 atm O_2 the time required is still less.

It has been suggested that growth of the resulting thin continuous

* Limited perhaps by the electron tunneling distance[2] below which electrons can penetrate the metal-oxide interface barrier without first acquiring the usual activation energy.

[5] F. Gronlund, *J. Chim. Phys.*, **53**, 660 (1956).

[6] J. Bénard, F. Gronlund, J. Oudar, and M. Duret, *Z. Elektrochem.*, **63**, 799 (1959).

oxide film is controlled by transfer of electrons from metal to oxide,[7] or in some instances by migration of metal ions in a strong electric field set up by negatively charged oxygen adsorbed on the oxide surface.[8] When the continuous film reaches a thickness in the order of several thousand angstroms, diffusion of ions through the oxide may become rate-controlling instead. The latter situation prevails so long as the oxide remains continuous. Eventually, at a critical thickness, the stresses set up in the oxide may cause it to crack (called spalling) and the oxidation rate increases irregularly.

PROTECTIVE AND NONPROTECTIVE SCALES

Rate of reaction in the later stages of oxidation depends on whether the thick film or scale remains continuous and protective as it grows, or whether it contains cracks and pores and is relatively nonprotective. Because reaction-product films are usually brittle and lack ductility, the initiation of cracks depends in some measure on whether the surface film is formed in tension favoring fracture or in compression favoring protection. This situation in turn depends on whether the volume of reaction product is greater or less than the volume of metal from which the product forms.[9] If $Md/nmD > 1$, a protective scale forms, where M is the molecular weight and D the density of scale, m and d are the atomic weight and density of metal, and n is the number of metal atoms in a molecular formula of scale substance ($n = 2$ for Al_2O_3). On the other hand, when this ratio is less than unity, the scale is formed in tension and tends to be nonprotective. For calcium and magnesium which tend to form nonprotective oxides, the ratios are 0.64 and 0.79, respectively, whereas for aluminum and chromium, which tend to form protective oxides, the ratios are 1.3 and 2.0. Tungsten has a ratio of 3.6 and hence the oxide, WO_3, is normally expected to be protective, except at high temperatures (approximately $>800°C$), where it volatilizes.

Three Equations of Oxidation

The three main equations that express thickness, y, of film or scale forming on any metal within time, t, are (1) the linear, (2) the parabolic, (3) the logarithmic. The particular equation that applies depends in part on the ratio Md/nmD and the relative thickness of the film or scale.

[7] H. H. Uhlig, *Acta Met.*, **4**, 541 (1956).

[8] N. Cabrera and N. Mott, *Reports Progr. Phys.*, **12**, 163 (1949).

[9] N. Pilling and R. Bedworth, *J. Inst. Metals*, **29**, 534 (1923).

For the linear equation, the rate of oxidation is constant, or $dy/dt = k$ and $y = kt +$ const., where k is a constant. Hence, the thickness of scale, y, plotted with time t is linear (Fig. 2). This equation holds whenever the reaction rate is constant at an interface as, for example, when the environment reaches the metal surface through cracks or pores in the reaction-product scale. Hence, for such metals, the ratio

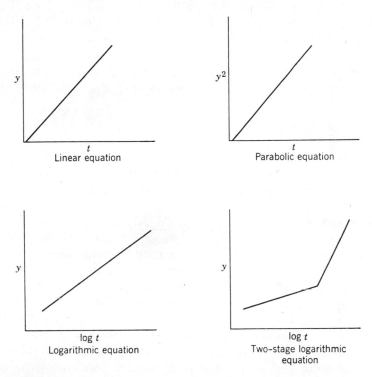

Fig. 2. Various equations expressing growth of film thickness, y, as a function of time of oxidation, t.

Md/nmD is usually less than unity. In special cases the linear equation may also hold even though the latter ratio is greater than unity, such as when the controlling reaction rate is constant at an inner or outer phase boundary of the reaction-product scale. For example, tungsten first oxidizes at 700 to 1000°C in accord with the parabolic equation, forming an outer porous WO_3 layer and an inner compact oxide scale of unknown composition.[10] When the rate of formation of the outer scale becomes equal to that of the inner scale, the linear equation is obeyed.

[10] W. Webb, J. Norton, and C. Wagner, *J. Electrochem. Soc.*, **103**, 107 (1956).

For the parabolic equation, the diffusion of ions or migration of electrons through the scale is controlling and the rate, therefore, is inversely proportional to scale thickness.

$$\frac{dy}{dt} = \frac{k'}{y} \quad \text{or} \quad y^2 = 2k't + \text{const.}$$

Accordingly, if y^2 is plotted with t, a linear relation is obtained (Fig. 2). This equation holds for protective scales corresponding to $Md/nmD > 1$ and is applicable to the oxidation of many metals at elevated temperatures such as copper, nickel, iron, chromium, and cobalt.

For relatively thin protective films, as when metals oxidize initially or when oxidation occurs at low temperatures, it is found that

$$\frac{dy}{dt} = \frac{k''}{t} \quad \text{or} \quad y = k'' \ln\left(\frac{t}{\text{const.}} + 1\right)$$

This is called the logarithmic equation. Correspondingly, if y is plotted with $\log(t + \text{const.})$, or with $\log t$ for $t \gg \text{const.}$, a linear relation is obtained. First reported by Tammann and Köster,[11] the logarithmic equation has been found to express the initial oxidation behavior of many metals, including Cu, Fe, Zn, Ni, Pb, Cd, Sn, Mn, Al, Ti, and Ta. Its validity was questioned at first; later several attempts were made to derive the equation on the basis of assumed diffusion blocks in oxides, ion-concentration gradients, or other special properties of oxides. These assumptions have not received experimental support. It has been shown, on the other hand, that the logarithmic equation can be derived on the condition that the oxidation rate is controlled by transfer of electrons from metal to reaction-product film[7, 12] when the latter is electrically charged; that is, when it contains a space charge throughout its volume. The preponderance of electric charge of usually negative sign in oxides near the metal surface, similar to the electrical double layer in aqueous electrolytes, has been demonstrated experimentally. Therefore, any factor changing the work function of the metal (energy required to extract an electron) such as grain orientation, lattice transition, or magnetic transition (Curie temperature) changes the oxidation rate, as has been observed. When the thickness of the film exceeds the maximum thickness of the space-charge layer, diffusion or migration through the film then usually becomes controlling, the parabolic equation applies, and the factors just mentioned, such as grain orientation, no longer affect

[11] G. Tammann and W. Köster, *Z. Anorg. Allg. Chem.*, **123**, 196 (1922).

[12] E. C. Williams and P. Hayfield, in *Vacancies and Other Point Defects in Metals and Alloys,* Inst. of Metals, Report 23, p. 131, London, 1958.

the rate. On this basis, metals forming protective films first follow the logarithmic equation and then the parabolic (or linear) equation.

If for thin film behavior, migration of ions controls the rate and the prevailing electric field within the film is considered to be set up by gaseous ions adsorbed on the outer surface, the rate of ion migration is an exponential function of the field strength, and the inverse logarithmic equation[8] results

$$\frac{1}{y} = \text{const.} - k \ln t$$

This relation has been reported to hold for copper and iron oxidized at low temperatures.[13-15] It is often difficult to distinguish between the logarithmic and the inverse logarithmic equations because of the limited range of time over which data can be accumulated for thin film behavior, either equation applying equally well. This situation also makes it difficult to evaluate other types of equations that have been proposed, such as the cubic equation $y^3 = kt + \text{const.}$ Data obeying this equation can also be represented in many cases by a two-stage logarithmic equation where an initial lower rate is followed by a final higher rate[16] (Fig. 2). The higher rate has been ascribed to formation of a diffuse space charge layer overlying an initially constant charge-density layer.[7]

The oxidation rate for thin or thick film conditions increases with temperature, obeying the usual Arrhenius equation

$$\log (\text{reaction rate constant}) = A \exp\left(\frac{-\Delta E'}{RT}\right)$$

where $\Delta E'$ is the activation energy, R is the gas constant, T is the absolute temperature, and A is a constant.

WAGNER THEORY OF OXIDATION

When certain metals oxidize (e.g., copper, zinc, nickel), it is found that metal ions migrate through the oxide to the outer oxide surface, reacting there with oxygen. For these metals, outward diffusion of metal ions occurs in preference to diffusion of the larger oxygen ions inward. Reaction occurring at the outer rather than the inner oxide surface was first reported by Pfeil,[17] who noticed that when an iron surface was painted with a slurry of Cr_2O_3, the green-colored layer after oxidation

[13] F. Young, J. Cathcart, and A. Gwathmey, *Acta Met.*, **4**, 145 (1956).

[14] T. Rhodin, *J. Amer. Chem. Soc.*, **72**, 5102 (1950); **73**, 3143 (1951).

[15] D. Gilray and J. Mayne, *Corros. Sci.*, **5**, 55 (1965).

[16] H. Uhlig, J. MacNairn, and J. Pickett, *Acta Met.*, **7**, 111, (1959).

[17] L. Pfeil, *J. Iron Steel Inst.*, **119**, 520 (1929).

was buried in the middle or lower layers of iron oxide. Iron ions, in other words, diffused through the Cr_2O_3 marker and reacted with oxygen at the gas-oxide interface. Similarly, Wagner showed by quantitative studies that Ag^+ and not S^{--} ions migrate through Ag_2S.[18] By placing two weighed pellets of Ag_2S over the silver, on top of which was molten sulfur (Fig. 3), he showed that after 1 hr at 220°C the bottom pellet

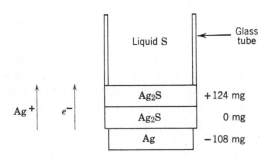

Fig. 3. Formation of silver sulfide from silver and liquid sulfur, 1 hr, 220°C (Wagner).

next to the silver neither gained nor lost weight, whereas the top pellet in contact with the sulfur gained weight chemically equivalent to the loss in weight of metallic silver. Wagner also showed that by assuming independent migration of Ag^+ and electrons, the observed reaction rate could be calculated from independent physical chemical data. He[19] derived an expression for the parabolic rate constant, which has the following simplified version:[20]

$$k = \frac{En_3(n_1 + n_2)\kappa}{F} \tag{1}$$

where E is the emf of the operating cell derived either from potential measurement or free energy data; n_1, n_2, and n_3 are the mean cation, anion, and electron transference numbers, respectively, within the reaction product film; κ is the mean specific conductivity of film substance;

[18] C. Wagner, *Z. Physik. Chem.*, **21B**, 25 (1933).

[19] C. Wagner, *Z. Physik. Chem.*, **21B**, 25 (1933); *Ibid.,* **32B**, 447 (1936); *Ibid.,* **40B**, 455 (1938); with K. Grünewald, *Trans. Faraday Soc.*, **34**, 851 (1938); *Atom Movements*, pp. 153–73, American Society for Metals, Cleveland, Ohio, 1951.

[20] T. P. Hoar and L. E. Price, *Trans. Faraday Soc.*, **34**, 867 (1938). (See also *Corrosion Handbook*, p. 17 and Per Kofstad, *High Temperature-Oxidation of Metals*, pp. 113–127, Wiley, New York, 1966.)

and F is the Faraday. The constant k appears in the relation

$$\frac{dw}{dt} = k\frac{A}{y}$$

where dw/dt is the rate of formation of film substance in equivalents/second, A is the area, and y is the thickness. The calculated value of k for silver reacting with sulfur at 220°C is 2 to 4×10^{-6} compared to the observed value of 1.6×10^{-6} eq/cm-sec. For copper oxidizing at 1000°C and 100-mm oxygen pressure, the calculated and observed values of k are 6×10^{-9} and 7×10^{-9} eq/cm-sec, respectively. The excellent agreement between theory and observation confirms the validity of the model suggested by Wagner for the oxidation process within the region for which the parabolic equation applies.

Equation 1 holds for either diffusion of cations or anions or both in the reaction product scale. Some metals, for example titanium and zirconium, oxidize in part by migration of oxygen ions, O^{--}, via anion vacancies in the corresponding outer oxides.

OXIDE PROPERTIES AND OXIDATION

Metal oxides are commonly in the class of electrical conductors called semiconductors; that is, their conductivity lies between insulators and metallic conductors. Conductivity increases with a slight shift from stoichiometric proportions of metal and oxygen and with an increase of temperature. There are two types of semiconducting oxides, namely, p- or n-types (p = positive carrier, n = negative carrier). In the p-type, shift of stoichiometric proportions takes the form of a certain number of missing metal ions in the oxide lattice called cation vacancies, represented by $\square$. At the same time, to maintain electrical neutrality, an equivalent number of positive holes, $\oplus$, form (i.e., sites where electrons are missing). A cupric ion, Cu^{++}, in a Cu_2O lattice is an example of a positive hole. Oxides of p-type are Cu_2O, NiO, FeO, CoO, Bi_2O_3, and Cr_2O_3. A model for the Cu_2O lattice is shown in Fig. 4. During the oxidation of copper, cation vacancies and positive holes are formed at the outer O_2-oxide surface. These migrate to the metal surface, a process which is equivalent to the reverse migration of Cu^+ and electrons.

For n-type oxides, excess metal ions exist in interstitial positions of the oxide lattice, and it is these which migrate during oxidation together with electrons (Fig. 4) to the outer oxide surface. Examples of n-type oxides are ZnO, CdO, TiO_2, and Al_2O_3. Wagner showed that the law of mass action can be applied to the concentration of interstitial ions and electrons, and also to cation vacancies and positive holes. Hence for

$$
\begin{array}{ccccc}
\text{Cu}^{++} & \text{Cu}^{+} & \text{Cu}^{+} & \text{Cu}^{+} & \text{Cu}^{+} \\
& \text{O}^{--} \quad\ \text{O}^{--} \quad\ \text{O}^{--} \quad\ \text{O}^{--} & & & \\
\text{Cu}^{+} & \square & \text{Cu}^{+} & \text{Cu}^{++} & \text{Cu}^{+} \\
& \text{O}^{--} \quad\ \text{O}^{--} \quad\ \text{O}^{--} \quad\ \text{O}^{--} & & & \\
\text{Cu}^{+} & \text{Cu}^{+} & \text{Cu}^{+} & \square & \text{Cu}^{+}
\end{array}
$$

Lattice Defects in Cu_2O

$$
\begin{array}{cccccc}
\text{Zn}^{++} & \text{O}^{--} & \text{Zn}^{++} & \text{O}^{--} & \text{Zn}^{++} & \text{O}^{--} \\
& e^{-} & & e^{-} & \text{Zn}^{++} & \\
\text{O}^{--} & \text{Zn}^{++} & \text{O}^{--} & \text{Zn}^{++} & \text{O}^{--} & \text{Zn}^{++} \\
\text{Zn}^{++} & & & e^{-} & e^{-} & \\
\text{Zn}^{++} & \text{O}^{--} & \text{Zn}^{++} & \text{O}^{--} & \text{Zn}^{++} & \text{O}^{--}
\end{array}
$$

Lattice Defects in ZnO

Fig. 4.

Cu_2O the equilibrium relations are

$$\tfrac{1}{2}O_2 \rightleftharpoons Cu_2O + 2Cu_{\square}^{-} + 2\oplus \tag{2}$$

and

$$C_{Cu_{\square}^{-}}^{2} \cdot C_{\oplus}^{2} = \text{const.} \cdot p_{O_2}^{\frac{1}{2}} \tag{3}$$

For ZnO the corresponding equilibrium is

$$ZnO \leftrightharpoons Zn_{int}^{++} + 2e^{-} + \tfrac{1}{2}O_2 \tag{4}$$

and

$$C_{Zn^{++}(Int.)} \cdot C_{e^{-}}^{2} = \frac{\text{const}}{p_{O_2}^{\frac{1}{2}}}. \tag{5}$$

(It has also been suggested that interstitial zinc ions in ZnO may be monovalent instead of divalent.) These relations lead to interesting predictions regarding the effect of impurities in the oxide lattice on oxidation rate. For example, if a few singly charged Li^+ ions are substituted for doubly charged Ni^{++} ions in the NiO lattice, the concentration of positive holes must increase in order to maintain electrical neutrality. Hence, to maintain equilibrium for the reaction

$$\tfrac{1}{2}O_2 \rightleftharpoons NiO + Ni_{\square}^{--} + 2\oplus$$

as expressed by

$$C_{Ni_{\square}^{--}} \cdot C_{\oplus}^{2} = \text{const.} \ p_{O_2}^{\frac{1}{2}} \tag{6}$$

the concentration of cation vacancies must decrease. This is accompanied by a decrease in oxidation rate,[21] because the rate is controlled

[21] K. Hauffe and H. Pfeiffer, *Z. Elektrochem.*, **56**, 390 (1952).

ox rate ∝ [cation vacancies]

by migration of cation vacancies. On the other hand, if small amounts of Cr^{+3} are added to NiO, the concentration of positive holes decreases, the concentration of cation vacancies correspondingly increases, and the oxidation rate increases. Data of Wagner and Zimens[22] showing the effect of alloyed chromium on oxidation of nickel are given in Table 1.

TABLE 1
Oxidation of Nickel Alloyed with Chromium, 1000°C, 1-atm O_2

Wt. % Cr	Parabolic Rate Constant k $g^2 cm^{-4} sec^{-1}$
0	3.1×10^{-10}
0.3	14×10^{-10}
1.0	26×10^{-10}
3.0	31×10^{-10}
10.0	1.5×10^{-10}

At 10% Cr the rate decreases again, possibly because a scale composed of Cr_2O_3 forms instead of NiO,* which alters the rate of ion migration apart from factors described previously.

In a sulfur atmosphere, for reasons paralleling those pertaining in the situation with O_2, up to 2% Cr alloyed with nickel is reported to accelerate reaction at 600 to 900°C.[23] Outward diffusion of Ni^{++} occurs through cation vacancies in $Ni_{1-x}S$ where x connotes a number between 0 and 1, and incorporation of Cr^{+3} increases cation vacancy concentration. In >40% Cr-Ni alloys, outward diffusion of Cr^{+3} occurs in a scale composed of Cr_2S_3. Incorporation of Ni^{++} ions into the Cr_2S_3 scale decreases cation vacancy concentration, thereby decreasing the reaction rate to a value below that for pure chromium. At intermediate chromium compositions, the scale is heterogeneous, consisting of both nickel and chromium sulfides, with the rate of sulfidation being less than that of pure chromium for >20% Cr-Ni.

From (3) it is apparent that higher partial pressures of oxygen for p-type semiconductors must be accompanied by higher concentrations of vacancies and holes at the O_2-oxide interface. Hence copper oxidizes at higher rates the higher the oxygen pressure, in accord with prediction.[24]

* C. Wagner and H. Rickert, private communication.

[22] C. Wagner and K. Zimens, *Acta Chem. Scand.*, **1**, 547 (1947).

[23] S. Mrowec, T. Werber, and N. Zastawnik, *Corros. Sci.*, **6**, 47 (1966).

[24] C. Wagner and K. Grünewald, *Z. Physik. Chem.*, **40B**, 455 (1938).

If small amounts of Li^+ are added to ZnO, which is an n-type semiconductor, the electron concentration decreases in order to preserve neutrality and the concentration of interstitial zinc ions increases in accord with the law of mass action (Eq. 5). This increase in concentration facilitates the diffusion of Zn^{++}_{int}, and hence Li^+ increases the oxidation rate of zinc, contrary to its effect in NiO. By similar reasoning, Al^{+3} decreases the rate. Data showing these effects are presented in Table 2.[25]

TABLE 2
Oxidation of Zinc, 390°C, 1-atm O_2

	Parabolic Rate Constant k $g^2cm^{-4}hr^{-1}$
Zn Pure	0.8×10^{-9}
Zn + 0.1% Al	1×10^{-11}
Zn + 0.4% Li	2×10^{-7}

The oxidation rate of zinc is almost independent of oxygen pressure because the concentration of interstitial zinc ions is already low at the O_2-oxide interface and any further decrease brought about by increasing oxygen pressure has but little effect on their concentration gradient referred to the metal surface where Zn^{++}_{int} concentration is highest.

Traces of impurities, therefore, which play a major role in semiconductor properties, also appreciably affect rates of oxidation of metals protected by semiconductor films. On the other hand, alloying elements present in large percentages (e.g., >10% Cr-Ni) affect the oxidation rate by a gross alteration of the actual composition and structure of the film, in addition to any effects on semiconducting properties.

GALVANIC EFFECTS AND ELECTROLYSIS OF OXIDES

The electrochemical nature of oxidation at elevated temperatures suggests that galvanic coupling of dissimilar metals should affect the rate, and in fact such effects are found.[26] The reaction of silver with gaseous iodine at 174°C, for example, is accelerated by contact of the silver with tantalum, platinum, or graphite. AgI, which is mainly an ionic conductor, forms on silver at a rate limited by transport of electrons

---

[25] C. Gensch and K. Hauffe, *Z. Physik. Chem.*, **196**, 427 (1950).
[26] C. Ilschner-Gensch and C. Wagner, *J. Electrochem. Soc.*, **105**, 198, 635 (1958).

across the AgI layer. When silver is coupled to tantalum, Ag⁺ ions diffuse over the surface of the tantalum, the latter supplying electrons that hasten conversion of silver to AgI. In addition, therefore, to the usual film of AgI on the silver, the compound spreads progressively over the tantalum surface (Fig. 5). Analogously, it was found[27] that

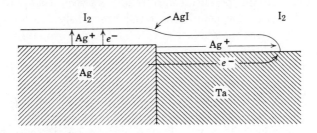

Fig. 5. Effect of coupling tantalum to silver on reaction of iodine vapor with silver (Ilschner-Gensch and Wagner).

when silver coated with a porous gold electroplate is exposed to sulfur vapor at 60°C, Ag_2S forms a tightly adherent coating over the gold surface.

A second example is given by nickel completely immersed in molten Na-K borax to a depth of about 0.3 cm at 780°C in 1 atm O_2 (Fig. 6). The rate of oxidation under these conditions is low because of limited access of oxygen from the gaseous phase. If nickel is coupled to

[27] T. Egan and A. Mendizza, *J. Electrochem. Soc.,* **107,** 353 (1960). For reasons still unknown, similar effects were not observed when rhodium and palladium were substituted for gold.

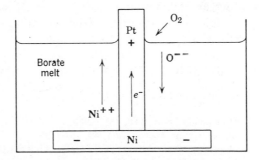

Fig. 6. Galvanic cell: Pt; O_2, borate melt; Ni, illustrating accelerated oxidation of nickel through contact with platinum.

platinum or silver gauze, the latter reaching above the liquid borax surface, corrosion of the nickel is very much accelerated (by a factor of 35 to 175 for 1-hr exposure). Nickel under these conditions corrodes even more rapidly than if exposed to pure oxygen at the same temperature, because a protective NiO scale does not form. Instead, Ni^{++} dissolves in the borax electrolyte and the platinum acts as an oxygen electrode; the open-circuit potential difference between the platinum and nickel is 0.7 V under the conditions cited previously. Addition of 1% FeO to borax increases the oxidation rate still more, probably by supplying Fe^{++} ions, which are oxidized to Fe^{+3} by oxygen near the electrolyte surface, the trivalent ions then being reduced again either at the cathode or by local cell action at the nickel anode.

The dependence of oxidation on ion migration within a reaction product layer suggests that the oxidation rate should be affected by an applied electric current. This was first shown to be the case by Schein et al.[28] By wrapping a platinum wire around oxidized iron and passing a current of 1.5 A/cm^2, these investigators successfully decreased the oxidation rate at 880°C when iron was made cathodic, and accelerated the rate when iron was made anodic. Similar relations were shown by Jorgensen for the oxidation of zinc in oxygen at 375°C.[29]

CATASTROPHIC OXIDATION

An 8% Al-Cu alloy, oxidizing in air 750°C in presence of MoO_3 vapor from an adjoining molybdenum wire not in contact, was found to react at an abnormally high rate.[30] Similar accelerated oxidation in air was reported for stainless steels containing a few percent molybdenum or vanadium,[31, 32] or as little as 0.04% boron.[33] The oxidation products were voluminous and porous.

An analogous accelerated damage is found in boiler tubes or gas turbine blades operating at high temperatures in contact with combustion gases of crude oils high in vanadium.[34] High-vanadium oils are found only in certain wells, some of which, for example, are located in South America. The vanadium is present in the oil as a soluble organic complex and hence is not easily removed. The solid residual ash of such

[28] F. Schein, B. LeBoucher, and P. LaCombe, *Compt. Rend. Acad. Sci. (Paris)*, **252,** 4157 (1961).

[29] P. Jorgensen, *J. Electrochem. Soc.,* **110,** 461 (1963).

[30] G. Rathenau and J. Meijering, *Metallurgia,* **42,** 167 (1950).

[31] W. Leslie and M. Fontana, *Trans. Amer. Soc. Metals,* **41,** 1213 (1949).

[32] A. Brasunas and N. Grant, *Ibid.,* **44,** 1117 (1952).

[33] A. Fry, *Tech. Mitt. Krupp,* **1,** 1 (1933).

[34] F. Monkman and N. Grant, *Corrosion,* **9,** 460 (1953).

oils may reach 65% V_2O_5 or higher, and the damage caused by the ash is not different from that observed when vanadium is alloyed with heat-resistant materials.

Damage of this kind is called catastrophic or accelerated oxidation. It has been suggested that the cause is a low-melting oxide phase, which acts as a flux to dislodge or dissolve the protective scale. The melting points of V_2O_5, MoO_3, and B_2O_3, for example, are 658, 795, and 294°C, respectively, compared to 1527°C for Fe_3O_4. Eutectic combinations of oxides reduce the melting point still further. In the case of vanadium-containing oils, the presence of sulfur and sodium compounds in the oil and the catalytic effect of V_2O_5 for converting SO_2 to SO_3 result in a scale containing Na_2SO_4 and various metal oxides, the melting point of which has been found to be as low as 500°C. Additions to oil of substances (e.g., calcium or magnesium soaps, powdered dolomite, or powdered Mg) which raise the melting point of the ash by forming CaO (m.p. 2570°C) or MgO (m.p. 2800°C) are found, accordingly, to be beneficial. Also, operating at temperatures below the melting point of the liquid phase avoids catastrophic oxidation. Similarly, alloys high in nickel are more resistant because of the high melting point of NiO (1990°C).

It was plausibly suggested by C. Ilschner-Gensch[26] that the mechanism is not one of melting point alone, but very likely involves galvanic effects of the kind discussed earlier. A spongy, porous network of an electronically conducting oxide such as Fe_3O_4 filled with a liquid electrolyte reproduces the cell described previously containing a platinum cathode and a nickel anode in contact with molten borax. The Fe_3O_4 sponge acts as an oxygen electrode of large area, and the base metal acts as anode. Supplied with a liquid electrolyte in which oxygen and metal ions rapidly migrate, such a cell accounts for an accelerated oxidation process far exceeding the rate for a metal reacting directly with gaseous oxygen through a continuous oxide scale.

OXIDATION OF COPPER

Copper oxidizes in air at low temperatures ($<260°C$) in accord with the two-stage logarithmic equation, forming a film of Cu_2O. The rate varies with crystal face, decreasing in the order (100) > (111) > (110). Heat treatment of polycrystalline copper with hydrogen at 300 to 450°C decreases the oxidation rate in oxygen at 200°C because submicroscopic surface facets are formed by adsorbed hydrogen, presumably favoring the (111) orientation. On the other hand, heat treatment with nitrogen or helium increases the rate because adsorbed oxygen (traces from gas or metal) favors submicroscopic facets of predominantly (100) orientation.[35, 36]

Between about 260° and 1025°C, the Cu_2O film is overlaid by a superficial film of CuO. Oxidation changes from logarithmic to parabolic behavior above 400 to 500°C. Only Cu_2O forms in air above 1025°C. Copper oxidizes at a rate slightly higher than that for iron, and much more rapidly than that for nickel or the heat-resistant Cr-Fe alloys. This is shown by the following temperatures[37] below which the scaling losses in air are found to be less than approximately 20 to 40 mg/dm²-hr: Cu, 450°C; Fe, 500°C; Ni, 800°C; 8 to 10% Cr-Fe (0.1% C), 750°C; 25 to 30% Cr-Fe (0.1% C), 1050 to 1100°C.

Alloying elements that are particularly effective for improving oxidation resistance at high temperatures are aluminum, beryllium, and magnesium. At 256°C, for example, a 2% Be-Cu alloy oxidizes in 1 hr at $\frac{1}{14}$ the rate of copper.[38] Maximum improvement by aluminum additions comes at about 8%.[39]

Internal Oxidation

When alloyed with small percentages of certain metals (e.g., aluminum, beryllium, iron, silicon, manganese, tin, titanium, and zinc), copper oxidizes with precipitation of oxide particles within the body of the metal as well as forming an outer oxide scale. Oxidation within the metal is called subscale formation or internal oxidation. (See Fig. 15, p. 1110, *Corrosion Handbook*.) Similar behavior is found for many silver alloys, but without formation of an outer scale. Internal oxidation is not observed in general with cadmium-, lead-, tin-, or zinc-base alloys. A few exceptions have been noted such as for alloys of sodium-lead, aluminum-tin, and magnesium-tin.[40] Internal oxidation is usually not pronounced for any of the iron alloys.

The mechanism is apparently one of oxygen diffusing into the alloy and reacting with alloying components of higher oxygen affinity than that of the base metal before the alloying components can diffuse to the surface. Above a critical concentration of the alloying component, a compact protective layer of the component oxide tends to be formed

[35] W. Bradley and H. Uhlig, *J. Electrochem. Soc.*, **114**, 669 (1967).

[36] A. Swanson and H. Uhlig, J. Electrochem. Soc., in press.

[37] B. Lustman, *Metal Progr.*, November 1946, p. 850.

[38] W. Campbell and U. Thomas, *Trans. Electrochem. Soc.*, **91**, 623 (1947).

[39] H. Nishimura, quoted by O. Kubaschewski and B. Hopkins, *Oxidation of Metals and Alloys*, 2nd ed., p. 251, Academic Press, New York, 1962.

[40] F. N. Rhines (with W. Johnson and W. Anderson), *Trans. Amer. Inst. Mining Met. Eng.*, **147**, 205 (1942); (with A. Grobe), *Ibid.*, p. 318.

at the external surface which thereafter suppresses internal oxidation. In accord with a diffusion-controlled mechanism, the depth of subscale grows in accord with the parabolic equation.[40] The subject was reviewed by Rapp.[41]

Reaction with Hydrogen ("Hydrogen Disease")

The tendency of copper to dissolve oxygen when the metal is heated in air leads to rupture of the metal along grain boundaries by formation of steam when the metal is subsequently heated in hydrogen. Cast tough-pitch copper containing free Cu_2O is very sensitive to this type of damage. Instances are on record where damage has been caused by hydrogen at temperatures as low as 400°C (750°F). So-called oxygen-free coppers are not susceptible. They may become moderately so, however, should they be heated at any time in oxygen or in air.

Silver similarly dissolves oxygen when heated at elevated temperatures in air, and becomes blistered or loses ductility if later heated in hydrogen above 500°C (925°F). The mechanism is the same as that applying to copper. Oxygen-free silver heated in hydrogen for 1 hr at 850°C (1550°F) is not embrittled or damaged. However, when it is heated immediately afterward in air at the same temperature, loss of ductility occurs that is similar to, but not as severe as, the characteristic loss when silver containing oxygen is heated in hydrogen.[42] Some dissolved hydrogen undoubtedly escapes before oxygen can diffuse into the silver, hence diminishing subsequent damage. Gold and platinum dissolve little or no oxygen and consequently are not subject to similar damage when heated in hydrogen.

OXIDATION OF IRON AND IRON ALLOYS

In the low temperature region (ca. 250°C) the oxidation rate of iron is sensitive to crystal face, reportedly decreasing in the order (100) > (111) > (110).[43] The oxide nuclei apparently consist of Fe_3O_4 which then grow to form a uniform film of oxide. Subsequently, αFe_2O_3 nucleates and covers the Fe_3O_4 layer.[44-46]

Oxidation of iron in the parabolic range is complicated by formation

[41] R. Rapp, *Corrosion*, **21**, 382 (1965).

[42] D. Martin and E. Parker, *Trans. Amer. Inst. Mining Eng.*, **152**, 269 (1943).

[43] J. Wagner, Jr., K. Lawless, and A. Gwathmey, *Trans. Met. Soc. AIME*, **221**, 257 (1961).

[44] R. Grauer and W. Feitknecht, *Corros. Sci.*, **6**, 301 (1966).

[45] W. Boggs, R. Kachik, and G. Pellisier, *J. Electrochem. Soc.*, **112**, 539 (1965).

[46] P. Sewell and M. Cohen, *J. Electrochem. Soc.*, **111**, 501 (1964).

of as many as three distinct layers of iron oxide, and the proportions of these layers change as the temperature or oxygen-partial pressure changes. Data reported by various investigators are not in good agreement, probably, for one reason, because of variations in the purity of iron used for oxidation tests, particularly with respect to carbon content.

At 600°C in 1-atm O_2 for 100 min, it is reported that the scale is composed of two layers: an inner FeO layer equal in weight to an outer Fe_3O_4 layer.[47] At 900°C for 100 min, a three-layer scale is composed of 90% FeO, 9% Fe_3O_4, and less than 1% Fe_2O_3. Below 570°C (1058°F), FeO is unstable and if any is formed above this temperature it decomposes at room temperature into Fe_3O_4 plus Fe.

The most efficient alloying elements for improving oxidation resistance of iron in air are chromium and aluminum. Use of these elements with additional alloyed nickel and silicon is especially effective. An 8% Al-Fe alloy is reported to have the same oxidation resistance as a 20% Cr-80% Ni alloy.[48] Unfortunately, the poor mechanical properties of aluminum-iron alloys, the sensitivity of their protective oxide scales to damage, and the tendency to form aluminum nitride which causes embrittlement, have combined to limit their application as oxidation resistant materials. In combination with chromium, some of these drawbacks of aluminum-iron alloys are overcome.

[47] M. Davies, M. Simnad, and C. Birchenall, *J. Metals*, **3**, 889 (1951).
[48] N. Ziegler, *Trans. Amer. Inst. Mining Met. Eng.*, **100**, 267 (1932).

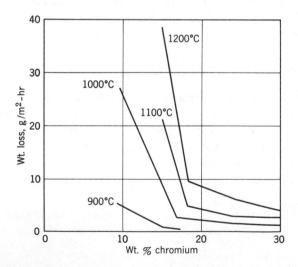

Fig. 7. Effect of alloyed chromium on oxidation of steels containing 0.5 % C, 220 hr (E. Houdremont, *Handbuch der Sonderstahlkunde*, vol. I, p. 815, Springer, Berlin, 1956).

The good oxidation resistance of the chromium-iron alloys combined with acceptable mechanical properties and ease of fabrication account for their wide commercial application. Typical oxidation behavior is shown in Fig. 7.

Improvement in oxidation resistance of iron by alloying with aluminum or chromium probably results from a marked enrichment of the innermost oxide scale with respect to aluminum or chromium. The middle oxide scales are known by chemical analysis to be so enriched, and electron-microprobe analyses confirm marked enrichment of chromium in the oxide adjacent to the metal phase in the case of chromium-iron alloys.[49, 50] These inner oxides resist ion and electron migration better than does FeO. For chromium-iron alloys, the enriched oxide scale is accompanied by depletion of chromium in the alloy surface immediately below the scale. This situation accounts for occasional rusting and otherwise poor corrosion resistance of hot-rolled stainless steels that have not been adequately pickled following high-temperature oxidation.

LIFE TEST FOR OXIDATION-RESISTANT WIRES

The merit of a particular alloy for resisting high-temperature environments, especially on long exposures, depends not only on the diffusion-barrier properties of reaction-product scales but also on the continuing adherence of such scales to the metal. Scales otherwise protective often spall (become detached) during cooling and heating cycles because their coefficient of expansion differs from that of the metal. Hence, an American Society for Testing and Materials accelerated test[51] for oxidation resistance of wires calls for a cyclic heating period of 2 min at a specific temperature followed by a cooling period of 2 min. Alternate heating and cooling results in much shorter life than would have followed if the wire had been heated continuously. The life of a wire in this test is measured as the time to failure or the time to reach a 10% increase in electrical resistance. An equation related to the Arrhenius reaction-rate equation expresses the dependence of life on temperature.[52]

$$\log \text{life (hr)} = \frac{\Delta E'}{2.3RT} + \text{const.}$$

Typical test results for life of several alloys are given in Fig. 8. The

[49] D. Lai, R. Borg, M. Brabers, J. MacKenzie, and C. Birchenall, *Corrosion*, **17**, 357t (1961).

[50] G. Wood and D. Melford, *J. Iron Steel Inst.*, **198**, 142 (1961).

[51] *Amer. Soc. Testing Mat. Std.*, B76-65, pt. 8, pp. 154–162 (1968).

[52] A. deS. Brasunas and H. H. Uhlig, Amer. Soc. Testing Mat. Bull. No. 182, p. 71, May, 1952.

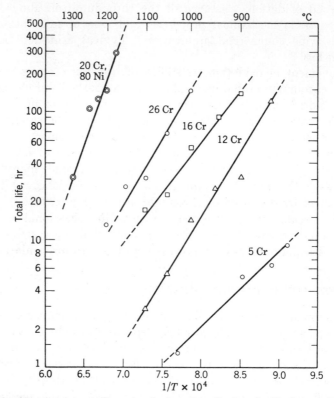

Fig. 8. Life of heat-resistant alloy wires in American Society for Testing and Materials life test as a function of temperature (20% Cr-Ni and 5-26% Cr-Fe), in air of 100% relative humidity at 25°C (Brasunas and Uhlig).

life of the most resistant alloys is more sensitive to temperature change (higher activation energy, $\Delta E'$) than for alloys with inherently shorter life.

OXIDATION-RESISTANT ALLOYS

Chromium-Iron Alloys

The 4 to 9% Cr alloys are used widely for oxidation resistance in oil-refinery construction. The 12% Cr-Fe alloy is used for steam turbine blades because of excellent oxidation resistance and good physical properties. The 9 to 30% Cr alloys are used for furnace parts and burners; when combined with silicon and nickel, and sometimes other alloying elements, they are used for valves of internal combustion engines. The approximate upper temperature limits for exposure to air are:

| | Max. Temperature in Air | |
% Cr in Cr-Fe Alloys	(°C)	(°F)
4–6	650	1200
9	750	1375
13	750–800	1375–1475
17	850–900	1550–1650
27	1050–1100	1925–2000

It is reported[53, 54] that addition of 1% yttrium to a 25% Cr-Fe alloy extends the upper limit of oxidation resistance to about 1375°C (2500°F). Rare earth metal additions in general are beneficial to oxidation resistance of chromium and chromium alloys, including gas turbine alloys.[55] The mechanism accounting for their behavior owes in part to improved resistance of the oxide scale to spalling.

Chromium-Aluminum-Iron Alloys

The chromium-aluminum-iron alloys have exceptional oxidation resistance, combining as they do the oxidation-resistant properties imparted by chromium and aluminum. For example, the 30% Cr, 5% Al, 0.5% Si alloy* resists oxidation in air up to at least 1300°C. Similarly the 24% Cr, 5.5% Al, 2% Co alloy† is resistant up to 1300°C (2375°F). They are used, among other applications, for furnace windings and parts, and for electric-resistance elements. Drawbacks in their properties are poor high-temperature strength and tendency toward embrittlement at room temperature after prolonged heating in air, caused in part by aluminum nitride formation. For this reason furnace windings must be well supported and are usually corrugated in order to allow for expansion and contraction.

* Trade name: Megapyr.
† Trade name: Kanthal A.
[53] J. Fox and J. McGurty, in *Refractory Metals and Alloys,* p. 207, edited by M. Semchyshen and J. Harwood, Interscience, New York, 1961.
[54] E. Fellen, *J. Electrochem. Soc.,* **108,** 490 (1961).
[55] R. Viswanathan, *Corrosion,* **24,** 359 (1968).

Nickel and Nickel Alloys

The good oxidation resistance of nickel is improved by adding 20% Cr, this alloy being resistant in air to a maximum temperature of about 1150°C (2100°F). It represents one of the best heat-resistant alloys available, combining excellent oxidation resistance with good physical properties at both low and elevated temperatures.* Oxidation resistance of the commercial alloy is considerably improved by the addition of calcium metal deoxidizer during the melting process which is said to avoid oxidation of the alloy along grain boundaries. Small amounts of zirconium, thorium, and rare earth metals (e.g., cerium) are also beneficial, probably in part because they decrease the tendency of protective oxides to spall. This explanation was also mentioned earlier as applying to the beneficial effect of the rare earth yttrium when alloyed with the chromium-iron alloys.

The 16% Cr, 7% Fe, 76% Ni alloy† is slightly less resistant to oxidation than the 20% Cr-Ni alloy, but similarly has excellent physical properties, is readily fabricated or welded, and can be used in air up to a maximum temperature of about 1100°C (2000°F). Electrical heating units for some stoves are fabricated of this alloy in tubular form. An electric heating wire of the 20% Cr-Ni alloy inside the tube is insulated from the outer sheath by powdered MgO. Because of its high nickel content and good strength (nickel does not readily form a carbide or nitride), this alloy is often used for construction of carburizing or nitriding furnaces.

Some heat-resistant thermocouple wires consist of nickel alloys. The 10% Cr-Ni alloy (Chromel P) and 2% Al, 2% Mn, 1% Si, bal. Ni alloy (Alumel) can be exposed to air at a maximum temperature of about 1100°C (2000°F).

Nickel and high-nickel alloys tend to oxidize along grain boundaries when subject to alternate oxidation and reduction. Alloying with chromium reduces this tendency. Also, in contact with sulfur or sulfur atmospheres at elevated temperatures, nickel and high-nickel alloys are subject to intergranular attack. Consequently, nickel is not usefully resistant to such atmospheres above about 315°C (600°F). For best resistance to sulfur-containing environments, iron-base alloys should contain high chromium and low nickel.

Gas turbine blades are essentially nickel-base alloys containing 10 to 20% chromium and various age-hardening additions to increase strength. At operating temperatures above about 750°C (1400°F) they

---

* One U. S. trade name is Nichrome V.

† Trade name: Inconel, Alloy 600.

undergo a so-called *sulfidation attack* in which grain boundaries are converted in part to Ni_3S_2; these areas in turn are subject to accelerated oxidation in air. The sulfur may originate either from fuels or from the atmosphere (e.g., sulfates from entrained sea salt). The attack is not dependent on presence of fuel ash or reducing conditions; it was shown, for example, that pure nickel encapsulated in an evacuated silica tube in presence of pure Na_2SO_4 heated to 800 to 1000°C (1475 to 1825°F) results in the formation of sulfides along the grain boundaries.[56] This type of attack on turbine blades is reduced by increasing the chromium content or by applying coatings of aluminum or Al + Cr. Alloying additions of cobalt raise the temperature at which such attack occurs.

Furnace Windings

Twenty percent chromium-nickel and various chromium-aluminum-iron alloys are in common use as furnace windings. To achieve higher temperatures in air, a 10% Rh-Pt alloy can be used up to at least 1400°C (2550°F). The alloy performs better than pure platinum because of higher strength and a lower rate of grain growth. A single crystal of the same dimensions as the resistance wire cross section tends to shear easily and cause failure.

Molybdenum-wound furnaces are operated to at least 1500°C (2725°F). Because molybdenum oxidizes in air, such furnace windings are blanketed in a protective atmosphere of hydrogen.

GENERAL REFERENCES

O. Kubaschewski and B. Hopkins, *Oxidation of Metals and Alloys,* 2nd ed., Academic Press, New York, 1962.

Per Kofstad, *High Temperature Oxidation of Metals,* Wiley, New York, 1966.

Karl Hauffe, *Oxidation of Metals,* Plenum Press, New York, 1965.

J. Bénard, "The Oxidation of Metals and Alloys," *Met. Rev.,* 9, 473–503 (1964).

K. Hauffe, in *Progress in Metal Physics,* edited by B. Chalmers, vol. 4, p. 71, Pergamon Press, New York, 1953.

Hot Corrosion Problems Associated with Gas Turbines, Spec. Tech. Publ. No. 421, Amer. Soc. Testing Mat., Philadelphia, Pa., 1967.

Corrosion and Deposits on Boilers and Gas Turbines, Amer. Soc. Mech. Engrs., New York, 1959.

R. Tylecoat, review, "Oxidation of Copper," *Metallurgist,* 2, 32 (1962); A. Rönnquist and H. Fischmeister, *J. Inst. Metals,* 89, 65 (1960–1961).

---

[56] A. Seybolt and A. Beltran, in *Hot Corrosion Problems Associated with Gas Turbines,* p. 21, Spec. Tech. Publ. No. 421, Amer. Soc. Testing Mat., Philadelphia, Pa., 1967.

chapter 11
Stray-current corrosion

Stray electric currents are those that follow paths other than the intended circuit, or they may be any extraneous currents in the earth. If currents of this kind enter a metal structure, they cause corrosion at areas where the currents leave again to enter the soil or water. Usually, natural earth currents are not important from a corrosion standpoint either because their magnitude is small or because their duration is short. Also, damage by alternating current (a-c) is less than by direct current (d-c), the resultant corrosion usually being greater for lower frequency and less for higher frequency currents. It is estimated that for metals like steel, lead, and copper, 60-cycle a-c current causes only about 1% or less the damage caused by an equivalent d-c current.[1, 2]

The effect of a-c current on metals that are passive (by def. 1, Chapter 5) can be greater. It is reported that stainless steels subjected to a-c current electrolysis are corroded;[3] similarly, aluminum in dilute salt solutions at 1.4 A/ft² (0.15 A/dm²) suffers 5%, and at 9.3 A/ft² (1.0 A/dm²) as much as 31% of the corrosion damage caused by equivalent d-c current densities. In a study of 1-Volt, 54-cycle a-c current superimposed on d-c current, Feller and Rückert[3] found that the passive region of potentiostatic polarization curves for nickel in $1N$ H_2SO_4 completely disappeared and that high anodic current densities persisted throughout the noble potential region. This result suggests that passivity lost during the cathodic cycle, if not restored during the anodic cycle, can account for unexpectedly high corosion rates of otherwise passive metals in aqueous media or in the soil.

[1] Scott P. Ewing, in *Corrosion Handbook,* pp. 601–606.
[2] F. Hewes, *Materials Prot.,* 8, 67 (Sept., 1969); J. Williams, *Ibid.,* 5, 52 (February 1966).
[3] H. Feller and J. Rückert, *Z. Metallk.,* 58, 635 (1967).

SOURCES OF STRAY CURRENTS

Sources of d-c stray currents are commonly electric railways, grounded electric d-c power, electric welding machines, cathodic protection systems, and electroplating plants. Sources of a-c stray currents are usually grounded a-c power lines. An example of stray current from an electric street railway system in which the steel rails are used for current return to the generating station is shown in Fig. 1. Because of poor

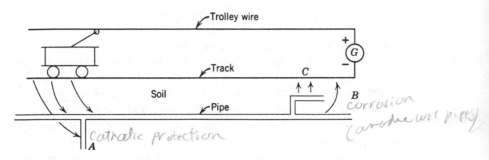

Fig. 1. Stray-current corrosion of buried pipe.

bonding between rails, combined with poor insulation of rails to the earth, some of the return current enters the soil and finds a low-resistance path such as a buried gas or water main. The owner of a household water service pipe at *A* benefits by cathodic protection and experiences no corrosion difficulty; but owner *B*, to the contrary, is harassed by corrosion failures because the service pipe of his house is anodic with respect to the rails. If *B* coats the pipe, which is an understandable layman's reaction to any corrosion problem, matters are made worse, because all stray currents now leave the pipe at defects in the insulating coating at high-current densities, accelerating penetration of the pipe.

Street railways have now in large part been replaced by other forms of transportation, so that they do not cause stray-current corrosion as frequently as in the past. However, cathodically protected structures requiring high currents, when located in the neighborhood of an unprotected pipe line, can produce damage similar to that by the railway illustrated in Fig. 1.

Another example of stray-current corrosion is illustrated in Fig. 2. A welding motor-generator located on shore with grounded d-c lines to a ship under repair can cause serious damage to the hull of the ship by current returning in part from the welding electrodes through the ship and through the water to the shore installation. In this case, it

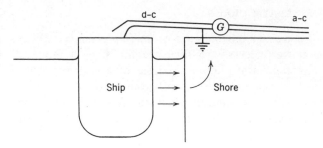

Fig. 2. Stray-current damage to ship by welding generator.

is better to place the generator on board ship and bring a-c power leads to the generator, since a-c currents leaking to ground cause less stray-current damage.

Current flowing along a water pipe (e.g., used as an electric ground) usually causes no damage *inside* the pipe because of the high electrical conductivity of steel or copper compared to water. For example, since the resistance of any conductor per unit length equals ρ/A where ρ is the resistivity and A is the cross-sectional area, then the ratio of current carried by a metal pipe compared to that carried by the water it contains is equal to $\rho_W A_M/\rho_M A_W$, where subscripts $_W$ and $_M$ refer to water and metal, respectively. For iron, $\rho_M = 10^{-5}$ Ω-cm and for a potable water ρ_W may be 10^4 Ω-cm. Assuming that the cross-sectional area of water is 10 times that of the steel pipe, it is seen that if current through the pipe is 1 A, only about 10^{-8} A flows through the water. This small current leaving the pipe and entering the water causes negligible corrosion. If seawater is transported instead, with $\rho_W = 20$ Ω-cm, the ratio of currents is 2×10^5, indicating that even in this case most of the current is carried by the metallic pipe and there is very little stray-current corrosion on the inner surface. However, where such currents leave the pipe and enter the soil, stray-current corrosion of the *outer* surface may be appreciable.

If insulating joints are installed in the above-mentioned pipe used for a ground connection, corrosion can now be serious on the water side of the joint where current leaves the pipe to enter the water. Or if a high-resistance joint exists between two sections of a buried pipe, corrosion may be high on the side where current enters the soil (Fig. 3).

QUANTITATIVE DAMAGE BY STRAY CURRENTS

In general, the amount of metal corroding at anodic areas because of stray current can be calculated using Faraday's law. Weight losses

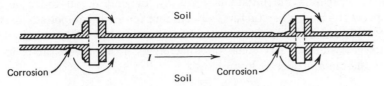

Fig. 3. Effect of current flowing along a buried pipeline on corrosion near insulated couplings.

of typical metals for the equivalent of 1 A flowing for one year are listed in Table 1.

At low current densities, local-action corrosion supplements stray-current corrosion. At high current densities in some environments, oxygen may be evolved, which reduces the amount of metal corroding per Faraday of electricity.

TABLE 1
Weight Loss of Metals by Stray-Current Corrosion

Metal	Equivalent Weight	Weight of Metal Corroded per Ampere-year	
Fe	$\dfrac{55.85}{2}$	20.1 lb	9.1 kg
Cu	$\dfrac{63.57}{2}$	22.8	10.4
Pb	$\dfrac{207.2}{2}$	74.5	33.8
Zn	$\dfrac{65.38}{2}$	23.6	10.7
Al	$\dfrac{26.98}{3}$	6.5	2.9

Amphoteric metals (e.g., lead, aluminum, tin, zinc) corrode in alkalies as well as in acids and hence may be damaged at cathodic areas where alkalies accumulate by electrolysis. This damage is in addition to damage at anodic areas. The amount of cathodic corrosion is not readily estimated. Plumbites ($NaHPbO_2$), aluminates (NaH_2AlO_3), stannates (Na_2SnO_3), or zincates (Na_2ZnO_2), which are all soluble in excess alkali, form in variable amounts per Faraday depending on diffusion rates, temperature, and other factors. Formulas for such compounds may also vary with conditions of formation. In general, the compounds hydrolyze at lower values of pH some distance away from the cathode to form insoluble metal oxides or hydroxides.

DETECTION OF STRAY CURRENTS

Stray currents may fluctuate over short or long intervals of time parallel to the varying load of the power source. This is in contrast to galvanic or cathodic protection currents, which are relatively steady. Hence by recording the potential of a corroding system with respect to a reference electrode over a 24-hr period, stray currents can sometimes be detected and their origin traced to the generator source whose load, day or night, varies in a pattern similar to that of the measured potential change. Thus if stray currents as indicated by potential measurements are larger at 7 to 9 A.M., and again at 4 to 6 P.M., a street railway is suspected. If interference by a cathodic protection system is suspected, the protective current can be turned off and on briefly at regular intervals, observing whether the potential of the corroded structure also fluctuates at the same frequency.

The magnitude of current leaving (or entering) a buried pipe from whatever source can be calculated by measuring the potential difference between a position on the soil surface directly over the pipe and a position on the soil surface some distance away and at right angles to the pipe. If $\Delta\phi$ is the measured potential difference, ρ the resistivity of soil, h the depth of pipe below the surface, and y the distance along the soil surface over which the potential difference is measured

$$\Delta\phi = \frac{\rho j}{2\pi} \ln \frac{y^2 + h^2}{h^2} \tag{1}$$

where j is the total current entering or leaving the pipe surface per unit length (for derivation, see Appendix, p. 401). If y is chosen equal to $10h$

$$\Delta\phi = 0.734\rho j \tag{2}$$

An outer reference electrode (+) with respect to a (−) reference electrode directly over the pipe corresponds to current entering the pipe.

SOIL-RESISTIVITY MEASUREMENT

The resistivity of a soil can be measured by the four-electrode method with each electrode arranged in a straight line and each separated by the distance a (Fig. 4). Steady current, I, from a battery is passed

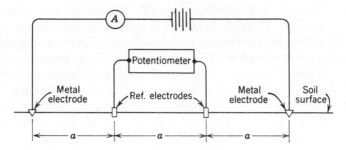

Fig. 4. Four-electrode method for measuring soil resistivity.

through the two outer metal electrodes, and the potential difference, $\Delta\phi$, of the two inner reference electrodes (e.g., Cu-CuSO$_4$) is measured simultaneously. The measurement is usually repeated with current direction reversed in order to cancel out any stray currents. Then

$$\rho = \frac{2\pi a \, \Delta\phi}{I} \tag{3}$$

where ρ is the resistivity of a uniform soil measured to an approximate depth equal to a (for derivation, see Appendix, p. 403).

MEANS FOR REDUCING STRAY-CURRENT CORROSION

1. *Bonding.* In Fig. 1, stray-current corrosion is completely avoided by placing a low-resistance metallic conductor between service pipe B and the rails at C. This is called bonding the rail and pipe systems. In the event of a cathodic protection system causing the damage, the bond may include a resistor just sufficient to avoid large change of potential of the unprotected system when the cathodic protection current is turned on and off. The resistor avoids major damage to the unprotected system. At the same time it avoids the necessity of a large increase in protective current caused by current flowing through the bond, which in effect serves to protect the adjoining system in addition to the protected structure.

2. *Intentional Anodes, Cathodic Protection.* If a bond from B to C in Fig. 1 is not feasible, an intentional anode of scrap iron may be buried in the direction of the rails and attached by a copper conductor to point B. Stray currents then cause corrosion only of the intentional anode, which is easily replaced at low cost. If a source of d-c current is inserted between the intentional anode and the pipe such that current

flows in the soil in a direction opposite to that of the stray current, the arrangement is equivalent to cathodically protecting the pipe. Cathodic protection is installed whenever the intentional anode is not sufficient to overcome all corrosion caused by stray currents.

3. *Insulating Couplings.* By installing one or more insulating couplings, the pipeline in Fig. 1 becomes a less favorable path for stray currents. Such couplings are often useful for minimizing stray-current damage. They are less useful if voltages are so large that current is induced to flow around the insulating joint, causing corrosion as depicted in Fig. 3. It is usual practice to coat the pipe for a distance of about 50 pipe diameters from the insulating joint on either side in order to reduce such localized corrosion on the soil side.[4]

[4] O. Mudd, *Corrosion,* 1, 42 (1945).

chapter 12

Cathodic protection

Cathodic protection is perhaps the most important of all approaches to corrosion control. By means of an externally applied electric current, corrosion is reduced virtually to zero, and a metal surface can be maintained in a corrosive environment without deterioration for an indefinite time.

As discussed in Chapter 4, p. 58, the mechanism of protection depends on external current polarizing cathodic elements of local action cells to the open-circuit potential of the anodes.[1] The surface becomes equipotential (cathode and anode potentials become equal), and corrosion currents no longer flow. Or looked at another way, at a high enough value of external current density, a net positive current enters the metal at all regions of the metal surface (including anodic areas), and hence there is no tendency for metal ions to enter into solution.

Cathodic protection can be applied in practice to protect metals such as steel, copper, lead, and brass against corrosion in all soils and in almost all aqueous media. Pitting corrosion can be prevented in passive metals like the stainless steels or aluminum. It can be used effectively to eliminate stress corrosion cracking (e.g., of brass, mild steel, stainless steels, magnesium, aluminum), corrosion fatigue of most metals (but not fatigue), intergranular corrosion (e.g., of Duralumin, 18-8 stainless steel), or dezincification of brass. It *cannot* be used to avoid corrosion above the water line (e.g., of water tanks) because the impressed current cannot reach metal areas that are out of contact with the electrolyte. Nor does the protective current enter electrically screened areas such as the interior of water condenser tubes (unless the auxiliary anode enters the tubes), even though the water box may be adequately protected.

[1] R. Mears and R. Brown, *Trans. Electrochem. Soc.* **74**, 519 (1938); *Ibid.*, **81**, 455 (1942); T. Hoar, *J. Electrodepositors Tech. Soc.*, **14**, 33 (1938).

BRIEF HISTORY

As a result of laboratory experiments in salt water, Sir Humphry Davy[2] reported in 1824 that copper could be successfully protected against corrosion by coupling it to iron or zinc. He recommended cathodic protection of copper-sheathed ships, employing sacrificial blocks of iron attached to the hull in the ratio of iron to copper surface of about 1:100. In practice, the corrosion rate of copper sheathing was appreciably reduced, as Davy had predicted, but unfortunately cathodically protected copper is subject to fouling by marine organisms, contrary to the behavior of unprotected copper which supplies a sufficient concentration of copper ions to poison fouling organisms (see p. 85). Since fouling reduced the speed of ships under sail, the British Admiralty decided against the idea. After Davy's death in 1829, his cousin, Edmund Davy (Professor of Chemistry at the Royal Dublin University), successfully protected the iron work of buoys by attaching zinc blocks, and Robert Mallet in 1840 produced a zinc alloy particularly suited as a sacrificial anode. When wooden hulls were replaced by steel, the fitting of zinc slabs became traditional on all Admiralty vessels.* These slabs provided localized protection, especially against the galvanic effects of the bronze propeller, but the overall cathodic protection of seagoing ships was not explored again until about 1950, this time by the Canadian Navy.[3] By proper use of antifouling paints in combination with anticorrosion paints, it was shown that cathodic protection of ships is feasible and can save appreciably in maintenance costs.

Cathodic protection was incidental to the mechanism of protecting steel sheet coated by dipping into molten zinc (galvanizing, see p. 234), a method first patented in France in 1836 and in England in 1837.[4] However, the practice of zinc coating of steel was apparently described in France as early as 1742.[5] The first application of impressed electric current for protection of underground structures took place in England and in the U. S., about 1910–1912.[4] Since then the general use of cathodic protection has spread rapidly, and now thousands of miles of buried pipeline and cables are effectively protected against corrosion by this means. Cathodic protection is also applied to canal gates,

* The author is indebted to Dr. W. H. J. Vernon and to Mr. L. Kenworthy for bringing these facts to his attention.

[2] H. Davy, *Phil. Trans. Roy. Soc.*, **114**, 151–158, 242–246, 328–346 (1824–1825).

[3] K. Barnard (with G. Christie) *Corrosion*, **6**, 232 (1950); *Ibid.*, **7**, 114 (1951); (with G. Christie and J. Greenblatt), *Ibid.*, **9**, 246 (1953).

[4] Wilson Lynes, *J. Electrochem. Soc.*, **98**, 3c (1951).

[5] R. Burns and W. Bradley, in *Protective Coatings for Metals*, 3rd edition, pp. 104–106, Reinhold, New York, 1967.

condensers, submarines, water tanks, marine piling, and chemical equipment.

HOW APPLIED

Cathodic protection requires a source of direct current and an auxiliary electrode (anode) usually of iron or graphite located some distance away from the protected structure. The d-c source is connected with its positive terminal to the auxiliary electrode and its negative terminal to the structure to be protected; in this way current flows from the electrode through the electrolyte to the structure. The applied voltage is not critical—it need only be sufficient to supply an adequate current density to all parts of the protected structure. In soils or waters of high resistivity, the applied voltage must be higher than in environments of low resistivity. Or when the extremities of a long pipeline are to be protected by a single anode, the voltage must be raised. A sketch of anode with respect to a protected buried pipeline is shown in Fig. 1.

The source of current is usually a rectifier supplying low-voltage d-c of several amperes. Motor generators have been used, although maintenance is troublesome. Windmill generators are employed in areas where prevailing winds are fairly dependable. Even in periods of calm, some degree of protection for steel persists temporarily because of the inhibiting effect of alkaline electrolysis products at the cathode surface.

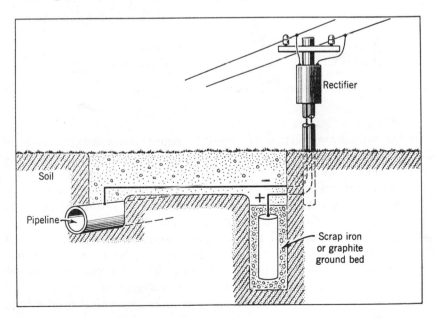

Fig. 1. Sketch of cathodically protected pipe, auxiliary anode, and rectifier.

Sacrificial Anodes

If the auxiliary anode is composed of a metal more active in the Galvanic Series than the metal to be protected, a galvanic cell is set up with current direction exactly as described in the previous section. The impressed source of current (i.e., the rectifier) can then be omitted and the electrode is called a sacrificial anode (Fig. 2). Sacrificial metals

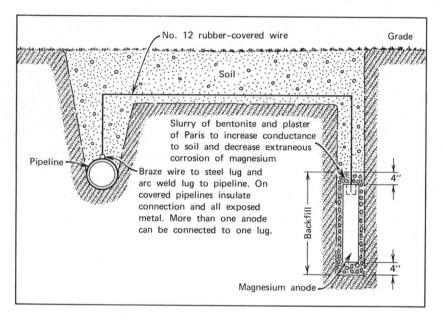

Fig. 2. Cathodically protected pipe employing sacrificial anode.

used for cathodic protection are mostly magnesium or magnesium-base alloys, and to a lesser extent, zinc and aluminum. Sacrificial anodes serve essentially as sources of portable electrical energy. They are useful particularly when electric power is not readily available, or in situations where it is not convenient or economical to install power lines for the purpose. The open-circuit potential difference of magnesium with respect to steel is about 1 V, (ϕ_H for magnesium in seawater $= -1.3$ V), so that only a limited length of pipeline can be protected by one anode, particularly in high-resistivity soils. This low voltage is sometimes an advantage over higher impressed voltages in that danger of overprotection to some portions of the system is less, and since the total current per anode is limited, the danger of stray-current damage (interference problems) to adjoining metal structures is reduced.

The potential of zinc is less than for magnesium (ϕ_H in seawater = -0.8 V) and hence current output per anode is also less. High purity zinc is usually specified in order to avoid significant anodic polarization with resultant reduction of current output caused by accumulation of adherent insulating zinc reaction products on commercial zinc. This tendency is less pronounced in zinc of high purity.

Aluminum operates theoretically at a voltage between magnesium and zinc. It tends to become passive in water or in soils with accompanying shift of potential to a value approaching that of steel, whereupon it ceases to function as a sacrificial electrode. A special chemical environment high in chlorides surrounding the electrode can be provided in order to avoid passivity; however, such an environment, called backfill, is only a temporary measure. In seawater, passivity is best avoided by alloying additions. For example, alloying aluminum with 0.1% Sn followed by heat treatment at 620°C for 16 hr and water quenching to retain the tin in solid solution very much decreases anodic polarization in chloride solutions.[6] The corrosion potential of the 0.1% Sn alloy in 0.1N NaCl is -1.2 V (s.h.e.) compared to -0.5 V for pure aluminum. Some sacrificial aluminum anodes contain about 0.1% Sn and 5% Zn.[7, 8] Another composition containing 0.6% Zn, 0.04% Hg, and 0.06% Fe, when tested in seawater for 254 days operated at a current efficiency of 94% (1270 A-hr/lb). Such alloys in the form of sacrificial anodes now account for about 5 million pounds of aluminum consumed each year in the United States.[9]

Most sacrificial anodes in use are of magnesium; about 15 million pounds are consumed annually in the United States for this purpose.[9] Magnesium anodes are often alloyed with 6% Al, 3% Zn to reduce pitting-type attack and to increase current efficiency. By using high-purity magnesium containing about 1% Mn,[10] the advantage of a higher potential (with higher current output per anode) is obtained. This alloy operates at a current efficiency in seawater similar to the first-mentioned alloy, but at somewhat lower efficiency in normal soils. The observed efficiency of magnesium anodes averages about 500 A-hr/lb compared to a theoretical efficiency of 1000 A-hr/lb.

[6] D. Keir, M. Pryor, and P. Sperry, *J. Electrochem. Soc.,* **114,** 777 (1967); *Ibid.,* **116,** 319 (1969).

[7] J. Burgbacher, *Materials Prot.,* **7,** 26 (April 1968).

[8] T. Lennox, M. Peterson, and R. Groover, *Materials Prot.,* **7,** 33 (February 1968).

[9] J. Robinson, Dow Chemical Co., private communication. 1970.

[10] P. George, J. Newport, and J. Nichols, *Corrosion,* **12,** 627t (1956).

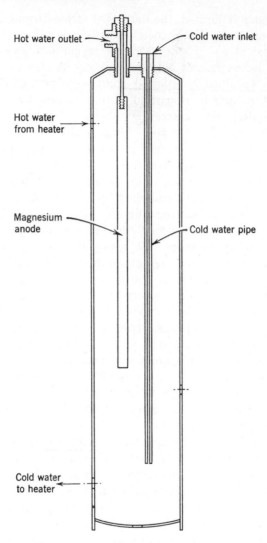

Fig. 3. Cathodically protected hot-water tank that uses magnesium anode.

A sketch of a magnesium anode rod installed in a steel hot-water tank is shown in Fig. 3. Such rods may increase the life of a steel tank by several years, particularly if the rod is renewed as required. The degree of protection is greater in waters of high conductivity, where the currents naturally set up by the magnesium-iron couple reach higher values than in waters of low conductivity (soft waters).

COMBINED USE WITH COATINGS

The distribution of current in a cathodically protected steel water tank is not ideal—too much current may flow to the sides and not enough to the top and bottom. Better distribution is accomplished by using an insulating coating (e.g., an organic coating at ordinary temperatures or a glass coating at elevated temperatures). The insulating coating need not be pore-free, since the protective current flows preferably to exposed metal areas, wherever located, which are precisely the areas needing protection. Also, since the total required current is less than for an uncoated tank, the magnesium anode lasts longer.

In hard waters, a partially protective coating may form on steel that consists largely of $CaCO_3$ precipitated by alkalies generated as reaction products at the cathode surface. A similar coating forms gradually on cathodically protected surfaces exposed to seawater (more rapidly at high current densities). Such coatings, if adherent, are also useful in distributing the protective current and in reducing total current requirements.

In the general application of cathodic protection using either impressed current or sacrificial anodes, it is expedient to use an insulating coating of some kind, and this combination is the accepted practice today. For example, the distribution of current to a coated pipeline is much improved over that to a bare pipeline, the total current and required number of anodes are less, and the total length of pipeline protected by one anode is much greater. Since the earth taken as a whole is a good electrical conductor, and the resistivity of the soil is localized only within the region of the pipeline or the electrodes, one magnesium anode can protect as much as 8 km (5 miles) of a coated pipeline. For a bare pipeline the corresponding distance might be only 30 m (100 ft). Using an impressed current at higher voltages, one anode might protect as much as 80 km (50 miles) of a coated pipeline. The limiting length of pipe protected per anode is imposed not by resistance of the soil but by the metallic resistance of the pipeline itself.

The potential decay, E_x, along an infinite pipeline measured from the point of attachment to the d-c source having potential, E_A, is expressed as an exponential relation with respect to distance x along the pipeline in accord with

$$E_x = E_A \exp\left[- \left(\frac{2\pi r R_L}{kz}\right)^{\frac{1}{2}} x \right] \tag{1}$$

where both E_x and E_A represent differences between polarized potential with current flowing and corrosion potential in absence of current, R_L is the resistance of pipe of radius r per unit length, k is a constant, and z is the resistance of pipe coating per unit area. (For derivation, see Appendix, p. 396.) This equation is derived by assuming that polarization of the cathodically protected surface is a linear function of current density. Note that E_x becomes zero at $x = \infty$.

Considering a finite pipeline for which $a/2$ is half the distance to the next point of bonding, and potential E_x at $a/2 = E_B$

$$E_x = E_B \cosh \left[\left(\frac{2\pi r R_L}{kz} \right)^{1/2} \left(x - \frac{a}{2} \right) \right] \tag{2}$$

Since cathodic protection is optimum within a specific potential range (see p. 224), it is obvious that the length of pipeline protected by one anode increases as the metallic pipe resistance, R_L, decreases and coating resistance, z, increases.

MAGNITUDE OF CURRENT REQUIRED

The current density required for complete protection depends on the metal and on the environment. It can be seen from Fig. 11, p. 59, that the applied current density must always exceed the current density equivalent to the measured corrosion rate in the same environment. Hence the greater the corrosion rate, the higher must be the impressed current density for protection.

When the corrosion rate is cathodically controlled and the corrosion potential approaches the open-circuit anode potential, the required current density is only slightly greater than the equivalent corrosion current. But for mixed control, the required current can be considerably greater than the corrosion current, and it is still greater for corrosion reactions that are anodically controlled.

If the protective current induces precipitation of an inorganic scale on the cathode surface, such as in hard waters or in seawater, the total required current, as described earlier, falls off as the scale builds up. However, at exposed areas of metal the current density remains the same as before scale formation; only the total current density per apparent unit area is less.

The precise requirements of current density for complete protection can be determined in several ways; the most important is potential measurement of the protected structure (see p. 224). In the absence of such measurements, only orders of magnitude can be given. These are listed in Table 1 for steel exposed to various environments.

ANODE MATERIALS AND BACKFILL

Auxiliary anodes for use with impressed current are usually composed of scrap iron or graphite. Scrap iron is consumed at the rate of 15

TABLE 1
Orders of Magnitude of Current Density Required for Cathodic Protection of Steel

Environment	Amp/dm^2	Amp/ft^2
Sulfuric acid pickle (hot)	4.	35.
Soils	0.0001–0.005	0.001–0.05
Moving sea water	{0.0003 (final) 0.0015 (initial)	0.003–0.015
Air-sat'd water (hot)	0.0015	0.015
Moving fresh water	0.0005	0.005

to 20 lb/A-year and must be renewed periodically. Graphite anodes are consumed at a lower rate not exceeding perhaps 2 lb/A-year. Graphite costs more than scrap iron, both initially and in subsequent higher electrical power costs because of the noble potential and accompanying high overvoltage for oxygen (or Cl_2) evolution on graphite compared to an active potential and lower overvoltage for the reaction $Fe \rightarrow Fe^{++} + 2e^-$. Graphite is also fragile compared to scrap iron and must be installed with greater care. The advantages and disadvantages of graphite apply in similar measure to 14% Si-Fe alloy anodes, which have also been recommended as auxiliary anodes.

For protection of structures in seawater, platinum-clad, 2% Ag-Pb, or platinized titanium electrodes have been recommended as corrosion-resistant anodes using impressed current.[11-13] Whereas sacrificial magnesium anodes require replacement approximately every two years, the 2% Ag-Pb anodes are estimated to last more than 10 years, and the 90% Pt-10% Ir anodes still longer.[12] In fresh waters, aluminum anodes are sometimes used for impressed current systems.

Because the effective resistivity of soil surrounding an anode is confined to the immediate region of the electrode, it is common practice to reduce local resistance by so-called backfill.* For impressed current systems this consists of surrounding the anode with a thick bed of coke, and adding a mixture of perhaps 3 or 4 parts gypsum ($CaSO_4 \cdot H_2O$) to 1 part NaCl. The coke backfill, being a conductor, carries part of the current, reducing in some measure consumption of the anode itself.

Whereas auxiliary anodes need not be consumed in order to fulfill

---

*Sometimes the anode is immersed in a river bed, lake, or the ocean, and then backfill is not required.

[11] H. Preiser and B. Tytell, *Corrosion*, 15, 596t (1959).

[12] K. Barnard, G. Christie, and D. Gage, *Corrosion*, 15, 581t (1959).

[13] R. Benedict, *Materials Prot.* 4, 36 (December 1965).

their purpose, sacrificial anodes are consumed not less than is required by Faraday's law in order to supply an equivalent electric current. In general, the observed rate of consumption is greater than the theoretical. For zinc the difference is not large, but for magnesium it is appreciable, the cause being ascribed to local-action currents on the metal surface, to formation of colloidal metal particles,[14, 15] or, perhaps more important, to initial formation of univalent magnesium ions.[16-18] The latter ions are unstable and react in part with water in accord with

$$2Mg^+ + 2H_2O \rightarrow Mg(OH)_2 + Mg^{++} + H_2$$

Hence in dilute sodium chloride about half the magnesium corroding anodically appears as $Mg(OH)_2$ and half as $MgCl_2$, accompanied by hydrogen evolution in about the amount expected according to this reaction.[16] Additional lesser side reactions may also take place at the same time. Accordingly, the observed yield of a magnesium anode is only about one-half the 1000 A-hr/lb calculated on the basis of Mg^{++} formation.

For magnesium anodes, backfill has the advantage of reducing resistance of insulating corrosion-product films [e.g., $Mg(OH)_2$], as well as increasing conductivity of the immediate environment. A suitable backfill may consist of approximately 20% bentonite (an inorganic colloid used for retention of moisture) 75% gypsum, and 5% Na_2SO_4. Sometimes the backfill is packaged beforehand in a bag surrounding the anode so that anode and backfill can be placed simultaneously into position in the soil.

Overprotection

Moderate overprotection of steel structures usually does no harm. The main disadvantages are waste of electric power and increased consumption of auxiliary anodes. In the extreme, additional disadvantages result if so much hydrogen is generated at the protected structure that blistering of organic coatings, hydrogen embrittlement of the steel (loss of ductility through absorption of hydrogen), or hydrogen cracking (see p. 142) is caused. Damage to steel by hydrogen absorption is particu-

[14] G. Marsh and E. Schashl, *J. Electrochem. Soc.*, **107**, 960 (1960).

[15] G. Hoey and M. Cohen, *Ibid.*, **105**, 245 (1958).

[16] J. Greenblatt, *J. Electrochem. Soc.*, **103**, 539 (1956); (with E. Zinck), *Corrosion*, **15**, 76t (1959); *Ibid.*, **18**, 125t (1962).

[17] R. Petty, A. Davidson, and J. Kleinberg, *J. Amer. Chem. Soc.*, **76**, 363 (1954).

[18] M. Rausch, W. McEwan, and J. Kleinberg, *Chem. Rev.*, **57**, 417 (1957).

larly apt to occur in environments containing sulfides[19] for reasons that are discussed on page 48.

In the case of amphoteric metals (e.g., aluminum, zinc, lead, tin) excess alkalies generated at the surface of overprotected systems damage the metals by causing increased attack rather than reduction of corrosion. It was shown in the case of lead that cathodic protection continues into the alkaline range of pH, but the critical potential for complete protection (see below) shifts to more active values.[20] Aluminum can be cathodically protected by coupling it to zinc[21] employed as a sacrificial anode, but if coupled to magnesium, overprotection may result with consequent damage to the aluminum.

CRITERIA OF PROTECTION

The effectiveness of cathodic protection in practice can be established in more than one way, and several criteria have been used in the past to prove whether protection is complete. For example, the observed number of leaks in an old buried pipeline are plotted against time, noting that leaks per year drop to a small number or to zero after cathodic protection is installed. Or the hull of a ship can be inspected at regular intervals for depth of pits.

It is also possible to check effectiveness of protection by short-time tests including the following measures:

1. *Coupon Tests.* A weighed metal coupon shaped to conform to the outside surface of a buried pipe is attached by a brazed connecting cable, and both the cable and the surface between coupon and pipe are overlaid with coal tar. After exposure to the soil for a period of weeks or months, the weight loss, if any, of the properly cleaned coupon is a measure of whether cathodic protection of the pipeline is complete.

2. *Colorimetric Tests.* A section of buried pipeline is cleaned, exposing bare metal. A piece of absorbent paper soaked in potassium ferricyanide solution is brought into contact, and the soil returned in place. After a relatively short time, examination of the paper for the blue color of ferrous ferricyanide indicates that cathodic protection is incomplete, whereas absence of blue indicates satisfactory protection.

Both the coupon and the colorimetric tests are qualitative and do not provide information about whether just enough or more than enough current is being supplied.

[19] W. Bruckner and K. Myles, *Corrosion,* **15,** 591t (1959).

[20] W. Bruckner and O. Jansson, *Corrosion,* **15,** 389t (1959).

[21] R. Mears and H. Fahrney, *Trans. Amer. Inst. Chem. Eng.,* **37,** 911 (1941).

Potential Measurements

A criterion that indicates degree of protection, including overprotection, is obtained through measuring the potential of the protected structure. This measurement is of greatest importance in practice, and it is the criterion generally accepted and used by corrosion engineers. It is based on the fundamental concept that optimum cathodic protection is achieved when the protected structure is polarized to the open-circuit anode potential of local-action cells. This potential for steel, as determined empirically, is equal to -0.85 V versus the copper-saturated $CuSO_4$ half-cell, or -0.53 V (s.h.e.).

The theoretical open-circuit anode potential for iron can be calculated assuming that the activity of Fe^{++} in equilibrium is determined by the solubility of a covering layer of $Fe(OH)_2$, in accord with previous discussions of oxide film composition on iron exposed to aqueous media (see p. 93). Employing the Nernst equation.

$$E_H = 0.44 - \frac{0.059}{2} \log (Fe^{++}) \qquad \text{where } (Fe^{++}) = \frac{\text{solubility product}}{(OH^-)^2}$$

The value for (OH^-) can be estimated assuming that its concentration at equilibrium is twice that of (Fe^{++}) in accord with $Fe(OH)_2 \rightarrow Fe^{++} + 2(OH^-)$. The oxidation potential so calculated is 0.59 V (s.h.e.), equivalent to a potential difference of 0.91 V versus copper-saturated $CuSO_4$.

The empirical value for lead, known only approximately, is about 0.78 V versus copper-saturated $CuSO_4$[22] compared to the calculated value for a $Pb(OH)_2$ film of lead equal to 0.59 V. In alkaline media, with formation of plumbites, the calculated value comes closer to the empirical value. Other calculated values are listed in Table 2.

For passive metals the criterion of protection differs from that just described. Since passive metals corrode uniformly at low rates but by pitting corrosion at high rates, cathodic protection of metals like aluminum or 18-8 stainless steel depends on polarizing them not to the usual open-circuit anode potential but only to a value more active than the critical potential at which pitting initiates (see p. 76). The latter potential lies within the passive range and is less noble the higher the Cl^- concentration; in 3% NaCl the value for aluminum is -0.45 V (s.h.e.). Hence, iron with a corrosion potential in seawater of about -0.4 V is not suited as a sacrificial anode for cathodically protecting aluminum, unlike zinc, which has a more favorable corrosion potential of about -0.8 V.

[22] D. Werner, *Corrosion*, **13**, 68 (1957).

TABLE 2
Calculated Minimum Potential, ϕ, for Cathodic Protection

Metal	$E°$ (V)	Solubility Prod., M(OH)$_2$	ϕ_H (V)	ϕ versus Cu-CuSO$_4$ Reference Electrode (V)
Iron	0.440	1.8×10^{-15}	-0.59	-0.91
Copper	-0.337	1.6×10^{-19}	0.16	-0.16
Zinc	0.763	4.5×10^{-17}	-0.93	-1.25
Lead	0.126	4.2×10^{-15}	-0.27	-0.59

For 18-8 stainless steel, the critical potential in 3% NaCl is 0.21 V; for nickel it is about 0.23 V. Coupling of the latter metals to a suitable area of either iron or zinc, therefore, can effectively protect them cathodically in seawater against pitting corrosion. Practical structures (e.g., ships and offshore oil-drilling platforms) are sometimes designed to take advantage of galvanic couples of this kind.

Doubtful Criteria

Criteria have sometimes been suggested based on empirical rules—for example, polarizing a steel structure 0.3 V more active than the corrosion potential. This criterion is not exact and in the average situation leads to under- or overprotection. It has also been suggested that polarization of the structure should proceed to breaks in slope of the voltage versus current plot. Such breaks may in principle occur in some environments when the applied current is just equal to or slightly greater than the corrosion current (e.g., oxygen depolarization control), but in other environments breaks may occur when concentration polarization or IR effects through partially protective surface films become appreciable. As Stern and Geary showed,[23] breaks of this kind in polarization measurements have varied causes and are of doubtful value as criteria of cathodic protection. Contrary to what is sometimes erroneously supposed, discontinuities in slopes of polarization curves have no general relation to the anode or cathode open-circuit potentials of the corroding system.

Position of Reference Electrode

The potential of a cathodically protected structure is determined ideally by placing the reference electrode as close as possible to the

[23] M. Stern and A. Geary, *J. Electrochem. Soc.*, **104**, 56 (1957).

structure to avoid an error caused by IR drop through the soil. Any IR drop through corrosion-product films or insulating coatings will persist, of course, even where adequate precautions are exercised otherwise, tending to make the measured potential more active than the actual potential at the metal surface. In practice, for buried pipelines, a compromise position is chosen at the soil surface located directly over the buried pipe. This is chosen because cathodic protection currents flow mostly to the lower surface and are minimum to the upper surface of a pipe buried a few feet below the soil surface.

The reference electrode is sometimes located at a remote position from the pipeline, recommended because currents do not penetrate remote areas and hence IR drop effects are avoided. Actually the potential measured at a remote position is a compromise potential at some value between that of the polarized structure and the polarized auxiliary or sacrificial anode. These potentials differ by the IR drop through the soil and through coatings. The potential measured at a remote location, therefore, tends to be more active than the true potential of the structure, resulting in a structure that may be underprotected.

ECONOMICS OF CATHODIC PROTECTION

For buried pipelines, the cost of cathodic protection is far less than for any other means offering equal assurance of protection. The guarantee that no leaks will develop on the soil side of a cathodically protected buried pipeline has made it economically feasible, for example, to transport oil and high-pressure natural gas across half the American continent.

Actually, lack of corrosion on the soil side makes it possible to specify thinner wall pipe adequate to withstand internal pressures and to avoid any extra thickness as a safety factor against corrosion. This saving alone has paid for capital equipment and installation costs of cathodic protection in some instances.[24]

The Panama Canal gates are protected by using impressed current, the initial costs of installation being less than 0.5% the cost of replacing the gates. One important advantage is that the gates can continue to operate without the necessity of periodic long shutdowns for repairs caused by corrosion. Similarly a ship cathodically protected can operate in principle for longer periods between dry-docking, thereby effecting the saving of thousands of dollars annually. The further economic advantages in other instances through avoidance of stress corrosion cracking, corrosion fatigue, and pitting of various structural metals have only begun to be appreciated.

[24] J. Stirling, *Corrosion*, 1, 19 (1945).

ANODIC PROTECTION

As mentioned briefly in Chapter 5, p. 68, some metals (e.g., iron and stainless steels) can also be protected effectively by making them anodic, and shifting their potential into the passive region of the anodic polarization curve (see Fig. 1, p. 62). The passive potential is automatically maintained, usually electronically, by an instrument called the potentiostat. Practical application of anodic protection and use of the potentiostat for this purpose were first suggested by Edeleanu.[25]

Anodic protection has found application in handling particularly sulfuric acid,[26] but the method is also applicable to other acids (e.g., phosphoric acid) and to alkalies and some salt solutions. It has been shown to be effective for increasing the resistance to corrosion fatigue of various stainless steels in $0.5M$ Na_2SO_4,[27] in 10% H_2SO_4 or 10% NH_4NO_3, and of 0.19% C steel in 10% NH_4NO_3.[28] Practical installations have been described for anodically protecting mild steel against uniform corrosion in NH_4NO_3 fertilizer mixtures,[29] carbon steel in 86% spent sulfuric acid at temperatures up to 60°C (140°F),[30] and carbon steel in 0.1 to $0.7M$ oxalic acid at temperatures up to 50°C (120°F).[31]

Since passivity of iron and the stainless steels is destroyed by halide ions, anodic protection of these metals is not possible in hydrochloric acid or in acid chloride solutions for which the current density in the otherwise passive region is very high. Also, if Cl^- should contaminate the electrolyte, the possible danger of pitting becomes a consideration even if the passive current density remains acceptably low. In the latter case, however, it is only necessary to operate in the potential range below the critical pitting potential for the mixed electrolyte.* Titanium, which has a very noble critical pitting potential over a wide range of Cl^- concentration and temperature, is passive in presence of Cl^- (low

---

*Stress corrosion cracking of type 304 stainless steel, which occurs at room temperature in $10N$ $H_2SO_4 + 0.5N$ NaCl, is prevented by anodically polarizing the alloy to 0.7 V (s.h.e.). S. Acello and N. Greene, Corrosion **18**, 286t (1962); J. Harston and J. Sully, *Ibid*, **25** 493 (1969).

[25] C. Edeleanu, *Nature*, **173**, 739 (1954); *Metallurgia*, **50**, 113 (1954).

[26] J. Sudbury, O. Riggs, and D. Shock, *Corrosion*, **16**, 47t (1960); D. Shock, O. Riggs, and J. Sudbury, *Ibid.*, 55t; O. Riggs, M. Hutchison, and N. Conger, *Ibid.*, 58t.

[27] H. Spähn, *Metalloberfläche* **16**, 369 (1962); *Z. Physik. Chem. (Leipzig)*, **234**, 1 (1967).

[28] W. Cowley, F. Robinson, and J. Kerrich, *Brit. Corros. J.*, **3**, 223 (1968).

[29] W. Banks and M. Hutchison, *Materials Prot.*, **8**, 31 (February 1969).

[30] L. Hays, *Materials Prot.*, **5**, 46 (September 1966).

[31] L. Perrigo, *Materials Prot.*, **5**, 73 (March 1966).

$i_{passive}$) and can be anodically protected without danger of pitting even in solutions of hydrochloric acid.

Anodic protection is applicable only to metals and alloys (mostly transition metals) which are readily passivated when anodically polarized and for which $i_{passive}$ is very low. It is not applicable, for example, to zinc, magnesium, cadmium, silver, copper, or copper-base alloys. Anodic protection of aluminum exposed to high-temperature water has been shown to be feasible (see p. 338).

Current densities to initiate passivity, $i_{critical}$, are relatively high, 0.6 mA/cm² being required for type 316 stainless steel in 67% H_2SO_4 at 24°C (75°F). But currents for maintaining passivity are usually low, the orders of magnitude being 0.1 μA/cm² for type 316 stainless steel to 15 μA/cm² for mild steel,[26] both in 67% H_2SO_4. Corrosion rates corresponding to these current densities are 0.2 to 25 mdd.

It is typical of anodic protection that corrosion rates, although small, are not reduced to zero, contrary to the situation for cathodic protection of steel. On the other hand, the required current densities in corrosive acids are usually much lower than for cathodic protection, since for cathodic protection the current cannot be less than the normal equivalent corrosion current in the same environment. For stainless steels this value of current density corresponds to the rather high corrosion rate for the active state of the alloys.

For anodic protection, it has been reported[26] that unusual throwing power (protection at distances remote from the cathode or in electrically screened areas) is obtained, far exceeding similar throwing power obtained in cathodic protection. The cause has been ascribed to high electrical resistance of the passive film, but this is probably not correct, because measurements have shown that such resistances are typically low. The cause instead may be related to the corrosion-inhibiting properties of anodic corrosion products released by stainless steels in small amounts (e.g., $S_2O_8^{--}$, $Cr_2O_7^{--}$, Fe^{+3}), which shift the corrosion potential into the passive region in the absence of an applied current.

GENERAL REFERENCES

Cathodic Protection, J. H. Morgan, Macmillan, New York, 1960.
Cathodic Protection, A Symposium, Nat. Assoc. Corros. Engrs. and Electrochem. Soc., Houston, Texas, 1949.
O. Mudd, "Catholic Protection," *Corrosion,* **1,** 192–218 (1945); *Ibid.,* **2,** 25–58 (1946).

chapter 13
Metallic coatings

METHODS OF APPLICATION

The majority of metal coatings are applied either by brief immersion in a molten bath of metal, called *hot dipping,* or by *electroplating* from an aqueous electrolyte. To a lesser extent coatings are also applied by other methods. *Metal spraying* employs a gun that simultaneously melts and propels small droplets of metal, usually by means of an air blast, onto the surface to be coated. The resultant coatings are porous but can be made adherent and of almost any desired thickness. One advantage is that such coatings can be applied on already fabricated structures. Sometime the pores are filled with a thermoplasic resin in order to increase corrosion protection. *Cementation* consists of tumbling the work in a mixture of metal powder and an appropriate flux at elevated temperatures, allowing the metal to diffuse into the base metal. Aluminum and zinc coatings can be prepared in this way. Diffusion coatings of chromium, nickel, titanium, aluminum, etc., can also be prepared by immersing metal parts under an inert atmosphere in a bath of molten calcium containing some of the coating metal in solution.[1]

Coatings are also sometimes produced by *gas-phase reaction.* For example, $CrCl_2$ when volatilized and passed over steel at about 1000°C, (1800°F) results in formation of a chromium-iron alloy surface containing up to 30% Cr in accord with the reaction:

$$\tfrac{3}{2}CrCl_2 + Fe \rightarrow FeCl_3 + \tfrac{3}{2}Cr \text{ (alloyed with Fe)}$$

Similar surface alloys of silicon-iron containing up to 19% Si can

[1] G. Carter, *Metal Progr.,* **93,** 117 (1968).

be prepared by reaction of iron with $SiCl_4$ at 800 to 900°C (1475 to 1650°F). Coatings are also produced by chemical reduction of metal-salt solutions, the precipitated metal forming an adherent overlay on the base metal. Nickel coatings of this kind are called *electroless* nickel plate.

CLASSIFICATION OF COATINGS

All commercially prepared metal coatings are porous in some degree. Furthermore coatings tend to become damaged during shipment or in use. Galvanic action at the base of a pore or scratch becomes an important factor, therefore, in determining coating performance. From the corrosion standpoint, metal coatings can be divided into two classes, namely: (1) noble (2) sacrificial.

As the names imply, noble coatings (e.g., nickel, silver, copper, lead, or chromium) on steel, are noble in the Galvanic Series with respect to the base metal. At exposed pores, the direction of galvanic current accelerates attack of the base metal and eventually undermines the coating (Figs. 1, 2). Consequently, it is important that noble coatings always be prepared with a minimum number of pores and that any existing pores be as small as possible, to delay access of water to the underlying metal. This usually means increasing the thickness of coating. Sometimes the pores are filled with an organic lacquer, or a second lower

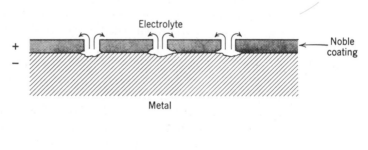

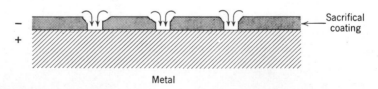

Fig. 1. Sketch of current flow at defects in noble and sacrificial coatings.

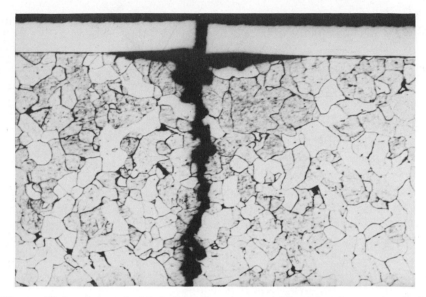

Fig. 2. Undermining of nickel electrodeposit on steel by galvanic corrosion in 3 % NaCl solution (100X). Crack resulted from cyclic stressing in a corrosion fatigue test. [H. Spähn and K. Fässler, *Werkst Korros.*, **17**, 321 (1966)].

melting metal is diffused into the coating at elevated temperatures (e.g., zinc or tin into nickel).

For sacrificial coatings (e.g., zinc, cadmium) and, in certain environments, also aluminum and tin on steel, the direction of galvanic current through the electrolyte is from coating to base metal; as a result of this, the base metal is cathodically protected (Fig. 1). So long as adequate current flows and the coating remains in electrical contact, corrosion of the base metal does not occur. The degree of porosity of sacrificial coatings, therefore, is not of great importance, contrary to the situation for noble coatings. In general, of course, the thicker the coating, the longer does cathodic protection continue.

The area of base metal over which cathodic protection extends depends on the conductivity of the environment. For zinc coatings on steel in waters of low conductivity, such as distilled or soft waters, a coating defect about 3 mm ($\frac{1}{8}$ in.) wide may begin to show rust at the center. However, in seawater, which is a good conductor, zinc protects steel several decimeters or feet removed from the zinc. This difference in behavior results from adequate current densities for cathodic protection extending over a considerable distance in waters of high conductivity whereas in waters of low conductivity, cathodic current densities fall off rapidly with distance from the anode.

SPECIFIC METAL COATINGS

Nickel Coatings

Nickel coatings are prepared usually by electroplating. The metal is plated either directly on steel or sometimes over an intermediate coating of copper. The copper underlayer is employed to facilitate buffing of the surface on which nickel is plated, by reason of copper being softer than steel, and also to reduce the required thickness of nickel (which costs more than copper) for obtaining a coating of minimum porosity. In an industrial atmosphere, too thin a coating of nickel may corrode more rapidly when plated over copper than over steel, presumably because copper corrosion products leaching out of pores in the nickel coating accelerate attack of nickel.[2] This situation would not necessarily apply to other types of atmospheres.

Because porosity is an important factor in the life and appearance of nickel coatings, a certain minimum thickness is recommended depending on exposure conditions. For indoor exposures, a coating 0.3 to 0.5 mil (0.008–0.013 mm) is considered adequate for many applications. For out-of-doors, a coating 0.8 to 1.5 mil (0.02–0.04 mm) may be specified. The thicker coatings in this range (or still thicker) are desirable (e.g., for auto bumpers) for use near the sea coast or in industrial environments, but thinner coatings are usually adequate for dry or unpolluted atmospheres. For use in the chemical industry, nickel coatings commonly range from 1 to 10 mils (0.025–0.25 mm) thick.

Nickel is sensitive to attack, particularly by industrial atmospheres. In the phenomenon called fogging, coatings tend to lose their specular reflectivity, owing to formation of a film of basic nickel sulfate which decreases surface brightness.[3] To minimize fogging, a very thin (0.01–0.03 mil; 0.0003–0.0008 mm) coating of chromium is electrodeposited over the nickel. This thin chromium overlayer has led to the term "chrome plate," although in reality such coatings are composed mostly of nickel. Optimum protection is obtained with "microcracked" chromium, electrodeposited with many inherent microscopic cracks obtained by suitable additions to the plating bath. This usually overlies a thin nickel coating formed from a plating bath containing additives (usually sulfur compounds) which improve surface brightness (bright nickel).

[2] W. Wesley, in *Corrosion Handbook,* p. 821.
[3] W. Vernon, *J. Inst. Metals,* **48,** 121 (1932); *Chem. Ind.,* **1934, 314.**

This thin coat in turn overlies a matte coating two to three times thicker, electrodeposited from a conventional nickel plating bath. The many cracks in chromium favor initiation of corrosion at numerous sites, thereby decreasing depth of penetration resulting otherwise from corresponding attack at fewer sites. The bright nickel, which incorporates in the deposit a small amount of sulfur, is anodic to the underlying nickel coat of lower sulfur content, so that the bright nickel acts as a sacrificial coating. Any pit beginning underneath the chromium coating tends to grow sidewise rather than through the nickel layers. In this way, appearance of rust is delayed and the multiple coating system is more protective for a given thickness of chromium and nickel.[4]

Electroless nickel plate is produced largely for the chemical industry. Nickel salts are reduced to the metal by sodium hypophosphite solutions at or near the boiling point. A typical solution is the following:

$NiCl_2 \cdot 6H_2O$	30 g/liter
Sodium hypophosphite	10 g/liter
Sodium hydroxyacetate	50 g/liter
pH	4–6

This particular formulation deposits nickel at the rate of about 0.6 mil (0.015 mm)/hr in the form of a nickel-phosphorus alloy.[5] The usual range of phosphorus content in coatings of this kind is 7 to 9%. Various metal surfaces including nickel act as catalysts for the reaction so that deposits can be built up to relatively heavy thicknesses. Specific additions are made to the commercial solution in order to increase rate of nickel deposition and also for coating glass and plastics, should this be desirable. Metals on which the coating does not deposit include lead, tin solder, cadmium, bismuth, and antimony. The phosphorus content makes it possible to harden the coating appreciably by low-temperature heat treatment, e.g., 400°C (750°F). Corrosion resistance of the nickel-phosphorus alloy is stated to be comparable to electrolytic nickel in many environments.

Lead Coatings

Lead coatings on steel are usually formed either by hot dipping or by electrodeposition. In the hot-dipping operation, a few percent tin are usually incorporated to improve bonding with the underlying steel. When considerable tin is incorporated (e.g., 25%) the resulting coating

---

[4] F. LaQue, *Trans. Inst. Metal Finishing,* **41,** 127 (1964).
[5] A. Brenner and G. Riddell, *J. Res. Nat. Bur. Std. (U. S.),* **39,** 385 (1947).

is called terne plate.* Coatings of lead or lead-tin alloy are usefully resistant to atmospheric attack, the pores tending to fill with rust which stifles further reaction. Lead coatings are not very protective in the soil. Applications include roofing and the protection of the insides of automobile gasoline tanks from corrosion by entrained water. Lead coatings should not be used in contact with drinking water or food products because of the poisonous nature of small quantities of lead salts (see pp. 3–4).

Zinc Coatings

Coatings of zinc, whether hot dipped or electroplated, are called galvanized. Electrodeposited coatings are somewhat more ductile than hot-dipped coatings, the latter forming brittle intermetallic compounds of zinc and iron at the coating interface (alloy layer). The corrosion rates of the two coatings are comparable, with the exception that hot-dipped coatings compared to rolled zinc (and probably electrodeposited zinc as well) tend to pit less in hot or cold water[6] and also in soils.[7] This difference suggests either that specific potentials of the intermetallic compounds favor uniform corrosion or that the incidental iron content of hot-dipped zinc is beneficial. In this connection, it is reported that zinc alloyed with either 5 or 8% Fe pits less than does pure zinc in water.[6] Although heat treatment of zinc-coated steel resulting in >15% Fe in the coating provides longer life of specimens exposed to industrial atmospheres,[8] the extra expense of heat treatment is not necessarily justified by the improvement. Diffusion coatings of zinc, aluminum, and chromium are reviewed by Drewett.[9]

Zinc coatings are relatively resistant to rural atmospheres and also to marine atmospheres, except when seawater spray comes into direct contact with the surface. Tests conducted in the United States showed that a thin coating 1 mil (0.03 mm) thick† was found to last about 11 years or longer in rural or suburban locations and about 8 years

---

* Terne, which means dull, was originally used to differentiate such coatings from bright tin plate. See Burns and Bradley, *Protective Coatings for Metals*, p. 269, Reinhold, New York, 1967.

† Thickness of zinc coatings is often expressed in ounces per square foot. For coated sheet, 1 mil Zn = 1.18 oz/ft^2 (2 sides). However, for pipe, wire, angles, channels, etc., the weight is given in terms of one side only; hence for these shapes, 1 mil Zn = 0.59 oz/ft^2.

[6] L. Kenworthy and M. Smith, *J. Inst. Metals*, **70**, 463 (1944).

[7] K. Logan, in *Corrosion Handbook*, pp. 460–462; W. Blum and A. Brenner, *Ibid.*, pp. 814–815.

[8] H. Campbell, J. Stanners, and K. Watkins, *J. Iron Steel Inst.*, **203**, 248 (1965).

[9] R. Drewett, *Anti-Corrosion Methods Materials*, April, June, August (1969).

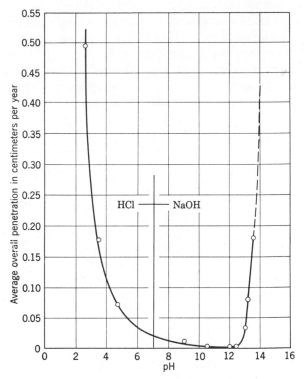

Fig. 3. Effect of pH on corrosion of zinc, aerated solutions, 30°C [B. Roetheli, G. Cox, and W. Littreal, *Metals and Alloys*, **3**, 73 (1932)].

in marine locations. In industrial atmospheres, on the other hand, average life was only four years, indicating sensitivity of zinc to attack by sulfuric acid of polluted atmospheres.[10]

In aqueous environments at room temperature the overall corrosion rate in short time tests is lowest within the pH range 7 to 12 (Fig. 3). In acid or very alkaline environments, the major form of attack is by hydrogen evolution. Above about pH 12.5, zinc reacts rapidly to form soluble zincates in accord with

$$Zn + OH^- + H_2O \rightarrow HZnO_2^- + H_2$$

In seawater, zinc coatings are effectively resistant for protecting steel against rusting, each mil (0.03 mm) of zinc equaling about one year of life. Thus a coating 5 mils (0.13 mm) thick will protect steel against appearance of rust for about five years.

In many aerated hot waters, *reversal of polarity* between zinc and

[10] W. Blum and A. Brenner, in *Corrosion Handbook,* pp. 803–813.

iron occurs at temperatures of about 60°C (140°F) or above.[11] This leads to zinc having the characteristics of a noble coating instead of a sacrificial coating, and hence a galvanized coating under these circumstances induces pitting of the base steel.

A 15-year service test on piping carrying Baltimore City water at a mean temperature of 46°C (115°F) and maximum of 80°C (176°F) confirmed that pitting of galvanized pipe was 1.2 to 2 times deeper than in black iron pipe (ungalvanized) of the same type, corresponding to shorter life of the galvanized pipe. In cold water, however, pits in galvanized pipe were only 0.4 to 0.7 as deep as those in black iron pipe, indicating in this case a beneficial effect of galvanizing.[12] It was found that waters high in carbonates and nitrates favor the reversal in polarity, whereas those high in chlorides and sulfates decrease the reversal tendency.[13]

The cause is apparently related to the formation of porous $Zn(OH)_2$ or basic zinc salts, which are insulators, under those conditions for which zinc is anodic to iron, but to formation of ZnO instead under conditions where the reverse polarity occurs.[14] The latter compound conducts electronically, being a so-called semiconductor. It can therefore perform in aerated waters as an oxygen electrode whose potential, like mill scale on steel, is noble to both zinc and iron. Accordingly, in deaerated hot or cold waters in which an oxygen electrode does not function because oxygen is absent, zinc is always anodic to iron, but this is not necessarily true in aerated waters. Apparently presence of HCO_3^- and NO_3^- aided by elevated temperatures stimulates formation of ZnO, whereas Cl^- and SO_4^{--} favor formation of hydrated reaction products instead.

At room temperature, in water or dilute sodium chloride, the current output of zinc as anode decreases gradually because of insulating corrosion products that form on its surface. In one series of tests, the current between a zinc-iron couple decreased to zero after 60 to 80 days and a slight reversal of polarity was reported.[15] This trend is less pronounced with high-purity zinc, on which insulating coatings have less tendency to form.

Cadmium Coatings

Cadmium coatings are produced almost exclusively by electrodeposition. The difference in potential between cadmium and iron is not so

[11] G. Schikorr, *Trans. Electrochem. Soc.*, **76**, 247 (1939).
[12] C. Bonilla, *Trans. Electrochem. Soc.*, **87**, 237 (1945).
[13] R. Hoxeng and C. Prutton, *Corrosion*, **5**, 330 (1949).
[14] P. Gilbert, *J. Electrochem. Soc.*, **99**, 16 (1952).
[15] H. Roters and F. Eisenstecken, *Arch. Eisenhüttenw*, **15**, 59 (1941).

large as that between zinc and iron, hence cathodic protection of steel by an overlayer of cadmium falls off more rapidly with size of coating defects. Cadmium is more expensive than zinc, but it retains a bright metallic appearance longer and is consequently often used for electronic equipment. Furthermore it is more resistant to attack by atmospheric condensate and by salt spray. Otherwise, however, cadmium coatings exposed to the atmosphere are not quite as resistant as zinc coatings of equal thickness.

In aqueous media cadmium, unlike zinc, resists attack by strong alkalies. Like zinc, it is corroded by dilute acids and by aqueous ammonia. Cadmium salts are toxic, and for this reason cadmium coatings should not come into contact with food products. Zinc salts are less toxic in this respect, and galvanized coatings are tolerated for drinking water, but they are also not recommended for contact with foods.

Tin Coatings

Several million tons of tin plate are produced each year, and most of it is used to manufacture many billions of food containers. Because electrodeposited tin is more uniform than hot-dipped tin and can be produced as thinner coatings, most sheet now available is so-called electrolytic tin plate. The nontoxic nature of tin salts makes tin plate ideal for handling beverages and foods.

The thickness of tin coatings is expressed in pounds per base box, where 1 lb/base box equals a coating 0.06 mil (0.0015 mm) thick. Hot-dipped tin ranges from 1 to 2.5 lb/base box. The usual thickness of electrolytic coatings for food containers is 0.5 lb/base box, but for corrosive foods the thickness may reach 1 lb/base box or more. Tin coatings so thin are naturally very porous, hence it is essential that the tin act as a sacrificial coating in order to avoid perforation by pitting of the thin-gage steel on which the tin is applied. This condition is generally observed.

Tin is cathodic to iron on the outside of a tinned container, in accord with the standard potential of tin equal to -0.136 V compared to -0.440 V for iron. On the inside, however, tin is almost always anodic to iron and hence tin cathodically protects the base steel. This fortunate reversal of potential occurs because stannous ions, Sn^{++}, are complexed by many food products which greatly reduce Sn^{++} activity, resulting in a change in corrosion potential of tin in the active direction. (See p. 28).

In addition to acids or alkalies as incidental and natural components, foods usually contain various organic substances some of which are complexing agents, as mentioned previously, and others act either as corrosion inhibitors or as cathodic depolarizers. Foods low in inhibitor and high in depolarizer substances may cause more rapid corrosion of food containers than, for example, highly acid foods. Because of the presence of organic depolarizers, corrosion of the inside tin coating of food containers commonly occurs with little or no hydrogen evolution. When the tin coating has all corroded, however, it is observed that subsequent corrosion usually occurs by hydrogen evolution. The reason for this behavior is not firmly established, but it may be related to the fact that Sn^{++} ions, which are known to inhibit corrosion of iron in acids, increase hydrogen overvoltage, thereby favoring reduction of organic substances at the iron cathode. Stannous ions are formed continuously at the iron surface during corrosion of the tin layer, but they are not present in sufficient concentration once the tin layer has dissolved. It is also possible that the potential difference of the iron-tin couple favors adsorption and reduction of organic depolarizers at the cathode, whereas these processes do not take place at lower potential differences. Eventual failure of containers is by so-called hydrogen swells, in which an appreciable pressure of hydrogen builds up within the can, making the contents suspect, since bacterial decomposition also causes gas accumulation.

The amount of hydrogen accumulated within the lifetime of the container is determined not only by the tin coating thickness, the temperature, and the chemical nature of food in contact, but most often by the composition and structure of the base steel. The rate of hydrogen evolution is increased by cold working of the steel (see p. 127), which is standard procedure for strengthening the container walls. Subsequent low-temperature heat treatment, incidental or intentional, may increase or decrease the rate (see Fig. 1, p. 128). Also a high phosphorus or sulfur content makes steel especially susceptible to acid attack, whereas a few tenths per cent copper in the presence of these elements can serve to decrease corrosion. The effect of copper is not always predictable, however, because of the presence, in certain foods, of organic depolarizers and inhibitors some of which may function only in the absence of alloyed copper.

PROPERTIES OF TIN

Tin is an amphoteric metal, reacting both with acids and alkalies, but it is relatively resistant to neutral or near-neutral media. It does not corrode in soft waters and has been used for many years as piping for distilled water. Only the scarcity and high price of tin account for its replacement for distilled water piping by other materials (e.g., aluminum).

Corrosion in dilute nonoxidizing acids (e.g., HCl or H_2SO_4) is determined largely by concentration of dissolved oxygen in the acid. The hydrogen overvoltage of tin is so high that corrosion accompanied by hydrogen evolution is limited. Deaerated acetic acid, for example, either hot or cold, corrodes tin slowly (15–60 mdd); NH_4OH and Na_2CO_3 have little effect, but in concentrated NaOH the rate of attack is rapid with formation of sodium stannate. As with acids, the rate is increased by aeration. In aerated fruit juices at room temperature, rates are only 1 to 25 mdd, but they become more than 10 times higher at the boiling point, indicating appreciable increase of attack with temperature.[16]

Chromium-Plated Steel for Containers

Because of a price advantage, a recent trend has been to coat steel with a combined thin chromium (0.3–0.4 μin.; 0.008–0.01 μ) followed by chromium oxide (5–40 mg Cr/m^2) and an organic top coat. About 10% of all plated steel sheet produced for containers is now chromium plated instead of tin plated, and the percentage is rapidly increasing.[17] Chromium ($\phi^\circ = -0.74$ V) is more active in the *Emf* Series than is iron ($\phi^\circ = -0.44$ V), but chromium has a strong tendency to become passive ($\phi^\circ_F = 0.2$ V). Hence, the potential of chromium in aqueous media is usually noble to that of steel. However, in galvanic couples of the two metals, especially in acid media, chromium is polarized below its Flade potential so that it exhibits an active potential. Hence, the corrosion potential of chromium-plated steel, which is always porous in some degree, is more active than that of either passive chromium or the underlying steel.[18] Under these conditions, chromium acts as a sacrificial coating in the same manner as does a tin coating but with dependence of its active potential on passive-active behavior rather than on formation of metal complexes.

A chromium oxide coating enhances adhesion of the organic top coating and diminishes undercutting by any corrosion reaction. The organic coating seals pores in the metal coating, increases resistance to the flow of electric current produced by galvanic cells, and reduces the amount of iron salts entering the contents of the container and affecting flavor or color.[19]

---

[16] B. Gonser and J. Strader, in *Corrosion Handbook,* p. 327.
[17] M. Vucich, National Steel Corp., private communication, 1970.
[18] G. Kamm, A. Willey, and N. Linde, *J. Electrochem. Soc.,* **116,** 1299 (1969).
[19] R. McKirahan and R. Ludwigsen, *Materials Prot.,* **7,** 29 (December 1968).

Aluminum Coatings

Aluminum coatings on steel are produced mostly by hot dipping or spraying, and to a lesser extent by cementation. Molten baths of aluminum for hot dipping usually contain dissolved silicon in order to retard formation of a brittle alloy layer. Hot-dipped coatings are used for oxidation resistance at moderately elevated temperatures such as for oven construction and for automobile mufflers. They are also used for protecting against atmospheric corrosion, an application that is limited by higher costs of aluminum compared to zinc coatings and by variable performance. In soft waters, aluminum exhibits a potential that is positive to steel and hence it acts as a noble coating. In seawater and in some fresh waters, especially those containing Cl^- or SO_4^{--}, the potential of aluminum becomes more active and the polarity of the aluminum-iron couple may reverse. An aluminum coating under these conditions is sacrificial and cathodically protects steel.

Sprayed coatings are commonly sealed with organic lacquers or paints to delay eventual formation of visible surface rust. The usual thickness of sprayed aluminum is 3 to 8 mils (0.08–0.2 mm). In one series of tests conducted in an industrial atmosphere, sprayed coatings 3 mils (0.08 mm) thick showed an average life of 12 years in comparison with 7 years for zinc, the latter coatings being sprayed, electrodeposited, or hot dipped.[20]

Cemented coatings ("Calorizing") are produced by tumbling the work in a mixture of aluminum powder, Al_2O_3, and a small amount of NH_4Cl as flux in a hydrogen atmosphere at about 1000°C (1800°F). An aluminum-iron surface alloy forms which imparts useful resistance to high-temperature oxidation in air [up to 850–950°C; (1550–1750°F)]; as well as resistance to sulfur-containing atmospheres such as are encountered in oil refining. Cemented aluminum coatings on steel are not useful for resisting aqueous environments. They are sometimes used to protect gas turbine blades (nickel-base alloys) from oxidation at elevated temperatures.

GENERAL REFERENCES

"Corrosion Resistance of Tin and Tin Alloys," S. Britton, Tin Res. Inst., Greenford, Middlesex, England, 1952.

Protective Coatings for Metals, R. Burns and W. Bradley, Reinhold, New York, 1967.

Corrosion Handbook, edited by H. H. Uhlig, pp. 803–857, Wiley, New York, 1948.

---

[20] J. Hudson, *Chem. Ind.* (*London*), **1961**, p. 3 (January 7).

chapter 14

Inorganic coatings

VITREOUS ENAMELS

Vitreous enamels, glass linings, or porcelain enamels are all essentially glass coatings of suitable coefficient of expansion fused on metals. Glass in powdered form (frits) is applied to a pickled or otherwise prepared metal surface, then heated in a furnace at a temperature that softens the glass and allows it to bond to the metal. Several coats may be applied. Vitreous enamel coatings are used mostly on steel, but some coatings are also possible on copper, brass, and aluminum.

In addition to decorative utility, vitreous enamels are usefully protective to base metals against corrosion by many environments. The glasses composed essentially of alkali borosilicates can be formulated to resist strong acids, mild alkalies, or both. Their highly protective quality results from virtual impenetrability to water and oxygen over relatively long exposure times, and from their durability at ordinary or above-room temperatures. Their use in cathodically protected hot-water tanks has already been mentioned. Although pores in the coating are permissible when cathodic protection supplements the glass coating, for other applications the coating must be perfect and without a single defect. This means that so-called glass-lined vessels for the food and chemical industries must be maintained free of cracks or other defects. Susceptibility to mechanical damage, or cracking by thermal shock, constitute the main weaknesses of glass coatings. If damage occurs, repairs can sometimes be made by tamping gold or tantalum foil into the voids.

Enameled steels exposed to the atmosphere last many years (gasoline pump casings, advertising signs, decorative building panels, etc.). Failure occurs eventually by formation of a network of cracks in the coating (crazing) through which rust appears. Vitreous enamels are also used

241

to protect against high-temperature gases (e.g., in airplane exhaust tubes), and they were found in Bureau of Standards tests to have long life exposed to soils.

PORTLAND CEMENT COATINGS

Portland cement coatings have the advantages of low cost, a coefficient of expansion ($1.0 \times 10^{-5}/°C$) approximating that of steel, ($1.2 \times 10^{-5}/°C$), and ease of application or repair. The coatings can be applied by centrifugal casting (as for the interior of piping), by trowelling or by spraying. Usual thickness ranges from 0.25 to more than 1 in. (0.5–2.5 cm); thick coatings are usually reinforced with wire mesh.

Portland cement coatings are used to protect cast iron or steel water pipe on the water or soil sides or both, with an excellent record of performance. Some coatings of this kind have been in use in New England for more than 60 years.[1] In addition, Portland cement coatings are used on the interior of hot- or cold-water tanks, oil tanks, and chemical storage tanks. They are also used to protect against seawater and mine waters. The coatings are usually cured for 8 to 10 days before exposure to nonaqueous media such as oils.

The disadvantages of Portland cement coatings lie in their sensitivity to damage by mechanical or thermal shock. Open tanks are easily repaired, however, by trowelling fresh cement into cracked areas. In cold-water piping, there is evidence that small cracks are automatically plugged with a protective reaction product of rust combining with alkaline products leached from the cement. In sulfate-rich waters, Portland cement may be attacked, but cement compositions are now available with improved resistance to such waters.

CHEMICAL CONVERSION COATINGS

These are protective coatings formed *in situ* by chemical reaction with the metal surface. They include special coatings such as $PbSO_4$ which forms when lead is exposed to sulfuric acid, or iron fluoride which forms when steel containers are filled with hydrofluoric acid ($>65\%$ HF).

*Phosphate coating*s on steel ("Parkerizing," "Bonderizing") are produced by brushing or spraying onto a clean surface of steel, a cold or hot dilute manganese or zinc acid orthophosphate solution (e.g., ZnH_2PO_4 plus H_3PO_4). The ensuing reaction produces a network of porous metal phosphate crystals firmly bonded to the steel surface (Fig. 1). Accelera-

[1] F. N. Speller, *Corrosion, Causes and Prevention,* pp. 370–376; 596–600, 3rd ed., McGraw-Hill, New York, 1951.

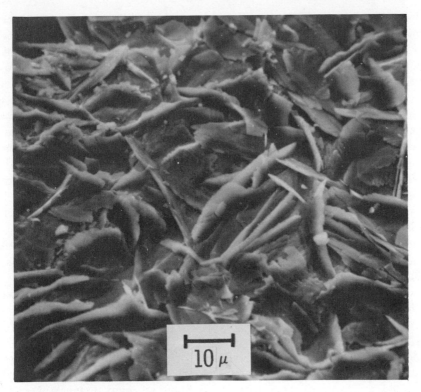

Fig. 1. Electron scanning photomicrograph of phosphated type 1010 mild steel; Acidic zinc phosphate + sodium nitrate accelerator applied at 65°C for 1 min (*Symposium on Interface Conversion for Polymer Coatings*, edited by P. Weiss and G. Cheever, Elsevier, New York, 1968).

tors are sometimes added to the phosphating solution (e.g., Cu^{++}, ClO_3^-, or NO_3^-) to speed the reaction.

Phosphate coatings do not provide appreciable corrosion protection in themselves. They are useful mainly as a base for paints, ensuring good adherence of paint to steel and decreasing the tendency for corrosion to undercut the paint film at scratches or other defects. Sometimes phosphate coats are impregnated with oils or waxes which may provide a degree of protection against rusting, especially if they contain corrosion inhibitors.

Oxide coatings on steel can be prepared by controlled high-temperature oxidation in air or, for example, by immersion in hot concentrated alkali solutions containing persulfates, nitrates, or chlorates. Such coatings, blue, brown, or black in color, consist mostly of Fe_3O_4 and, like phosphate coatings, are not protective against corrosion. When rubbed with inhibiting oils or waxes, as is often done with oxidized gun barrels, some protection is obtained.

Oxide coatings on aluminum are produced at room temperature by anodic oxidation of aluminum (called *anodizing*) in a suitable electrolyte, e.g., dilute sulfuric acid, at current densities of 1 or more A/dm^2. The resultant coating of Al_2O_3 may be 0.1 to 1 mil (0.0025–0.025 mm) thick. The oxide so formed must be hydrated to improve its protective qualities by exposing anodized articles to steam or hot water for several minutes, a process called sealing. Improved corrosion resistance is obtained if sealing is done in a hot dilute chromate solution. The oxide coating may be dyed various colors, either in the anodizing bath or afterward.

Anodizing provides aluminum with some degree of improved corrosion resistance, but the additional protection is not spectacular and it is certainly less than proportional to oxide thickness. Anodized coatings provide a good base for paints on aluminum, which is a metal otherwise difficult to paint without special surface preparation.

Coatings of MgF_2 on magnesium can be formed by anodizing the metal at 90 to 120 V in 10 to 30% NH_4HF_2 solution at room temperature. The coating is recommended for surface cleaning or as a base for finishing treatments.[2]

Chromate coatings are produced on zinc by immersing the cleaned metal for a few seconds in sodium dichromate solution (e.g., 200 g/liter) acidified with sulfuric acid (e.g., 8 ml/liter) at room temperature, and then rinsing and drying. A zinc chromate surface is produced which imparts a slight yellow color and which protects the metal against spotting and staining by condensed moisture. It also increases life of zinc to a modest degree on exposure to the atmosphere. Similar coatings have been recommended for cadmium.

GENERAL REFERENCES

"Glass Linings and Vitreous Enamels," Report of N.A.C.E. Committee T-6H; *Corrosion,* **16,** 107 (1960).
Corrosion Handbook, edited by H. H. Uhlig, pp. 857–878, Wiley, New York, 1948.
R. Burns and W. Bradley, *Protective Coatings for Metals;* Reinhold, New York, 1967.
High Temperature Inorganic Coatings, edited by J. Huminik, Jr., Reinhold, New York, 1963.
F. Sacchi, "Recent Research on Anodizing and its Practical Implications," *Trans. Inst. Metal Finishing,* **41,** 182 (1964).

---

[2] E. Emley, *Principles of Magnesium Technology,* Pergamon Press, New York, 1966.

chapter 15

Organic coatings

Paints are a mixture of insoluble particles of pigment suspended in a continuous organic or aqueous vehicle. Pigments usually consist of metallic oxides (e.g., TiO_2, Pb_3O_4, Fe_2O_3) or other compounds such as $ZnCrO_4$, $PbCO_3$, $BaSO_4$, clays, etc. The vehicle may be a natural oil— linseed or tung oil, for example. When these so-called drying oils are exposed to air, they oxidize and polymerize to solids, a process which can be hastened by small amounts of catalysts such as lead-, manganese-, or cobalt soaps. Synthetic resins are now often used as vehicles or components of vehicles, particularly for continuous contact with water or where resistance to acids, alkalies, or higher temperatures is required. These resins may dry by evaporation of the solvent in which they are dissolved, or they may polymerize through application of heat, or by addition of suitable catalysts. *Varnishes* consist generally of a mixture of drying oil, dissolved resins, and a volatile thinner. *Lacquers* consist of resins dissolved in a volatile thinner; they sometimes contain pigments as well.

Examples of synthetic *resins* include phenol-formaldehyde formulations, which withstand boiling water or slightly higher temperatures, and are used in the chemical industry in the form of multiple coats, baked on, for resisting a variety of corrosive media. Silicone and polyimide resins are useful at still higher temperatures. Alkyd resins, because of favorable cost, fast drying properties, and durability, have found wide application for protecting the metal surfaces of machinery and home appliances.

Vinyl resins have good resistance to penetration by water. Their resistance to alkalies makes them useful for painting of structures which are to be protected cathodically. Linseed- and tung-oil paints, by comparison, are quickly saponified and disintegrated by alkaline reaction products formed at a cathode, whether in waters or in the soil.

Epoxy resins are also resistant to alkalies and to many other chemical media, and have the distinguishing property of adhering well to metal surfaces. The latter property presumably derives from many available polar groups in the molecule. These resins are the basis of plastic mixtures that, on addition of a suitable catalyst, solidify in place within a short time. They are useful, for example, to seal leaks in ferrous or nonferrous piping.

Paints in general are not useful for protecting buried structures, one reason being that mechanical damage to thin coatings by contact with the soil is difficult to avoid. Tests have shown that for this purpose their life is relatively short. Less expensive thickly applied coal-tar coatings have been found to be far more practical. Similarly, the usual linseed-tung oil paints are not durable for metal structures totally immersed in water, except possibly for short periods of time in the order of one year or less. In hot water, life is still shorter. More adequate protection extending up to several years at ordinary temperatures can be obtained by applying four or five coats of a synthetic vehicle paint, as is done in the chemical industry. Because of the expense of a multiple-coat system of this kind, many applications for fresh water or seawater make use instead of thick coal-tar coatings. By far the majority of paints perform best for protecting metals against atmospheric corrosion, and this is their main function.

The total value of paints and lacquers produced in the United States amounts to about $2.5 billion annually, half of which is estimated to be used for corrosion protection. Added to this is an average cost for labor of application equaling two to three times the cost of the paint. Protection of the George Washington Bridge, for example, requires 13,000 man-days to remove rust, prime, and touch up the approximately 6 million ft^2 of surface.[1] For this type of painting, the ratio of labor to paint costs runs more nearly 5:1 or higher. Naval vessels require four coats of paint for exterior exposures and two coats for interior surfaces, each coat on a 1600-ton destroyer requiring 1.5 tons of paint.

REQUIREMENTS FOR CORROSION PROTECTION

To protect against corrosion, a good paint should meet the following requirements:

1. *Provide a Good Vapor Barrier.* All present-day paints are per-

---

[1] R. Burns and W. Bradley, *Protective Coatings for Metals,* p. 524, Reinhold, New York, 1967.

meable in some degree to water and oxygen. Some vehicles are less permeable than others, but their better performance as a diffusion barrier applies only to well-adhering multiple-coat applications that effectively seal up pores and other defects. The diffusion path through a paint film is normally increased by incorporating pigments. Particularly effective in this regard are pigments having the shape of flakes oriented parallel (e.g., by brushing) to the metal surface (e.g., micaceous or flaky hematite, aluminum powder).

2. *Inhibit Against Corrosion.* Pigments incorporated into the prime coat (the coat immediately adjacent to the metal) should be effective corrosion inhibitors. Water reaching the metal surface then dissolves a certain amount of pigment, making the water less corrosive. Corrosion-inhibiting pigments must be soluble enough to supply the minimum concentration of inhibiting ions necessary to reduce the corrosion rate, yet not soluble to a degree that they are soon leached out of the paint.

Among pigments that have been recommended for prime coats, only relatively few actually do the job that is required. Effective pigments whose performance has been established by many service tests include (1) red lead (Pb_3O_4) having the structure of plumbous ortho-plumbate (Pb_2PbO_4) and (2) zinc chromate ($ZnCrO_4$) and basic zinc chromate or zinc tetroxychromate. The inhibiting ion in the case of red lead is probably PbO_4^{-4}, which is released in just sufficient amounts to passivate steel, protecting it against rusting by water reaching the metal surface. It is likely that some other composition lead oxides and hydroxides also have inhibiting properties in this regard, but red lead appears to be the best of the lead compounds.

For zinc chromate, the inhibiting ion is CrO_4^{--}, solubility relations being just right to release at least the minimum concentration of the ion ($>10^{-4}$ mole/liter) for optimum inhibition of steel. The solubility of zinc tetroxychromate is reported to be 2×10^{-4} mole/liter.[2] Lead chromate, on the other hand, is not nearly soluble enough (solubility $= 1.4 \times 10^{-8}$ mole/liter) and acts only as an inert pigment. Commercial formulations of lead chromate sometimes contain lead oxides, present either inadvertently or intentionally, which may impart a degree of inhibition.

Zinc molybdate has been suggested as an inhibiting pigment for paints,[3] being white instead of the characteristic yellow of chromates. It is said to be less toxic than chromates.

The sulfate and chloride content of a commercial $ZnCrO_4$ (or $ZnMoO_4$)

[2] P. Nylen and E. Sunderland, *Modern Surface Coatings*, p. 689, Interscience, New York, 1965.

[3] *Chem. Eng. News,* p. 58, November 14 (1960).

pigment must be low so that it can perform effectively to passivate the metal surface. Because inhibiting pigments passivate steel, it is obvious that they are relatively ineffective for this purpose in presence of high concentrations of chlorides such as in seawater.

Paints pigmented with zinc dust (zinc-rich paints) are also useful as prime coats, the function of the zinc being to cathodically protect the steel in the same manner as galvanized coatings. Such paints are sometimes used over partly rusted galvanized surfaces because they also adhere well to zinc (rust should first be removed). It is reported[4, 5] that in order to ensure good electrical contact between zinc particles and with the base metal, the amount of pigment in the dried paint film should account for 95% of its weight. A coating of this kind protected steel in seawater against rusting at a scratch for one to two years, whereas with 86 and 91% Zn, rust appeared after 1 to 2, and 10 to 20 days, respectively. Recommended vehicles for zinc-rich paints to accommodate such a large fraction of pigment include chlorinated rubber, polystyrene, epoxy, and polyurethane.

3. *Provide Long Life at Low Cost.* A reasonable cost for paint should be gauged by its performance. A paint system lasting five years justifies double the cost for paint if the more expensive paint provides 35% longer life or lasts short of seven years (labor to paint cost ratio of 2:1).

The rate of deterioration of a paint depends on the particular atmosphere to which it is exposed, which in turn depends on amount of atmospheric pollution, and on the amount of rain and sunshine. The color of the top coat (i.e., its ability to reflect infrared and ultraviolet radiation) and the type of vehicle used play some part. Other things being equal, the performance of good quality paints used for corrosion protection is largely determined by the thickness of the final paint film. In achieving a given coating thickness, it is advantageous to apply several coats rather than one, probably because pores are better covered by several applications, and also because evaporation or dimensional changes during polymerization are better accommodated by thin films.

METAL SURFACE PREPARATION

Tests have shown[6] that the most important single factor influencing the life of a paint is the proper preparation of the metal surface. This

---

[4] J. Mayne and U. Evans, *Chem. Ind. (London)*, **1944**, 109.
[5] J. Mayne, *Brit. Corros. J.*, **5**, 106 (1970).
[6] J. Hudson, *J. Iron Steel Inst.*, **168**, 153 (1951); (with W. Johnson), *Ibid.*, 165.

factor is generally more important than the quality of the paint that is applied. In other words, a poor paint system on a properly prepared metal surface usually outperforms a better paint system on a poorly prepared surface.

Adequate surface preparation consists of two main processes.

1. CLEANING ALL DIRT, OILS, GREASES FROM SURFACE

Initial cleaning can be accomplished by using solvents or alkaline solutions.

Solvents. Mineral spirits, naphtha, alcohols, ethers, chlorinated solvents, etc., are applied by dipping in, brushing, or spraying. Perhaps the best in this category is Stoddard Solvent, which is a petroleum base mineral spirit whose flashpoint (40–55°C; 100–130°F) is sufficiently high to minimize fire hazard and which is not particularly toxic.

Chlorinated solvents, on the other hand, although nonflammable, are relatively toxic. In addition, they may leave chloride residues on the metal surface which can later initiate corrosive attack. They are used largely for vapor degreasing (tri- or perchlorethylene), in which the work is suspended in the vapor of the boiling solvent. Care must be exercised in the vapor degreasing of aluminum, ensuring that adequate chemical inhibitors are added and maintained in the chlorinated solvent in order to avoid catastrophic corrosion (see p. 340), or in the extreme, to avoid an explosive reaction.

Alkaline Solutions. Aqueous solutions of certain alkalies provide a method of removing oily surface contamination that is cheaper and less hazardous than the use of solvents. They are more efficient in this particular function than solvents, but perhaps less effective for removing heavy or carbonized oils. Suitable solutions contain one or more of the following substances: Na_3PO_4, $NaOH$, $Na_2O \cdot nSiO_2$, Na_2CO_3, borax, and sometimes sodium pyro- or metaphosphate and a wetting agent. Cleaning may be done by immersing the work in the hot solution [80°C (180°F) to b.p.] containing about 30 to 75 g/liter alkali (4–10 oz/gal). The hot solution in somewhat more dilute concentration may also be sprayed onto the surface. Electrolytic cleaning in alkaline solutions is also sometimes used; this process makes use of the mechanical action of hydrogen gas and the detergent effect of released OH^- at the surface of the work connected as cathode to a source of electric current.

If the metal is free of mill scale and rust, final rinse in water and in a dilute chromic-phosphoric acid ensures both removal of alkali from the metal surface, which would otherwise interfere with good bonding of paint, and also temporary protection against rusting.

2. COMPLETE REMOVAL OF RUST AND MILL SCALE

Rust and mill scale are best removed by either pickling or sandblasting.

Pickling. The metal, cleaned as described previously, is dipped into an acid (e.g., 3–10% H_2SO_4 by weight) containing a pickling inhibitor (see p. 265) at a temperature of 65 to 90°C (150–190°F) for an average of 5 to 20 min. Oxide next to the metal surface is dissolved, loosening the upper Fe_3O_4 scale. Sometimes sodium chloride is added to the sulfuric acid, or HCl alone is used at lower temperatures, or a 10 to 20% H_3PO_4 is employed at temperatures reaching 90°C (190°F). The latter acid is more costly but has the advantage of producing a phosphate film on the steel surface which is beneficial to paint adherence. Some pickling procedures in fact call for a final rinse in dilute H_3PO_4 in order to ensure removal from the metal surface of residual chlorides and sulfates which are damaging to the life of a paint coating. Sometimes the final dip is a dilute solution of chromic (30–45 g/liter; 4–6 oz/gal) or chromic-phosphoric acid, which serves to prevent rusting of the surface before the prime coat is applied.

Blasting. Using this procedure, scale is removed by high-velocity particles impelled by an air blast or by a high-velocity wheel. Blast materials usually consist of sand, or sometimes of steel grit, silicon carbide, alumina, refractory slag, or rock wool byproducts.

Other methods of removing mill scale include *flame cleaning* by which scale spalls off the surface through sudden heating of the surface with an oxyacetylene torch. *Weathering* for several weeks or months is also possible; the natural rusting of the surface dislodges scale, which can then be further removed by wire brushing. But these procedures are less satisfactory than complete removal of scale and rust by pickling or blasting.

The poor performance of paints on weathered steel exposed to an industrial atmosphere was shown in tests reported by Hudson[6] (Table 1). The relatively long life of paint shown for intact mill scale would probably not be achieved in actual service. It would, for example, be difficult to keep large areas and various shapes of mill scale from cracking before or after painting. Fragmentation of mill scale allows paint to become dislodged, particularly after electrolytic action has occurred between metal and scale by aqueous solutions that have penetrated to the metal surface.

APPLYING PAINT COATINGS

The prime coat should be applied to the dry metal surface as soon as possible after the metal is cleaned in order to achieve a good bond.

TABLE 1
Effect of Surface Preparation of Steel on Life of Paint Coatings

(Hudson)

Surface Preparation	Durability of Paint, years (Sheffield, England)	
	2 coats red lead + 2 coats red iron oxide paint	2 coats red iron oxide paint
Intact mill scale	8.2	3.0
Weathered and wire brushed	2.3	1.2
Pickled	9.5	4.6
Sandblasted	10.4	6.3

Better still, the metal should first be given a phosphate coat (see p. 242) in which case the prime coat, if necessary, can be delayed for a short while. The advantages of a phosphate coat lie in a better bond of paint to metal, and good resistance to undercutting of the paint film at scratches or other defects in the paint at which rust forms and otherwise progresses beneath the organic coating. It has been standard practice for many years to coat auto bodies and electric appliances with phosphate before painting.

Only in unusual cases should paint be applied over a damp or wet surface. Poor bonding of paint to steel results. A second prime coat can be applied after the first has dried, or a sequence of top coats can follow. A total of four coats with combined thickness of not less than about 5 mils (0.13 mm) is considered by some authorities to be the recommended minimum for steel that will be exposed to corrosive atmospheres.[7]

Wash Primer

A so-called wash primer, WP1, was developed during World War II in order to facilitate the painting of aluminum. Subsequently it was also found advantageous as a prime coat for steel and several other metals. It consists of approximately 9% polyvinyl butyral and 9% zinc tetroxychromate by weight in a mixture of isopropanol and butanol as one solution, which is then mixed just before using with a solution of 18 wt. % H_3PO_4 in isopropanol and water in the weight ratio of four of the pigmented solution to one of the latter.[8] The mixture must

[7] J. Hudson, *Chem. Ind.* (*London*), **1961**, 3 (January 7).
[8] L. Whiting, *Corrosion*, **15**, 311t (1959).

be used within 8 to 24 hr after mixing. It has the advantage of providing in one operation, instead of two, a phosphating treatment of the metal and the application of the prime coat. It has proved to be an effective prime coat on steel, zinc (galvanized steel), and aluminum.

Painting of Aluminum and Zinc

Paints do not adhere well to aluminum without special surface treatment, unless the wash primer is used which provides its own surface treatment. Otherwise phosphating or anodizing is suitable. The prime coat should in general contain zinc chromate as an inhibiting pigment. Red lead is not recommended because of the galvanic interaction of aluminum with lead deposited by replacement of lead compounds. Paints pigmented with zinc dust plus ZnO (zinc-rich paints) can also be used satisfactory as a prime coat, forming a good bond with the metal. In this case, Zn and ZnO apparently react beforehand with organic acids of the vehicle, ensuring that aluminum soaps and other compounds do not form at the paint-metal interface to weaken the attachment of paint to metal.

Zinc or galvanized surfaces are also difficult to paint and should be phosphated beforehand, or a wash primer should be used as a prime coat. As with aluminum, zinc chromate but not red lead is a suitable inhibiting pigment in the prime coat, or zinc-rich paints can be used.

FILIFORM CORROSION

Metals coated with organic substances may undergo a type of corrosion resulting in numerous meandering threadlike filaments of corrosion product. This is sometimes known as underfilm corrosion, and was called filiform corrosion by Sharmon[9] (Fig. 1). It has been described by several investigators and reproduced in the laboratory.[10-12] According to reported descriptions, the filaments or threads on steel are typically 0.1 to 0.5 mm wide. The thread itself is red in color, characteristic of Fe_2O_3, and the head is green or blue, corresponding to presence of ferrous ions. Each thread grows at a constant rate of about 0.4 mm/day in random directions, but threads never cross each other. If a head approaches another thread, it either glances off at an angle or stops growing.

Filiform corrosion occurs independent of light, metallurgical factors

[9] C. Sharmon, *Nature*, **153**, 621 (1944); *Chem. Ind. (London)*, **46**, 1126 (1952).

[10] M. Van Loo, D. Laiderman, and R. Bruhn, *Corrosion*, **9**, 277 (1953).

[11] W. Slabaugh and M. Grotheer, *Ind. Eng. Chem.*, **46**, 1014 (1954).

[12] H. Kaesche, *Werkst. Korros.*, **10**, 668 (1959).

Lacquered tin can 1×.

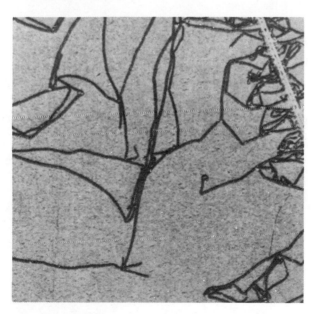

Clear varnish on steel, 10×(86% R.H., 840 hrs.) (van Loo, Laiderman, and Bruhn)

Fig. 1. Filiform corrosion.

in the steel, and bacteria. Although threads are visible only under lacquers or varnishes, they probably also occur with some frequency under opaque paint films. They have been observed under various types of paint vehicles, and on various metals including steel, zinc, aluminum, magnesium, and chromium-plated nickel. This type of corrosion takes place on steel only in air of high relative humidity (e.g., 65–95%). At 100% relative humidity, the threads may broaden to form blisters. They may not form at all if the film is relatively impermeable to water, as is stated to be the case for paraffin.[11] The mechanism appears to be a straightforward example of a differential aeration cell.

Theory of Filiform Corrosion

Various schematic views of a filiform thread or filament are shown in Fig. 2. The head is made up of a relatively concentrated solution

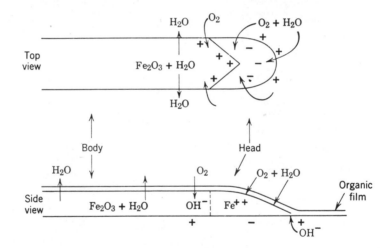

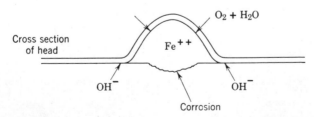

Fig. 2. Schematic views of filiform filament on iron showing details of differential aeration cell causing attack.

of ferrous salts, as was shown by analysis.[11] Hence water tends to be absorbed from the atmosphere in this region of the thread. Oxygen also diffuses through the film and reaches higher concentrations, so far as the metal surface is concerned, at the interface of head and body and at the periphery of the head, compared to lower concentrations in the center of the head. This sets up a differential aeration cell with cathode and accumulation of OH^- ions in all regions where the film makes contact with the metal, and also at the rear of the head. The anode is located in the central and forward portions of the head attended by formation of Fe^{++}.* The liberated OH^- ions probably play an important role in undermining the film in view of the well-known ability of alkalies to destroy the bond between paints and metals. In addition they diffuse toward the center of the head, reacting with Fe^{++} to form $FeO \cdot nH_2O$ which, in turn, is oxidized by O_2 to $Fe_2O_3 \cdot nH_2O$. The precipitated oxide assumes a typical V shape because more alkali is produced in the region between the head and the body (more O_2), compared to the periphery of the head. Behind the V-shaped interface, Fe_2O_3 exists predominantly, and since it is less hygroscopic than ferrous salt solutions, water again diffuses out through the film, leaving this portion relatively dry. Oxygen continues to diffuse through the film, serving to keep the main portion of the filament cathodic to the head.

If a head should approach another filament body, the previous participation of the film in filiform growth will have depleted the film of organic and inorganic anions necessary to the accumulation of high concentrations of ferrous salts in the head and also of cations necessary to build up a high pH at the periphery. This serves to discourage further growth of the filament in the direction of the old filament. Furthermore, and perhaps more important, the previous accumulation of OH^- added to that being formed, plus still greater abundance of oxygen, tends to ensure that the old filament body remains cathodic, encouraging the approaching anode to veer off in another direction. If the head should lose its electrolyte because of dislodged film at the old thread which it approaches, the filament would stop growing, a situation sometimes observed.

Phosphate surface treatments and chromate prime coats of paint serve to limit filiform corrosion, but apparently do not prevent its occurrence. Wholly adequate remedies have not yet been found.

---

* The accumulation of alkali at the periphery can be demonstrated by placing a large drop of dilute sodium chloride solution (1–5%), preferably deaerated, containing a few drops of phenolphthalein and about 0.1% $K_3Fe(CN)_6$ on an abraded surface of iron. Within a few minutes, the periphery turns pink and the center turns blue.

PLASTIC LININGS

Protection against acids, alkalies, and corrosive liquids and gases in general, can be obtained by bonding a thick sheet of plastic or rubber to a steel surface. Rubber, Neoprene, and vinylidene chloride (Saran) are examples of materials that are so used. A thickness of 3 mm ($\frac{1}{8}$ in.) or more ensures a relatively good diffusion barrier and protects the base metal against attack for a long time. The expense of such coatings usually excludes their use for any but severely corrosive environments such as are common to the chemical industry.

Plastic coatings of vinyl or polyethylene are also applied as an adhesive tape, particularly to protect buried metal structures. Such tape finds practical use for coating pipe and auxiliary equipment, including pipe connections and valves, exposed to the soil.

One of the most stable plastics in terms of resisting a wide variety of chemical media is *tetrafluorethylene* (Teflon). It successfully resists aqua regia and boiling concentrated acids, including HF, H_2SO_4, and HNO_3. It resists concentrated boiling alkalies, gaseous chlorine, and all organic solvents up to about 250°C (480°F). It is said to react only with elementary fluorine and with molten sodium. Slow decomposition into HF and fluorinated hydrocarbons begins at temperatures above 200°C (400°F), the mixture of gases being highly toxic.[13] Toxic gases may also be liberated by heat generated during machining operations.

As a plastic, Teflon is not strong, tending to creep readily under stress. Its extreme inertness makes bonding to any surface difficult. Use for this compound is found typically for linings, gasket material, and diaphragm valves.

GENERAL REFERENCES

Steel Structures Painting Manual, vol. 1, "Good Painting Practice"; vol. 2 "Systems and Specifications"; edited by J. Bigos, Steel Structures Painting Council, Pittsburgh, Pa., 1955.

R. Burns and W. Bradley, *Protective Coatings for Metals,* Reinhold, New York, 1967.

"Report on Surface Preparation of Steels for Organic and Other Protective Coatings," Committee TP-6G; *Corrosion,* **9,** 173 (1953).

A. G. Roberts, *Organic Coatings, Properties, Selection and Use,* Building Science Series 7, Govt. Printing Office, Washington, D.C., 1968.

J. Mayne, "The Mechanism of Protection by Organic Coatings," *Trans. Inst. Metal Finishing,* **41,** 121 (1964).

---

[13] *Bulletin,* M.I.T. Occupational Medical Service, December 1961.

chapter 16

Inhibitors and passivators

An inhibitor is a chemical substance that, when added in small concentration to an environment, effectively decreases the corrosion rate. There are several classes of inhibitors conveniently designated as follows: (1) passivators, (2) organic inhibitors, including slushing compounds and pickling inhibitors, and (3) vapor-phase inhibitors.

Passivators are usually inorganic oxidizing substances (e.g., chromates, nitrites, or molybdates) that passivate the metal and shift the corrosion potential several tenths volt in the noble direction. Nonpassivating inhibitors such as the pickling inhibitors are usually organic substances that have only slight effect on the corrosion potential, changing it either in the noble or active direction usually by not more than a few milli- or centivolts. In general, the passivating-type inhibitors reduce corrosion rates to very low values, being more efficient in this regard than most of the nonpassivating types. They represent, therefore, the best inhibitors available for certain metal-environment combinations. Most of the discussion in this chapter pertains to steel; only limited information is available in the literature for other metals.

PASSIVATORS

Mechanism of Passivation

The theory of passivators has already been dealt with in part, and should be referred to again in Chapter 5, pp. 65–67. Passivators in contact with a metal surface act as depolarizers, initiating high current densities at residual anodic areas which exceed $i_{critical}$ for passivation. Only those ions can serve as passivators which have both an oxidizing

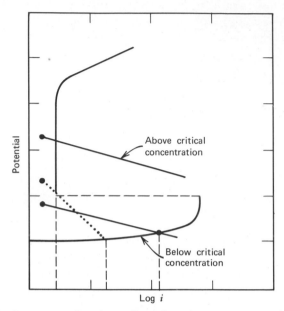

Fig. 1. Polarization curves that show effect of passivator concentration on corrosion of iron. An oxidizing substance that reduces sluggishly does not induce passivity (dotted cathodic polarization curve).

capacity in the thermodynamic sense (noble oxidation-reduction potential) and which are readily reduced (shallow cathodic polarization curve) (Fig. 1). Hence SO_4^{--} or ClO_4^- ions are not passivators for iron because they are not readily reduced, nor are NO_3^- ions compared to NO_2^-, because nitrates are reduced less rapidly than are nitrites, the former reducing too sluggishly to achieve the required high value of $i_{critical}$. The extent of chemical reduction on initial contact of a passivator with metal, according to this viewpoint, must be at least chemically equivalent to the amount of passive film formed as a result of such reduction. For the passive film on iron, as has been discussed earlier, this is in the order of 0.01 C/cm^2 of apparent surface. The total equivalents of chemically reduced chromate is found to be of this order and it is probably also the same for other passivators acting on iron. The amount of chromate reduced in the passivation process is arrived at from measurements[1-3] of residual radioactivity of a washed iron surface after exposure

[1] H. Uhlig and P. King, *J. Electrochem. Soc.,* **106**, 1 (1959).
[2] R. Powers and N. Hackerman, *Ibid.,* **100**, 314 (1953).
[3] M. Cohen and A. Beck, *Z. Elektrochem.,* **62**, 696 (1958).

to a chromate solution containing radioactive ^{51}Cr. The following reaction applies, assuming, as measurements appear to substantiate,[4] that all reduced chromate (or dichromate) remains on the metal surface as adsorbed Cr^{+3} or as hydrated Cr_2O_3:

$$Cr_2O_7^{--} + 8H^+ + Fe \text{ surface} \rightarrow 2Cr^{+3} + 4H_2O + O_2 \cdot O_{ads. \text{ on } Fe}$$
$$\Delta G^\circ = -97 \text{ kcal}$$

Residual radioactivity accounts for 3×10^{16} Cr atoms/cm^2 (1.5×10^{-7} eq or 0.015 C passive film substance/cm^2). The equation assumes an adsorbed passive film structure, but the same reasoning applies whatever the structure.

Reduction of passivator continues at a low rate after passivity is achieved, equivalent in absence of dissolved oxygen to the value of $i_{passive}$ (<0.3 μA/cm^2 based on observed corrosion rates of iron in chromate solutions). Iron oxide and chromate reduction products slowly accumulate. The rate of reduction increases with factors that increase $i_{passive}$, such as higher H^+ activity, higher temperatures, and the presence of Cl$^-$. It is found in practice that less chromate is consumed as exposure time continues, partially because the secondary film of oxides eventually exposes less of the metal surface at which the passive film requires repair.

For optimum inhibition, the concentration of passivator must exceed a certain critical value. Below this concentration, passivators behave as active depolarizers and increase the corrosion rate at localized areas (pits). Lower concentrations of passivator correspond to more active values of the oxidation-reduction potential, and eventually the cathodic polarization curve intersects the anodic curve in the active region instead of in the passive region alone (Fig. 1).

The critical concentration for CrO_4^{--}, NO_2^-, MoO_4^{--}, or WO_4^{--} is about 10^{-3} to 10^{-4} M.[5-8] A concentration of 10^{-3} M Na$_2$CrO$_4$ is equivalent to 0.016% or 160 ppm. Chloride ions and elevated temperatures increase $i_{critical}$ as well as $i_{passive}$, which in effect raises the critical passivator concentration to higher values. At 70 to 90°C, for example, the critical concentration of CrO_4^- and NO_2^- is about 10^{-2} M.[7, 8] Should the passivator concentration fall below the critical value in stagnant areas (e.g., at threads of a pipe or at crevices), the active potential at such areas in galvanic contact with passive areas elsewhere

[4] D. Brasher and A. Kingsbury, *Trans. Faraday Soc.,* **54**, 1214 (1958).

[5] W. Robertson, *J. Electrochem. Soc.,* **98**, 94 (1951).

[6] S. Matsuda and H. Uhlig, *J. Electrochem. Soc.,* **111**, 156 (1964).

[7] A. Mercer and I. Jenkins, *Brit. Corros. J.,* **3**, 130 (1968).

[8] A. Mercer, I. Jenkins, and J. Rhoades-Brown, *Brit. Corros. J.,* **3**, 136 (1968).

of noble potential promotes corrosion (pitting) at the active areas (passive-active cells). For this reason it is important to maintain the concentration of passivators above the critical value at all portions of the inhibited system by use of stirring, rapid flow rates, and avoidance of crevices or of surface films of grease and dirt. Since consumption of passivators increases in the presence of chloride and sulfate ions, it is also essential to maintain as low a concentration of these ions as possible.

Some substances indirectly facilitate passivation of iron (and probably of some other metals, too) by making conditions more favorable for adsorption of oxygen. In this category are alkaline compounds (e.g., $NaOH$, Na_3PO_4, $Na_2O \cdot nSiO_2$, $Na_2B_4O_7$). These are all nonoxidizing substances requiring dissolved oxygen in order to inhibit corrosion, hence oxygen is properly considered the passivating substance. The mechanism of passivation is similar to that described in Chapter 6 under "Higher Partial Pressures of Oxygen," p. 94 and "Effect of pH," p. 98. High concentrations of OH^- displace H adsorbed on the metal surface, thereby decreasing the probability of a reaction between dissolved O_2 and adsorbed H. The excess oxygen is then available to adsorb instead, producing passivity. In addition to passive films of this kind, protection is supplemented by diffusion barrier films of iron silicate, iron phosphate, etc.

It was found that passivation of iron by molybdates and tungstates, both of which inhibit in the near neutral pH range, also requires dissolved oxygen,[9] contrary to the situation for chromates and nitrites. In this case, dissolved oxygen may help create just enough additional cathodic area to ensure anodic passivation of the remaining restricted anode surface at the prevailing rate of reduction of MoO_4^{--} or of WO_4^{--}, whereas in the absence of O_2, $i_{critical}$ is not achieved.

Sodium benzoate[9-11] (C_6H_5COONa), sodium cinnamate[12] ($C_6H_5 \cdot CH \cdot CH \cdot COONa$), and sodium polyphosphate[13-15] ($NaPO_3)_n$ (Fig. 2) are further examples of nonoxidizing compounds that effectively passivate iron in the near-neutral range, apparently through facilitating the adsorption of dissolved oxygen. As little as 5×10^{-4} M sodium benzoate (0.007%) effectively inhibits steel in aerated distilled water,[16]

[9] M. Pryor and M. Cohen, *J. Electrochem. Soc.,* **100,** 203 (1953).
[10] W. Vernon, *J. Soc. Chem. Ind.,* **66,** 137 (1947); F. Wormwell and A. Mercer, *J. Appl. Chem.,* **2,** 150 (1952).
[11] D. Brasher and A. Mercer, *Brit. Corros. J.,* 3, 120 (1968).
[12] F. Wormwell, *Chem. Ind. (London),* 1953, 556.
[13] G. Hatch and O. Rice, *Ind. Eng. Chem.,* **32,** 1572 (1940).
[14] H. Uhlig, D. Triadis, and M. Stern, *J. Electrochem. Soc.,* **102,** 59 (1955).
[15] G. Hatch, *Materials Prot.,* **8,** 31 (November 1969).
[16] S. Matsuda, D.Sc. thesis, Dept. of Metallurgy, M.I.T., 1960.

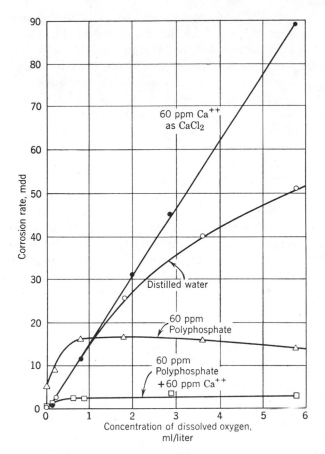

Fig. 2. Effect of oxygen concentration on sodium polyphosphate as a corrosion inhibitor of iron showing beneficial effect of dissolved O_2 and Ca^{++}, 48-hr test, 25°C (Uhlig, Triadis, and Stern).

but inhibition is not observed in deaerated water. The steady-state corrosion rate of iron in aerated $0.01M$ (0.14%) sodium benzoate, pH 6.8, is only 0.01 mdd, whereas in deaerated solution the rate is 0.73 mdd. Inhibition occurs only above about pH 5.5; below this value the hydrogen evolution reaction, for which benzoate ions have no inhibiting effect, presumably becomes dominant, and the passive film of oxygen is destroyed.

Gatos[17] found that optimum inhibition of steel in water of pH 7.5

[17] H. Gatos, in *Symp. sur les Inhibiteurs de Corrosion*, p. 257, University of Ferrara, Italy, 1961.

containing 17 ppm NaCl occurred at and above 0.05% sodium benzoate or 0.2% sodium cinnamate. By using radioactive ^{14}C as tracer, only 0.07, 0.12, and 0.16 monomolecular layer of benzoate (25 $Å^2$, roughness factor 3) was found on a steel surface exposed 24 hr to 0.1, 0.3, and 0.5% sodium benzoate solutions, respectively, and inhibition was observed in all the solutions. These values confirmed previous measurements on benzoates reported by Brasher and Stove[18] also using ^{14}C as tracer. To explain how so small an amount of benzoate on the metal surface can increase adsorption of oxygen, or alternatively in some manner decrease reduction of oxygen at cathodic areas, requires further study. The effect is specific to the cathodic areas of iron because iron continues to corrode in 0.5% sodium benzoate when coupled to gold, on which the reduction of oxygen is apparently not retarded.

The mechanism of inhibition in the case of sodium polyphosphate solutions may depend in part on the ability of polyphosphates to interfere with oxygen reduction on iron surfaces, making it easier for dissolved oxygen to adsorb and thereby to induce passivity. Other factors enter as well; there is for example evidence of protective film formation of the diffusion-barrier type on cathodic areas.[19, 20] Such diffusion-barrier films probably account for the observed inhibition of steel exposed to as high as 2.5% NaCl solutions containing several hundred ppm calcium polyphosphate.[21] In low concentrations of dissolved oxygen, corrosion of iron is accelerated by sodium polyphosphate because of its metal-ion-complexing properties (Fig. 2). Calcium, iron, and zinc polyphosphates are better inhibitors than the sodium compound, but the reasons are not yet fully understood. This and the benzoate system require further study before all the complexities of the mechanism can be spelled out.

In line with the theory of passivators just described, transition metals are those expected and found to be inhibited best by passivators; their anodic polarization curves have the shape depicted in Fig. 1, allowing passivity to be established and then maintained at low current densities. A lesser degree of inhibition can be obtained with the nontransition metals such as Mg, Cu, Zn, Pb, using for example chromates. Protection of these metals apparently results largely from formation of relatively thick diffusion barrier films of insoluble metal chromates mixed with oxides. There is also the possibility that adsorption of CrO_4^{--} on the metal surface contributes in some degree to the lower reaction rate by

[18] D. Brasher and E. Stove, *Chem. Ind. (London)*, **1952**, 171.

[19] G. Hatch, *Ind. Eng. Chem.*, **44**, 1774 (1952).

[20] J. Lamb and R. Eliassen, *J. Amer. Water Works Assoc.*, **46**, 445 (1954).

[21] G. Butler and D. Owen, *Corros. Sci.*, **9**, 603 (1969).

decreasing the exchange current density for the reaction $M \rightarrow M^{++} + 2e^-$ but this has not been proved.

An inhibiting mechanism similar to that for nontransition metals in contact with passivators probably also applies to steel in concentrated refrigerating brines (NaCl or $CaCl_2$) to which chromates are added as inhibitors (approximately 1.5 to 3.0 g $Na_2Cr_2O_7$/liter adjusted with NaOH to form CrO_4^{--}). In the presence of so large a Cl^- concentration, passivity of the kind discussed under Def. 1 (p. 61) does not take place. The reduction in corrosion rate is not as pronounced as when chlorides are absent (see Table 1), and any reduction that occurs apparently results

TABLE 1
Effect of Chromate Concentration, Chlorides, and Temperature on Corrosion of Mild Steel (Roetheli and Cox)

Velocity of spec.: 37 cm/sec.; 14-day tests

$Na_2Cr_2O_7 \cdot 2H_2O$, gm/l → 0			0.1	0.5	1.0
% NaCl	Temp. °C			Corrosion Rate, ipy	
0	20°	0.021	0.0001	0.0001	0.0000
	75	0.036	0.014*	0.0004	0.0002
	95	0.017	0.011*	0.0004	0.0000
0.002	20°	0.026	0.0006	0.0000	0.0000
	75	0.067	0.005*	0.0002	0.0000
	95	0.021	0.017*	0.005*	0.0003
0.05	20°	0.031	0.0012	0.0015	0.0008
	75	0.085	0.002	0.003	0.002
	95	0.023	0.007*	0.005*	0.002
3.5	20°	0.024	0.0017	0.0016	0.0015
22.0	20°	0.007	0.0009	0.0006	0.0013

* Pitted.

from formation of a surface diffusion barrier of chromate reduction products and iron oxides. It should be noted that chromates are not adequate inhibitors for the hot concentrated brine solutions that are sometimes mistakenly proposed as antifreeze solutions for engine cooling systems.

Applications of Passivators

Chromates are applied mostly as inhibitors for recirculating cooling waters (e.g., of internal combustion engines, rectifiers, and cooling towers). The concentration of Na_2CrO_4 used for this purpose is about 0.04 to 0.2%, the higher concentrations being employed at higher temperatures or in fresh waters of chloride concentration above 10 ppm. The pH is adjusted, if necessary, to 7.5 to 9.5 by addition of NaOH. Periodic analysis (colorimetric) is required to ensure that the concentration remains above the critical level (10^{-3} M or 0.016% Na_2CrO_4 at room temp.). Sometimes combinations of chromates with polyphosphates or other inhibitors permit the concentration of chromates to fall below the critical level. This results in some sacrifice of inhibiting efficiency, but there is adequate protection against pitting for the treatment of very large volumes of water employing cooling towers.[22]

It should be remembered that chromates are toxic and also cause a rash on prolonged contact with the skin.

Corrosion rates of mild steel as a function of chromate and chloride concentration at various temperatures are shown in Table 1.[23] Such data in the region of the critical chromate concentration are not readily reproduced because of erratic pitting behavior.

Nitrites find use as inhibitors for antifreeze cooling waters (see p. 279) because, unlike the chromates, they have little tendency to react with alcohols or ethylene glycol. They are not so well suited to cooling-tower waters because they are gradually decomposed by bacteria.[24] They are used to inhibit cutting oil-water emulsions employed in the machining of metals (0.1–0.2%). Inhibition of the internal surface of pipelines transporting gasoline and other petroleum products is also accomplished by continuously injecting a 5 to 30% $NaNO_2$ solution into the line.[25] In this connection, gasoline is corrosive to steel because, on reaching lower temperatures underground, it releases dissolved water which in contact with large quantities of oxygen dissolved in the gasoline (solubility of O_2 in gasoline is 6 times that in water) corrodes steel, forming voluminous rust products which clog the line. Sodium nitrite enters the water phase and effectively inhibits rusting. Chromates are also used for the same purpose but have the disadvantage that they tend to react with some constituents of the gasoline.

The corrosion rates of steel in contact with water-gasoline mixtures

[22] H. Kahler and P. Gaughan, *Ind. Eng. Chem.*, **44**, 1770 (1952).

[23] B. Roetheli and G. Cox, *Ind. Eng. Chem.*, **23**, 1084 (1931).

[24] T. P. Hoar, *Corrosion*, **14**, 103t (1958).

[25] A. Wachter and S. Smith, *Ind. Eng. Chem.*, **35**, 358 (1943).

TABLE 2
Corrosion Rates of Mild Steel in Sodium Nitrite Solutions Containing Gasoline

(Rotating Bottle Tests Using Pipeline Water, pH 9, and Regular Gasoline)
14-Day Exposure, Room Temperature (Wachter and Smith)

% $NaNO_2$	Corrosion Rate, mils/year
0.0	4.3
0.02	3.0
0.04	0.6
0.06	0.0
0.10	0.0

containing increasing amounts of $NaNO_2$ are listed in Table 2. The minimum amount of $NaNO_2$ for effective inhibition is 0.06% or 7×10^{-3} M which, because of impurities present in the water, is higher than the critical concentration in distilled water. Nitrites are inhibitors only above about pH 6.0. In more acid environments they decompose, forming volatile nitric oxide and nitrogen peroxide. In common with other passivators they tend to induce pitting at concentrations near the critical in presence of Cl^- or SO_4^{--} ions. In this regard nitrites are less sensitive to Cl^- than to SO_4^{--}, contrary to the situation for chromates[6-8, 26] (Table 3).

PICKLING INHIBITORS

Most pickling inhibitors function by forming an adsorbed layer on the metal surface, probably no more than a monolayer in thickness, which essentially blocks discharge of H^+ and dissolution of metal ions. Some inhibitors block the cathodic reaction (raise hydrogen overvoltage) more than the anodic reaction, or vice versa; but adsorption appears to be general over all the surface rather than at specific anodic or cathodic sites, and both reactions tend to be retarded. Hence on addition of an inhibitor to an acid, the corrosion potential of steel is not greatly altered (<0.1 V), although the corrosion rate may be appreciably reduced (Fig. 3).

Compounds serving as pickling inhibitors require, by and large, a favorable polar group or groups by which the molecule can attach itself to the metal surface. These include the organic N, amine, S, and OH groups. The size, orientation, shape, and electric charge of the molecule

[26] S. Sussman, O. Nowakowski, and J. Constantino, *Ind. Eng. Chem.*, **51**, 581 (1959).

TABLE 3
Critical Concentrations of NaCl or Na₂SO₄ above which Pitting of Armco Iron in Chromate or Nitrite Solutions Occurs*

5-Day Tests, 25°C, Stagnant Solutions (Matsuda)

	Critical Concentration	
	NaCl	Na₂SO₄
Na₂CrO₄, 200 ppm	12 ppm	55 ppm
500	30	120
NaNO₂ 50 ppm	210 ppm	20 ppm
100	460	55
500	> 2000	450

* See also Refs. 7 and 8.

play a part in the effectiveness of inhibition. Whether a compound adsorbs on a given metal and the relative strength of the adsorbed bond often depend on factors such as surface charge of the metal.[27] For inhibitors that adsorb better at increasingly active potentials as measured from the point of so-called zero surface charge, (potential

[27] A. Frumkin, *Z. Elektrochem.,* **59**, 807 (1955); *Trans. Faraday Soc.,* **55**, 156 (1959); *J. Electrochem. Soc.,* **107**, 461 (1960); "Kinetics of Electrode Processes and Null Points of Metals," L. Antropov, pp. 48–82, Council Sci. Ind. Res., New Delhi, 1960.

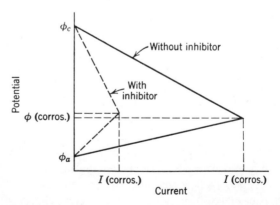

Fig. 3. Polarization diagram for steel corroding in pickling acid with and without inhibitor.

of minimum ionic adsorption) cathodic polarization in presence of the inhibitor provides better protection than either the equivalent cathodic protection or use of an inhibitor alone. This was demonstrated by Antropov[28] for iron and zinc in sulfuric acid containing various organic inhibitors.

The anion of the pickling acid may also take part in the adsorbed film or so-called double-layer structure, accounting for differing efficiencies of inhibition for the same compound in HCl compared to H_2SO_4. For example, the corrosion rate of steel at 20°C inhibited with 20 g/liter quinoline is 260 mdd in $2N$ H_2SO_4 but only 48 mdd in $2N$ HCl, whereas the corrosion rates in the absence of inhibitor are 360 and 240 mdd, respectively.[29] In addition, specific electronic interaction of polar groups with the metal (chemisorption) may account for a given compound being a good inhibitor for iron but not for zinc, or vice versa. The latter factor in certain cases may be more important than the steric factor (diffusion-barrier properties) of a closely packed oriented layer of high-molecular-weight molecules. This is shown by the outstanding inhibition provided by the simple molecule carbon monoxide, CO, dissolved in HCl to which 18-8 stainless steel is exposed,[30] (99.8% efficient* in $6.3N$ HCl, 25°C) or which is provided to iron by a small amount of iodide in dilute H_2SO_4.[31, 32] Both CO and iodide chemisorb on the metal surface, interfering mostly with the anodic reaction.[33] Kaesche[34] showed that $10^{-3}M$ KI is a much more effective inhibitor for iron in $0.5M$ Na_2SO_4 solution at pH 1 (89% efficient) compared to pH 2.5 (17% efficient), indicating that adsorption of iodide, particularly in this pH range, is pH dependent.

The interaction of factors just mentioned, plus perhaps still others, enter into accounting for the fact that some compounds (e.g., *o*-tolyl-thiourea[35] in 5% H_2SO_4) are better inhibitors at elevated temperatures

---

* % efficiency = [rate$_{no\ inhib}$ — rate$_{with\ inhib}$] × 100/rate$_{no\ inhib}$.

[28] L. Antropov, *Inhibitors of Metallic Corrosion and the phi-scale of Potentials*, First International Congress on Metallic Corrosion, Butterworths, London, 1961.

[29] I. Putilova, S. Balezin, and V. Barannik, *Metallic Corrosion Inhibitors*, transl. G. Ryback, p. 63, Pergamon Press, New York, 1960.

[30] H. Uhlig, *Ind. Eng. Chem.*, **32**, 1490 (1940).

[31] K. Hager and M. Rosenthal, *Ordnance*, **35**, 479 (1951); *Chem. Abstr*, **49**, 12254 (1955).

[32] S. Iofa and L. Medwedjewa, *Compt. Rend. Acad. Sci. URSS*, **69**, 213 (1949); S. Iofa, E. Ljachowezkeja, and K. Scharifow, *Ibid.*, **84**, 543 (1952); S. Iofa and G. Roshdesfwenskaja, *Ibid.*, **91**, 1159 (1953). See also Ref. 29.

[33] K. Heusler and G. Cartledge, *J. Electrochem. Soc.*, **108**, 732 (1961).

[34] H. Kaesche, in *Symp. sur les Inhibiteurs de Corrosion*, p. 137, University of Ferrara, Italy, 1961.

[35] T. Hoar and R. Holliday, *J. Appl. Chem.*, **3**, 502 (1953).

than at room temperature, presumably because adsorption increases or the film structure becomes more favorable at higher temperatures. Others (e.g., quinoline derivatives) are more efficient in the lower temperature range.[35]

Typical effective organic pickling inhibitors for steel are quinolin-ethiodide, o- and p-tolylthiourea, propyl sulfide, diamyl amine, formaldehyde, and p-thiocresol. Other compounds are listed in the *Corrosion Handbook*, pages 910 to 912.

Inhibitors containing sulfur, although effective otherwise, sometimes induce hydrogen embrittlement of steel in the event that the compound itself or hydrolysis products (e.g., H_2S) are formed which favor entrance of H atoms into the metal (see p. 48). In principle, inhibitors containing arsenic or phosphorus can behave similarly.

Applications of Pickling Inhibitors

Pickling inhibitors used in practice are seldom pure compounds. They are usually mixtures, which may be a byproduct, for example, of some industrial chemical process for which the active constituent is unknown. They are added to a pickling acid in small concentration, usually in the order of 0.01 to 0.1%. The typical effect of concentration of an inhibitor on the reaction between steel and 5% H_2SO_4 is shown in Fig. 4,[35] showing that above a relatively low concentration, presumably

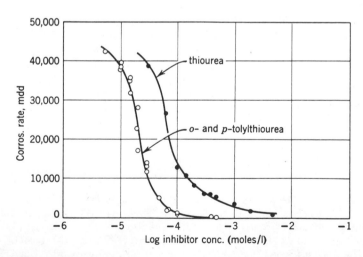

Fig. 4. Effect of inhibitor concentration on corrosion of 0.1% C steel in 5% H_2SO_4, 70°C (Hoar and Holliday).

that necessary to form an adsorbed monolayer, an additional amount of inhibitor has little effect on further reducing the rate.

Inhibitors are commonly used in the acid pickling of hot-rolled steel products in order to remove mill scale. The advantages of using an inhibitor for this purpose are (1) saving of steel, (2) saving of acid, (3) reduction of acid fumes caused by hydrogen evolution. Inhibited dilute sulfuric or hydrochloric acid is also used to clean out steel water pipes clogged with rust, to clean boiler tubes encrusted with $CaCO_3$ or iron oxide scales, and to activate oil wells underground, the inhibitor protecting steel oil-well tubing. For example, boiler scale can be removed by using 0.1% hexamethylene tetramine in 10% HCl at a maximum temperature of 70°C (160°F).[36]

SLUSHING COMPOUNDS

Slushing compounds are used to protect steel surfaces temporarily from rusting during shipment or storage. They consist of oils, greases, or waxes that contain small amounts of organic additives. The latter are polar compounds that adsorb on the metal surface in the form of a closely packed oriented layer. In this respect the mechanism of inhibition by organic additives is similar to that of inhibition by pickling inhibitors, except that for use as slushing compounds additives must suitably adsorb in the near-neutral range of pH, whereas pickling inhibitors adsorb best in the low pH range, calling for somewhat different properties.

Whether an oil or a wax is chosen as the vehicle depends on (1) the relative length of time protection is desired, the wax usually providing longer life and (2) the factor of ease of removal before the protected machine part is put into service, an oil being easier to wipe off or dissolve in solvents. The thickness of an applied coat varies from 5 to more than 100 mils (0.1 to more than 2.5 mm).

Suitable organic additives for use in slushing compounds include organic amines, zinc naphthenate, various oxidation products of petroleum, alkali and alkaline earth metal salts of sulfonated oils, and various other compounds.[37] A substance that has been used successfully for a long time is lanolin, obtained from wool scouring; its active constituents are various high molecular-weight fatty alcohols and acids. Sometimes lead soaps are added to slushing compounds, these soaps

[36] W. Cerna, *Proc. 7th Annual Water Conf.*, Pittsburgh, Pa., 1947.

[37] D. Atkins, H. Baker, C. Murphy, and W. Zisman, *Ind. Eng. Chem.*, **39**, 491 (1947); H. Baker and W. Zisman, *Ibid.*, **40**, 2338 (1948).

reacting to form relatively insoluble $PbCl_2$ with any NaCl from perspiration transferred to steel surfaces through handling.

VAPOR-PHASE INHIBITORS

Substances of low but significant vapor pressure, the vapor of which has corrosion-inhibiting properties, are called vapor-phase inhibitors. They are used to protect critical machine parts (e.g., ball bearings or other manufactured steel articles) temporarily against rusting by moisture during shipping or storage. They have the advantage over slushing compounds of easy application, and the possibility of immediate use of the protected article without first removing a residual oil or grease film. They have the disadvantage of accelerating the corrosion of some nonferrous metals, discoloring some plastics, and requiring relatively effective sealing of a package against loss of the inhibiting vapor. The latter requirement is relatively easily achieved, however, by using wrapping paper impregnated on the inside surface with the inhibitor and incorporating a vapor-barrier coating on the outside.

The mechanism of inhibition has not been studied in detail, but it appears to be one of adsorbed film formation on the metal surface which provides protection against water or oxygen, or both. In the case of volatile nitrites, the inhibitor may also supply a certain amount of NO_2^- which passivates the surface.

Detailed data have been presented for dicyclohexylammonium nitrite,[38] which is one of the most effective of the vapor-phase inhibitors. This substance is white, crystalline, almost odorless, and relatively nontoxic. It has a vapor pressure of 0.0001 mm Hg at 21°C (70°F), which is about one-tenth the vapor pressure of mercury itself.* One gram saturates about 550 m³ (20,000 ft³) of air, rendering the air relatively noncorrosive to steel. The compound decomposes slowly; nevertheless, in properly packaged paper containers at room temperature it effectively inhibits corrosion of steel over a period of years. However, it should be used with caution in contact with nonferrous metals. In particular, corrosion of zinc, magnesium, and cadmium is accelerated.

Cyclohexylamine carbonate has the somewhat higher vapor pressure of 0.4 mm Hg at 25°C and its vapor also effectively inhibits steel.[39] The

---

* I. Rosenfeld et al. (*Symp. sur les Inhibiteurs de Corrosion,* p. 344, University of Ferrara, Italy, 1961) report the lower value of 0.00001 mm Hg obtained for dicyclohexylammonium nitrite carefully purified by multiple recrystallization from alcohol.

[38] A. Wachter, T. Skei, and N. Stillman, *Corrosion,* 7, 284 (1951).

[39] E. Stroud and W. Vernon, *J. Appl. Chem.,* 2, 178 (1952).

higher vapor pressure provides more rapid inhibition of steel surfaces either during packaging or on opening and again closing a package, during which time concentration of vapor may fall below that required for protection. The vapor is stated to reduce corrosion of aluminum, solder, and zinc, but it has no inhibiting effect on cadmium and it increases corrosion of copper, brass, and magnesium.

Ethanolamine carbonate and various other compounds are described as vapor phase inhibitors in Refs. 29 and 39. A combination of urea and sodium nitrite has found practical application, including use in impregnated paper. The mixture probably reacts in the presence of moisture to form ammonium nitrite, which is volatile although unstable and conveys inhibiting nitrite ions to the metal surface.

Inhibitor to Reduce Tarnishing of Copper

Copper dipped in 0.25% benzotriazole in water at 60°C for 2 min forms a thin adsorbed or reaction-product film that protects the metal against subsequent tarnish in the atmosphere. The film, which probably acts as a diffusion-barrier layer although the precise mechanism is not yet known, is stated to be effective also for brasses and for nickel-silver (zinc-nickel alloys).[40]

GENERAL REFERENCES

Metallic Corrosion Inhibitors, I. Putilova, S. Balczin, and V. Barannik, transl. G. Ryback, Pergamon Press, New York, 1960. (The authors of this volume believe that the primary process of pickling inhibitors is the formation of protective metal reaction-product films formed by metal, inhibitor, and ions of the acid. Although a few isolated compounds may adhere to this pattern, evidence indicates that it cannot be followed in general.)

N. Hackerman and A. Makrides, "Action of Polar Organic Inhibitors," *Ind. Eng. Chem.,* **46**, 523 (1954).

R. Mears and G. Eldredge, "Inhibitors for Aluminum," *Trans. Electrochem. Soc.,* **83**, 403 (1943); *Ind. Eng. Chem.,* **37**, 736 (1945).

Symposium sur les inhibiteurs de Corrosion, University of Ferrara, Italy, 1961. (A collection of 35 papers, most of them in English, which were presented at an international symposium organized by Prof. Leo Cavallaro at the University of Ferrara.) See also, *Second Symposium,* Vols. I and II, 1966.

H. Baker, "Volatile Rust Inhibitors," *Ind. Eng. Chem.,* **46**, 2592 (1954).

---

[40] J. Cotton and I. Scholes, *Brit. Corros. J.,* **2**, 1 (1967).

chapter 17

Treatment of water and steam systems

DEAERATION AND DEACTIVATION

In accord with principles described in Chapter 6, pp. 92–94, corrosion of iron is negligible at ordinary temperatures in water that is free of dissolved oxygen. An effective practical means, consequently, for reducing corrosion of iron or steel in contact with fresh water or seawater is to reduce the dissolved oxygen content. In this way, corrosion of copper, brass, zinc, and lead is also minimized.

Removal of dissolved oxygen from water is accomplished by either chemically reacting oxygen beforehand, called *deactivation*, or by distilling it off in suitable equipment, called *deaeration*. Deactivation can be carried out in practice by slowly flowing hot water over a large surface of steel laths or sheet contained in a closed tank called a deactivator. The water remains in contact long enough to corrode the steel, and by this process most of the dissolved oxygen is consumed. Subsequent filtration removes suspended rust. Water so treated is much less corrosive to a metal pipe distribution system. Deactivators of this kind have been employed in some buildings, a description of which is given by Speller.[1] However, since use of deactivators requires regular attention in addition to periodic renewal of the steel sheet, this approach to oxygen removal is usually too cumbersome compared with the use of chemical inhibitors or deaeration.

Deactivation of industrial waters (not potable waters because the

---

[1] F. N. Speller, *Corrosion, Causes and Prevention,* pp. 418–424, McGraw-Hill, New York, 1951.

chemicals used are toxic) is possible by employing sodium sulfite in accord with the reaction

$$Na_2SO_3 + \tfrac{1}{2}O_2 \rightarrow Na_2SO_4 \qquad (1)$$

for which Na_2SO_3 reacts with oxygen in the weight ratio of 8:1 (8 lb Na_2SO_3 to 1 lb O_2). The reaction is relatively fast at elevated temperatures, but slow at ordinary temperatures. It can be speeded up by adding catalysts such as Cu^{++} or, still better, Co^{++} salts.[2, 3]

The rapid decrease with time of dissolved oxygen in San Joaquim, Calif., river water after treatment with 80 ppm Na_2SO_3 (0.67 lb/1000 gal) plus copper or cobalt salts is shown in Fig. 1. Water so treated

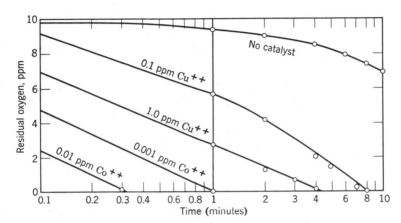

Fig. 1. Effect of cobalt and copper salts on reaction rate of sodium sulfite with dissolved oxygen at room temperature (Pye).

using $CoCl_2$ as catalyst was found by Pye[2] to be noncorrosive to a steel heat-exchanger system that without treatment had previously suffered serious corrosion and loss of heat transfer. Tests showed a reduction in corrosion rate from 0.008 ipy (pitting factor = 7.4) before treatment to 0.00016 ipy afterward.

Hydrazine (N_2H_4), supplied as a concentrated aqueous solution, also reacts with dissolved oxygen according to

$$N_2H_4 + O_2 \rightarrow N_2 + 2H_2O \qquad (2)$$

in the weight ratio of 1:1. The reaction is similarly slow at ordinary temperatures, but can be speeded by use of catalysts (e.g., activated

[2] D. Pye, *J. Amer. Water Works Assoc.*, **39**, 1121 (1947).
[3] S. Pirt, D. Callow, and W. Gillett, *Chem. Ind.* (*London*), **1957**, 730.

charcoal, metal oxides, alkaline solutions[4] of Cu^{++} and Mn^{++}) and by raising the temperature. The reaction, however, in absence of specific catalysts, is still slow at 175°C (345°F)[5] or even higher.[6] At elevated temperatures decomposition occurs slowly at 175°C (350°F) and more rapidly at 300°C (570°F), producing ammonia:[5]

$$3N_2H_4 \rightarrow N_2 + 4NH_3 \tag{3}$$

Note that the normal reaction products—nitrogen, water, and a small

[4] H. Gaunt and E. Wetton, *J. Appl. Chem.*, **16**, 171 (1966).

[5] M. Baker and V. Marcy, *Trans. Amer. Soc. Mech. Engrs.*, **78**, 299 (1956).

[6] G. Everitt, E. Potter, and R. Thompson, *Chem. Ind.* (*London*), **1962**, 609 (March 31).

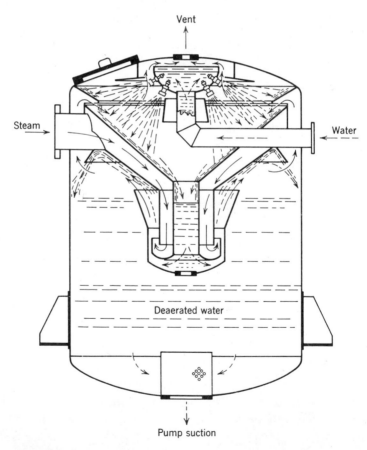

Fig. 2. Sketch of one type of steam deaerator.

TABLE 1
Approximate Allowable Oxygen Concentration in Deaerated Water for Corrosion Control in Steel Systems

	Maximum Oxygen Concentration	
Cold water	0.3 ppm	0.2 ml/liter
Hot water	0.1	0.07
Low pressure boilers (<250 psi)	0.03	0.02
High pressure boilers	0.005	0.0035

amount of NH_3—are all volatile, and that, unlike sulfite additions, no salts accumulate in the treated water.

Recently, special ion-exchange resins have been made available for reducing dissolved oxygen. These are resins that incorporate within the resin structure substances such as metal sulfites, ferrous hydroxide, or manganous hydroxide, which rapidly react with oxygen. The resins can be regenerated by suitable chemical treatment. In laboratory tests using a resin containing ferrous hydroxide, Potter[7] succeeded in continuously reducing the oxygen content of a water containing 8.8 ppm to below 0.002 ppm O_2.

Deaeration is accomplished by spraying water or flowing it over a large surface, countercurrent to steam. Oxygen distills off and also some dissolved carbon dioxide (Fig. 2). Water is heated in the process and is suitable, therefore, as feed water for boilers. Steam deaerators of this kind are standard equipment for all high-pressure stationary boilers. On the other hand, if the water is to be used cold, dissolved gases are distilled off by lowering the pressure, employing a mechanical pump or steam ejector instead of a countercurrent flow of steam. This is called vacuum deaeration. Equipment of this kind has been designed to deaerate several million gallons of water per day.

In principle it is more difficult and more expensive to remove the last traces of dissolved oxygen by distillation compared to the first 90 to 95%, and it is more difficult at low temperatures than at high temperatures. To achieve a low enough oxygen level in cold water, it is often necessary to use multiple-stage vacuum treatment. Fortunately acceptable levels of dissolved oxygen for corrosion control in cold water are higher than in hot water or in steam. The allowable levels established through experience[8] are given in Table 1.

[7] E. Potter and G. Whitehead, *J. Appl. Chem.*, **7**, 629 (1957).
[8] F. N. Speller, in *Corrosion Handbook*, p. 506.

HOT- AND COLD-WATER TREATMENT

1. *Hot-Water Heating Systems.* These are closed steel systems in which the initial corrosion of the system soon uses up dissolved oxygen; corrosion is negligible so far as the life of the metal equipment is concerned, once the dissolved oxygen is used up. A continuing minor reaction of steel with water produces hydrogen, with a characteristic odor due to traces of hydrocarbon gases originating from the reaction of carbides in steel with water. It is stated that the hydrogen reaction can be minimized by treating the water with NaOH (or Na_3PO_4) to a pH of 8.5.[9]

2. *Municipal Water Supplies.* Usually, hard waters of positive Saturation Index are relatively noncorrosive and do not require treatment of any kind for corrosion control. Soft waters, on the other hand, cause rapid accumulation of rust in ferrous piping, are readily contaminated with toxic quantities of lead salts on passing through lead piping, and cause blue staining of bathroom fixtures by copper salts originating from slight corrosion of copper or brass piping. Vacuum deaeration of such waters would be ideal as a corrosion control measure. The expense, however, runs high for treating the large quantities of water involved, and no practical installations have yet been constructed for community water supplies. The possibility nevertheless deserves consideration.

Chemical treatment of potable waters is limited to small concentrations of inexpensive, nontoxic chemicals, such as the addition of alkalies or lime. Some water supplies are treated with about 2 ppm sodium polyphosphate, which helps to reduce the red color originating from ferric salts or suspended rust in the water. This treatment also reduces the corrosion rate to a modest extent wherever the water moves with some velocity and is fully aerated. In stagnant areas of the distribution system, however, there is probably no practical benefit. Also in hot-water systems, the addition of small amounts of polyphosphates, apart from their rapid decomposition into orthophosphates which are less effective as inhibitors, provides no advantage in protecting the system against corrosion.

Raising of the Saturation Index, on the other hand, provides a potentially useful means for reducing the corrosion rate in either flowing or stagnant portions of the distribution system, and is also effective for reducing corrosion of hot-water systems. This treatment requires addition of lime [$Ca(OH)_2$] or both lime and soda ash (Na_2CO_3) to the water in amounts which raise the Saturation Index to about +0.5 (see

[9] F. Speller, in *Corrosion Handbook*, p. 498.

p. 115). For the treatment to be successful, the water must be low in colloidal matter and in dissolved solids other than calcium salts. Corrosion of copper, lead, and brass is also reduced by this treatment. In hot-water systems the possibility of excess deposition of $CaCO_3$, which causes scaling, must be taken into account in arriving at the proper proportions of added chemicals.

Sodium silicate treatment in the amount of about 4 to 15 ppm SiO_2 is sometimes used by individual owners of buildings in soft-water areas. This treatment reduces "red water" caused by suspended rust resulting from corrosion of ferrous piping, and also eliminates blue staining by water that has passed through copper or brass piping. At the same time, a practical reduction in the corrosion rate of steel in the order of 50 to 90% may be observed,[10, 11] but not for all waters.[12, 13]

The conditions under which protection exists or is optimum are not entirely understood, but it is clear that dissolved calcium and magnesium salts have an effect and that some protection may result alone from the alkaline properties of sodium silicate. In presence of silicate, passivity of iron may be observed at pH 10 with an accompanying reduction of the corrosion rate to 1 to 7 mdd.[13] Sodium hydroxide induces similar passivity and corresponding low corrosion rates at the somewhat higher pH range of 10 to 11.* Under other conditions (e.g., pH 8) a protective diffusion-barrier film is formed, apparently containing SiO_2 and consisting perhaps of an insoluble iron silicate. Laboratory tests in distilled water at 25°C showed a reduction in the corrosion rate of iron in the order of 85 to 90% when sodium silicate was added (5 ppm SiO_2) to bring the pH to 8.[13] However, no inhibition was obtained in the laboratory at the same SiO_2 content using Cambridge tap water (pH 8.3, 44 ppm Ca, 10 ppm Mg, 16 ppm Cl⁻). If larger quantities of sodium silicate were added to raise the pH of the water to 10 or 11, the range in which passivity of iron occurs, a marked decrease in the corrosion rate was observed.

Domestic or industrial hot-water heaters of galvanized steel through

---

* An increase in the corrosion rate is observed at pH 9.5 to 10, just before passivity is established, which appears to be associated with the presence of carbonates in NaOH. Corrosion rates of iron in Na_2CO_3 solutions show a similar maximum [E. Heyn and O. Bauer, *Mitt. Königl. Materialprüf.*, **26**, 84–85 (1908)].

[10] H. Shuldener and S. Sussman, *Corrosion,* **16**, 354t (1960).

[11] F. N. Speller, *Corrosion, Causes and Prevention,* pp. 389–395, McGraw-Hill, New York, 1951.

[12] R. Eliassen, R. Skrinde, and W. Davis, "Corrosion Control Studies by Manometric Techniques," Birmingham Regional Meeting, Amer. Iron and Steel Inst. (1958).

[13] S. Matsuda, Sc. D. thesis, Dept. of Metallurgy, M.I.T. (1960).

which hot aerated water passes continuously are not protected reliably in all types of water by nontoxic chemical additions such as silicates or polyphosphates. Adjustment of the Saturation Index to a more positive value, as discussed earlier, is sometimes helpful. Often cathodic protection or use of nonferrous metals [e.g., copper or 70% Ni-Cu (Monel)], is the best or only practical measure.

Cooling Waters

Once-through cooling waters usually cannot be treated chemically, both because of the large quantities of inhibitors required and because of the problem of water pollution. Sometimes additions of about 2 to 5 ppm sodium or calcium polyphosphate are made to help reduce corrosion of steel equipment. In such small concentrations polyphosphates are not toxic, but water disposal may continue to be a problem because of the need to avoid accumulation of phosphates in rivers and lakes. Adjustment of the Saturation Index to a more positive value is sometimes a practical possibility. Otherwise use must be made of a suitable protective coating or of metals more corrosion resistant than steel.

Recirculating cooling waters, as for engine-cooling systems, can be treated with sodium chromate (Na_2CrO_4) in the amount of 0.04 to 0.2% (or the equivalent amount of $Na_2Cr_2O_7 \cdot 2H_2O$ plus alkali to pH 8). Chromates inhibit corrosion of steel, copper, brass, aluminum, and soldered components of such systems. Since chromate is consumed slowly, additions must be made at long intervals in order to maintain the concentration above the critical. For diesel or other heavy-duty engines 2000-ppm sodium chromate (0.2%) is recommended in order to reduce damage by cavitation-erosion as well as by aqueous corrosion (see pp. 109–111).

Chromates cannot be used in the presence of antifreeze solutions because of their tendency to react with organic substances. Many proprietary inhibitor mixtures are on the market which are usually dissolved beforehand in methanol or in ethylene glycol in order to simplify the packaging problem, but this also limits the available number of suitable inhibitors. In the United States, borax ($Na_2B_4O_7 \cdot 10H_2O$) is a common ingredient. Sulfonated oils, which produce an oily protective coating, and mercaptobenzothiazole, which specifically inhibits corrosion of copper and at the same time removes the accelerating influence of dissolved Cu^{++} on corrosion of other portions of the system, are sometimes added to borax. One specification calls for a final concentration in the antifreeze solution of 1.7% borax, 0.1% mercaptobenzothiazole, and 0.06% Na_2HPO_4, the latter being added specifically to protect aluminum.

Borax, although an inhibitor of zinc in 50% ethylene glycol solution at elevated temperatures (80°C; 170°F), accelerates attack at room temperature or below.[14] Cadmium and magnesium (but not Al, Fe, or Cu) are affected similarly. In England, 0.1% sodium nitrite plus 1.5% sodium benzoate is used,[15] the latter substance being necessary in part to protect solder from accelerated attack by nitrite, and the nitrite being necessary to ensure the protection of cast iron. Ethanolamine phosphate is also used as an inhibitor for engine-cooling systems containing ethylene glycol.

For industrial waters cooled by recirculation through a spray or tray-type tower, ideal inhibitors are not yet available. Chromates are the most reliable from the standpoint of efficient inhibition. However, the critical concentration is relatively high, and as the sulfate and chloride concentrations build up through evaporation of the water, chromates tend to cause pitting or may cause increased galvanic effects at dissimilar metal couples. Windage losses (loss of spray by wind) must be carefully avoided because chromates are toxic. Toxicity also makes it difficult to dispose of chromate solutions whenever it becomes necessary to reduce the concentration of accumulated chlorides and sulfates.

Sodium polyphosphate is often used in a concentration of about 10 to 100 ppm, sometimes with added zinc salts to improve inhibition. The pH value is adjusted to 5 to 6 in order to minimize pitting and tubercle formation, as well as scale deposition. Polyphosphates decompose slowly into orthophosphates, which in the presence of calcium or magnesium ions precipitate insoluble calcium or magnesium orthophosphate, causing scale formation on the warmer parts of the system. Unlike chromates, they favor algae growth, which necessitates the addition of algicides to the water. Corrosion inhibition with polyphosphates is less effective than that by chromates, but polyphosphates in low concentration are not toxic and the required optimum amount of inhibitor is less than for chromates. Combining polyphosphates with chromates permits the chromate concentration to be reduced to values appreciably below the critical level without danger of pitting, although the inhibiting efficiency is not the same as that obtained above the critical concentration. Other combinations have been proposed and used with some stated advantage over single inhibitors employed alone.[16-18]

[14] D. Caplan and M. Cohen, *Corrosion*, **9**, 284 (1953).
[15] F. Wormwell and A. Mercer, *J. Appl. Chem.*, **3**, 22 (1953); (with H. Ison), *Ibid.*, 133.
[16] W. Palmer, *J. Iron Steel Inst.*, **163**, 42 (1949).
[17] H. Kahler and C. George, *Corrosion*, **6**, 331 (1950).
[18] H. Kahler and P. Gaughan, *Ind. Eng. Chem.*, **44**, 1770 (1952).

BOILER WATER TREATMENT

Boiler Corrosion

Steam boilers are constructed according to various designs, but they consist essentially of a low-carbon steel or low-alloy steel container for water which is heated by hot gases. The steam may afterward pass to a superheater of higher alloy steel at higher temperature than the boiler itself. For maximum heat transfer, a series of boiler tubes is usually incorporated, the hot gases passing either around the outer surface or, less frequently, through the inner surface of the tubes. The steam, after doing work or completing some other kind of service, eventually reaches a condenser constructed generally of copper-base alloy tubes. Steam is cooled on one side of the tubes by water passing along the opposite side of a quality ranging from fresh to polluted, brackish, or seawater. The condensed steam then returns to the boiler and the cycle repeats.

Some boilers are equipped with an *embrittlement detector* by means of which the chemical treatment of a water can be evaluated continuously in terms of its potential ability to induce stress corrosion cracking (Fig. 3).[19, 20] A specimen of plastically deformed boiler steel is stressed by setting a screw; adjustment of this screw regulates a slight leak of hot boiler water in the region where the specimen is subject to maximum tensile stress and where boiler water evaporates. A boiler water is considered to have no embrittling tendency if specimens do not crack within successive 30-, 60-, and 90-day tests. Observation of the detector is a worthwhile safety measure because the tendency toward cracking is more pronounced in the plastically deformed test piece than in any portion of the welded boiler, and hence water treatment can be corrected, if necessary, before the boiler is damaged.

Corrosion of boiler and superheater tubes is sometimes a problem on the hot combustion gas side, especially if vanadium-containing oils are used as fuel. This matter is discussed under *Catastrophic Oxidation* on p. 196. On the steam side, since modern boiler practice ensures removal of dissolved oxygen from the feed water, a reaction occurs between H_2O and Fe resulting in a protective film of magnetite (Fe_3O_4) as follows:

$$3Fe + 4H_2O \rightarrow Fe_3O_4 + 4H_2 \qquad (4)$$

[19] American Society for Testing and Materials Standard D 807-52, ASTM *Book of Standards,* Part 23, p. 68 (1970).

[20] W. Schroeder and A. Berk, *Intercrystalline Cracking of Boiler Steel and Its Prevention,* Bureau of Mines Bulletin 443, 1941.

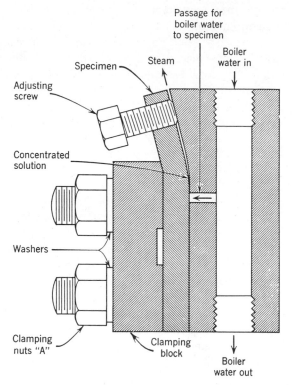

Fig. 3. Embrittlement detector which, when attached to an operating boiler, detects tendency of a boiler water to induce stress corrosion cracking.

The mechanism of this reaction, so far as it is understood, indicates that Fe_3O_4 is formed only below about 570°C (1060°F) and that above this temperature FeO forms instead. The latter then decomposes on cooling to a mixture of magnetite and iron in accord with

$$4FeO \rightarrow Fe_3O_4 + Fe \qquad (5)$$

Measurements of hydrogen accumulation in boilers as a function of time, as well as laboratory corrosion rate determinations, indicate that growth of the oxide obeys the parabolic equation.[21] Hence the rate is diffusion controlled, in line with the mechanism depending on ion and electron migration through solid reaction products as described under *Oxidation and Tarnish*, p. 183.

[21] E. Ulrich, in *Passivierende Filme und Deckschichten, Anlaufschichten,* edited by H. Fischer, K. Hauffe, and W. Wiederholt, p. 308, Springer, Berlin, 1956; see also data by G. Hawkins and H. Solberg in *Corrosion Handbook,* Fig. 1, p. 512.

At lower temperatures (e.g., room temperature to about 100°C) and probably at higher temperatures before a relatively thick surface film develops, experiments show that $Fe(OH)_2$ is the initial reaction product and not Fe_3O_4.[22, 23] The mechanism of corrosion in this temperature range follows that described for anode and cathode interaction on the plane of the metal surface in contact with an electrolyte. Ferrous hydroxide eventually decomposes, at a rate depending on temperature, into magnetite and hydrogen in accord with a reaction first described by Schikorr (Schikorr reaction)[24]

$$3Fe(OH)_2 \rightarrow Fe_3O_4 + H_2 + 2H_2O \qquad (6)$$

The reaction is inhibited by OH^- and is accelerated by Ni^{++}, Pt^{+4}, colloidal platinum, nickel powder, and copper powder, but not by Mn^{++}.[22, 25]

Any factors that disturb the protective magnetite layer on steel, either chemically or mechanically, induce a higher rate of reaction usually in a localized region, causing pitting or sometimes grooving of the boiler tubes. In this regard, the specific damaging chemical factor of excess OH^- concentration is discussed later; mechanical damage, on the other hand, may take place each time the boiler is cooled down. Differential contraction of the oxide and metal causes some degree of spalling of the oxide, thereby exposing fresh metal. Accordingly, it is observed that rate of hydrogen evolution is momentarily high after a boiler is started up again, hydrogen production then falling to normal values presumably after a reasonable thickness of oxide has again built up at damaged areas.

Conditions of boiler operation that lead to metal oxide or inorganic deposits (from the boiler itself or from condenser leakage) on the water side of boiler tubes cause local overheating accompanied by additional precipitation of solutes from the water. Pitting usually results, or so-called plug-type oxidation occurs, which accentuates local temperature rise, leading eventually to stress rupture of the tube. Furthermore, hydrogen resulting from the H_2O-Fe corrosion reaction may enter the steel causing decarburization, followed by microfissuring along grain boundaries and eventual blowout of the affected tube. The latter type of failure may take place without any major loss of tube wall thickness.

[22] U. Evans and J. Wanklyn, *Nature,* **162**, 27 (1948).

[23] V. Linnenbom, *J. Electrochem. Soc.,* **105**, 322 (1958).

[24] G. Schikorr, *Z. Elektrochem.,* **35**, 65 (1929); *Z. Anorg. Allgm. Chem.,* **212**, 33 (1933).

[25] F. Schipko and D. Douglas, *J. Phys. Chem.,* **60**, 1519 (1956).

In the absence of deposits within boiler tubes, these types of damage are not observed.[26]

Boiler Water Treatment for Corrosion Control

Feed waters of boilers are chemically treated both to reduce corrosion of the boiler and auxiliary equipment and to reduce formation of inorganic deposits on the boiler tubes (scaling), which interfere with heat transfer. If steam is used for power production, concentrations of silica and silicates in feed waters must also be reduced in order to minimize volatilization of SiO_2 with steam, causing formation of damaging deposits on turbine blades. Control of scale formation usually requires removing all calcium and magnesium salts by various means, including use of ion-exchange resins or adding substances to the water which favor precipitation of sludges rather than adherent continuous scales on the metal surface. Details are discussed in standard references on boiler-water treatment.

For corrosion control, the basic treatment consists of removal of dissolved gases, addition of alkali, and addition of inhibitors.

1. *Removal of Dissolved Oxygen and Carbon Dioxide.* For high-pressure boilers, any remaining dissolved oxygen in the feed water combines quantitatively with the metals of the boiler system, usually causing pitting of the boiler tubes and general attack elsewhere. Removal of oxygen is accomplished by steam deaeration of the water, followed by addition of a scavenger such as sodium sulfite or hydrazine (see pp. 272–275). Final oxygen concentration is usually held below values (<0.005 ppm O_2) analyzable by chemical methods of analysis, e.g., the Winkler method.

Deaeration is accompanied by some reduction of carbon dioxide content, particularly if the water is acidified before the deaeration process to liberate carbonic acid from the dissolved carbonates. Carbonic acid is corrosive to steel in the absence of dissolved oxygen and more so in its presence,[27] but addition of alkali to boiler water limits any corrosion caused by carbon dioxide to the boiler itself by converting dissolved carbon dioxide to carbonates. At prevailing boiler temperatures, however, carbonates dissociate as follows:

$$Na_2CO_3 + H_2O \rightarrow CO_2 + 2NaOH \tag{7}$$

bringing hot carbonic acid into contact with the condenser and return-line systems. Steel return-line systems suffer serious corrosion, there-

---

[26] P. Goldstein and C. Burton, *Trans. Amer. Soc. Mech. Engrs., Series A*, **91**, 75 (1969).

[27] G. Skaperdas and H. Uhlig, *Ind. Eng. Chem.*, **34**, 748 (1942).

fore, if the carbon dioxide content of boiler water is high. Soluble $FeCO_3$ is formed and it returns with the condensate to the boiler, where it decomposes into $Fe(OH)_2 + CO_2$, the carbon dioxide being again available for further corrosion. The copper alloy condenser system also suffers corrosion should dissolved oxygen be present together with carbon dioxide, but attack of copper-base alloys is negligible in the absence of oxygen. Since carbon dioxide is not used up in the corrosion process, it accumulates increasingly in the boiler with each addition of feed water unless an occasional blow down (intentional release of some boiler water) is arranged.

2. *Addition of Alkali.* Addition of NaOH to water reduces the rate of reaction (4) at 310°C (590°F), according to data of Berl and Van Taack[28] and at 100°C (100–150 hr tests) according to data of Thiel and Luckmann.[29] Berl and Van Taack used mild steel powder (0.11% C) in a bomb of the same material in short time tests (7.5 hr) (Fig. 4). Some doubt has been expressed whether these data apply to condi-

[28] E. Berl and F. van Taack, *Forschungsarbeiten aus dem Gebiete des Ingenieurwesens,* Heft 330, Berlin, (1930).

[29] A. Thiel and H. Luckmann, *Korros. Metallschutz,* **4**, 169 (1928).

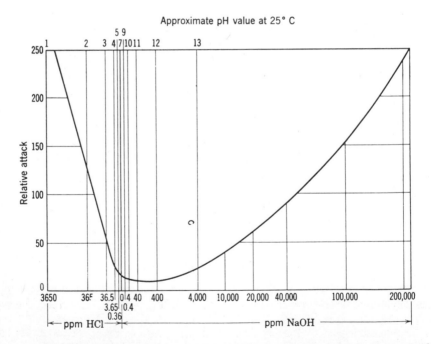

Fig. 4. Corrosion of iron by water at 310°C (590°F) at various values of pH measured at 25°C (Partridge and Hall, based on data of Berl and van Taack).

tions of boiler operation.* Nevertheless, alkali addition to boiler waters constitutes standard practice for most high-pressure boilers presently in operation in the United States and abroad.

Feed water for a high-pressure boiler is treated to a pH (measured at room temperature) of about 9.5 to 11.0. In low-pressure boilers (<250 psi) this value is commonly raised to 11 to 11.5. In some high-pressure boilers, NH_3 is used instead of $NaOH$ at correspondingly lower pH values (8.5–9.0).

The value of nonvolatile alkali additions under some conditions of boiler operation is borne out by statistical analysis of boiler experience by Potter.[30] Of a total of 513 boilers operating in England without alkali addition to the feed water, 29% showed evidence of boiler tube corrosion, whereas of 121 boilers operating with alkali addition, only 5% were corroded.

It is apparent from Fig. 4 that excess alkali can be damaging to a boiler in that the corrosion rate increases rapidly as pH is increased above 13. The danger is not so much that the initial pH of the boiler water may be too high as that accidental concentration of an alkaline boiler water at a crevice, such as is formed between riveted plates or in a cracked oxide scale, or at a hot spot on the tube surface, may reach OH^- concentration levels above the safe range. For this reason it has been held advisable to add buffer ions, such as PO_4^{-3} (Na_3PO_4), which limit the increase in pH a water can achieve no matter how concentrated it becomes. Such ions are also uesful in avoiding similar high OH^- concentrations, which lead to stress corrosion cracking of any portion of the boiler under high residual or applied stress. The minimum amount of PO_4^{-3} recommended for this purpose varies from 30 ppm at pH 10.5 to 90 ppm at pH 11. The amounts at other pH values are given in Fig. 9 on page 530 of the *Corrosion Handbook* in accord with data by Purcell and Whirl.[31] Goldstein and Burton[26] reported that 5 to 10 ppm phosphate at pH 9.5 to 10.0 was more effective in reducing corrosion of high-pressure boiler tubes under a variety of operating conditions than either $NaOH$ or NH_3 treatment.

3. *Addition of Inhibitors.* It is presently possible to add inhibitors for controlling two kinds of corrosion in boiler systems, namely, stress

* M. Bloom, for example, reported that the initial reaction of water with iron at 315°C (600°F) up to at least 100 days is increased at pH 10.6 compared to pH values near neutral. (21st Ann. Water Conf., Engrs. Soc. Western Pennsylvania, Pittsburgh, 1960). See also W. Fraser and M. Bloom, *Corrosion,* 18, 163t, (1962).

[30] E. Potter, *J. Inst. Fuel,* 32, 218 (1959).

[31] T. Purcell and S. Whirl, *Trans. Am Soc. Mech. Engrs.* 64, 397 (1942); *Trans. Electrochem. Soc.,* 83, 279 (1943).

corrosion cracking and return-line corrosion. The first can be minimized by addition of phosphates as mentioned previously. Tests using the embrittlement detector[20] have shown that tannins are also effective inhibitors for this purpose. Quebracho extract from the bark of a South American tree by the same name is one such tannin; it is also sometimes added to boiler waters to prevent scaling. Tests using the embrittlement detector have shown that nitrates are also effective when added as $NaNO_3$ equal to 20 to 30% of the sodium hydroxide alkalinity of the water.[20] This treatment has been used with success in treating feed waters for locomotive boilers, its use practically eliminating further incidence of stress corrosion cracking.

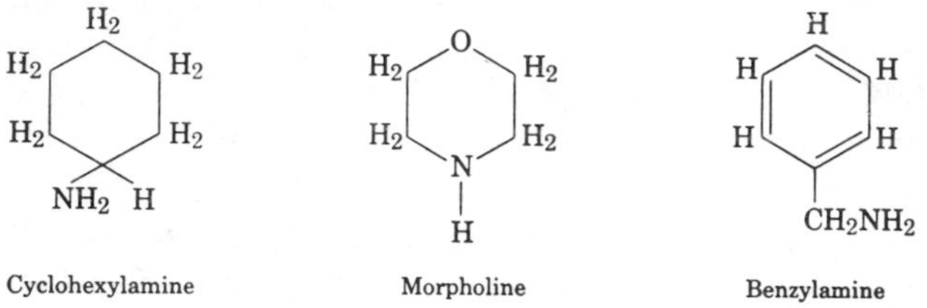

Fig. 5. Structural formulas for three neutralizing-type amines.

Corrosion caused by dissolved carbon dioxide in steam condensate is minimized by adding a volatile amine to the boiler water. There are two categories of volatile amines used for this purpose: (1) neutralizing amines and (2) filming amines. The first group includes cyclohexylamine, benzylamine, and morpholine (Fig. 5). When any of these is added to boiler water in sufficient amount, it neutralizes carbonic acid and raises the pH of steam condensate to an alkaline value, thereby making the condensate less corrosive. Operating by a different principle, volatile octadecylamine, hexadecylamine, and dioctadecylamine are typical of the filming-type inhibitors. They are stated to protect against corrosion by building up a protective organic film on the condenser surface. The filming amines more nearly fit the definition of an inhibitor, whereas the other amines are actually for the most part neutralizers.

MECHANISMS

Mechanisms of boiler corrosion, or the inhibition of such corrosion, are not well understood and many questions remain unanswered. Perhaps the most fundamental questions are whether any chemical additives are needed at all and whether a high-pressure boiler might operate more efficiently and suffer less corrosion damage by using pure deaerated water and nothing else. A few boilers are presently operating on this principle.[32, 33]

Removal of dissolved oxygen is generally agreed upon as a necessary step in all boiler water treatment. It is not likely, however, that any oxygen remaining in a boiler affects the overall uniform reaction of iron to form Fe_3O_4, because the Fe-O_2 reaction at elevated temperatures proceeds at a rate similar to the Fe-H_2O reaction.[34] At higher concentrations (540 ppm O_2) at 100 to 200°C, oxygen was found, in fact, to retard the overall reaction compared to the rate prevailing at less than 0.1 ppm O_2.[35] Instead, oxygen is damaging apparently because it initiates pitting, possibly through action of differential aeration cells. Economizers, in which water is heated before it enters the boiler, are especially susceptible to pitting if the water contains dissolved oxygen.[36]

It is not entirely clear from the available evidence to what extent last traces of O_2, at concentrations for which oxygen concentration cells no longer function, must be scrupulously removed. However, traces of O_2, even if not damaging to the steel boiler directly, are nevertheless effective in causing attack of the condenser system, especially if CO_2 or NH_3 is also present in the condensate. Such corrosion is sufficient to return small amounts of copper salts to the boiler. Although the condensers may or may not be damaged appreciably by such corrosion, the question remains whether pitting of the boiler is initiated by copper contamination of boiler waters. Such contamination would not have occurred had the oxygen content of the water been maintained at zero. There is no general agreement, however, whether copper deposits inside a boiler induce formation of pits around which such deposits are concentrated, or whether they damage the protective qualities of the Fe_3O_4 film in which copper happens to be found as a constituent. It is possible, in the opinion of some, that copper deposition is incidental to galvanic action in which Cu^{++} ions are deposited at cathodic areas instead of H^+. In support of this premise, it is pointed out that many boilers containing copper deposits are not damaged.

In practice, oxygen is not always removed from feed waters (e.g., many locomotive and ship boilers), but for boilers operating above about

---

[32] U. Evans, *The Corrosion and Oxidation of Metals*, p. 466, Ed. Arnold, London, 1960.

[33] D. Voyles, "Some Applications of the Pure Water Concept," presented at the Kansas City Convention of NACE, March 1962.

[34] O. Kubaschewski and B. Hopkins, *Oxidation of Metals and Alloys,* 2nd ed., p. 271, Academic Press, New York, 1962.

[35] W. Ruther and R. Hart, *Corrosion,* **19,** 127t (1963).

[36] *Corrosion Handbook,* p. 524.

350 lb/in.^{2} pressure, deaeration is standard procedure. For power boilers in England operating in the higher pressure range, Potter[30] showed that boiler waters containing less than 0.043 ppm oxygen were associated with only half the incidence of corrosion compared with boilers operating with dissolved oxygen above this level. He also cautioned that in interpreting these figures it should be remembered that of the 86 power stations operating below 0.043 ppm O_2, 27% had nevertheless suffered boiler tube corrosion. He concluded, in any case, that the data justify deaerating water to values below 0.05 ppm O_2, with the question remaining whether more efficient deaeration would reduce the incidence of corrosion still further.

Hence, whether a particular boiler is damaged by a given water treatment is not evidence in itself that the treatment in general is good or bad. Statistical analysis of many boilers, or a fundamental investigation of corrosion mechanisms, is required in order to obtain a final answer. There are probably many interacting factors of feed water composition, condenser leakage, boiler design, and boiler operation varying from one boiler to another which determine whether oxygen and copper contamination in specific instances are damaging. At present, the magnitude of all these interdependent factors has not been firmly established.

Alkali Additions. The mechanism by which alkali additions retard boiler corrosion is not well understood either. Evans[37] proposed that alkalies in presence of oxygen reduce the solubility of hydroxides and oxides of iron favoring precipitation of protective magnetite films on the metal surface rather than in the body of the water where any oxide formed is not protective. Such a mechanism, however, must refer to corrosion at defects in the oxide scale, because the high-temperature oxidation of iron is controlled by diffusion through an anhydrous oxide rather than by liberation of hydrated metal ions, which later react to become part of a corrosion product film. It has been proposed that the benficial effect of a dilute alkaline water is not so much to retard the Fe-H_2O reaction as to decrease appreciable transfer of the Fe_3O_4 reaction product by the flowing boiler water. In addition, in the event of condenser leakage leading to contamination of boiler water with salts like magnesium chloride which hydrolyze to acid corrosion products, sodium hydroxide additions prevent acid attack of the boiler tubes.[38] In this regard, NH_4OH is less effective than $NaOH$.

Excess alkali, on the other hand, is damaging because it may slowly

[37] U. Evans, *The Corrosion and Oxidation of Metals*, p. 438, Ed. Arnold, London, 1960.

[38] H. Masterson, J. Castle, and G. Mann, *Chem. Ind. (London)*, **1969, 1261.**

dissolve the magnetite film in accord with

$$Fe_3O_4 + 4NaOH \rightarrow 2NaFeO_2 + Na_2FeO_2 + 2H_2O \qquad (8)$$

forming sodium hypoferrite or sodium ferroate (Na_2FeO_2) and sodium ferrite ($NaFeO_2$), both of which are soluble in hot concentrated NaOH. In addition, concentrated alkali reacts directly and more rapidly with iron to form hydrogen and sodium ferroate

$$Fe + 2NaOH \rightarrow Na_2FeO_2 + H_2 \qquad (9)$$

Reactions of this kind account in part for pitting and grooving of boiler tubes and are those that account for the excessive corrosion rate of iron at high values of pH, as Fig. 4 shows. Concentration of NaOH is favored by crevices or by deposits of metal oxides and boiler scales at which liquid flow is impeded and heat transfer is restricted, resulting in overheating of the boiler tube and local evaporation of the otherwise mildly alkaline boiler water. In the absence of conditions that allow concentration of alkaline solutes, damage by corrosion is expected to be minimal. As mentioned earlier, the major function of phosphate additions is to avoid the high values of pH in concentrated boiler water which are necessary for reactions (8) and (9).

Oxygen Scavengers. The use of sodium sulfite as a scavenger in high-pressure boilers is frowned upon by some authorities because of the decomposition of sulfites at high temperatures into sulfides or SO_2. It is said to be satisfactory below 650 lb/in.^{2} (260°C, 500°F) steam pressure, but at 900 lb/in.^{2} (280°C, 535°F), for example, failure of a superheater tube was ascribed to acid decomposition products of Na_2SO_3.[39] It has been suggested that one possible path for decomposition is the following:

$$4Na_2SO_3 + 2H_2O \rightarrow 3Na_2SO_4 + 2NaOH + H_2S \qquad (10)$$

The use of hydrazine is not subject to similar objections, but its slow reaction with oxygen and partial decomposition into NH_3 should be taken into account in calculating proper dosages. Also NH_3 is a possible cause of stress corrosion cracking of copper-base alloys in the condenser system, should oxygen accidentally contaminate the condensate. Stress corrosion cracking of copper-base alloys and uniform attack of condensers by condensate containing NH_4OH, on the other hand, are not expected in absence of oxygen or other depolarizers.

Inhibitors, the Sulfate Ratio. The reasons that tannins inhibit against

[39] W. Stones, *Chem. Ind.* (*London*), **1957**, 120.

stress corrosion cracking are not likely to include competitive adsorption with OH⁻ at the prevailing high temperatures of a boiler. The characteristically weak bonding of organic molecules to a metal surface makes this improbable. It has been suggested that tannins act as scavengers of dissolved oxygen, but such a function would not necessarily affect stress corrosion cracking because no evidence exists to rule out this type of damage in NaOH solutions free of dissolved oxygen. It can be speculated that sodium compounds are formed between tannins and NaOH which act as buffer ions in the same way as do PO_4^{-3}, but this has not been proved. They may function in part to seal any leaks such as at welded sections of the boiler where boiler water otherwise concentrates.

The inhibiting action of nitrates at boiler temperatures is probably the result of a shift of potentials such that the corrosion potential of steel lies outside the critical potential range for stress corrosion cracking. At lower temperatures, near 100°C, the corresponding shift probably results instead in the corrosion potential falling within the critical potential range, thereby accelerating stress corrosion cracking (see p. 133).

At one time specific ratios of sulfate to hydroxide in boiler waters were specified in order to inhibit stress corrosion cracking. Such ratios, however, did not prevent cracking in tests employing the embrittlement detector,[20] and the use of sulfates for this purpose has now been largely abandoned. The original recommendation apparently stemmed from observations made in Illinois[40] to the effect that local waters high in $NaHCO_3$ and low in sulfates caused stress corrosion cracking of boilers. The damaging effect could be overcome by treatment of the waters with sulfuric acid. Laboratory tests, in addition, showed that stress corrosion cracking of steel, although not eliminated, occurred in longer times for increasing ratios of Na_2SO_4 to NaOH in the water. It was concluded that the practical beneficial effect of sulfuric acid treatment resulted from the inhibiting effect of sulfates. From the standpoint of present-day evidence, this conclusion was partially correct; but some additional benefit undoubtedly also accrued from neutralization of $NaHCO_3$, thereby avoiding accumulation of NaOH in the boilers by hydrolysis of bicarbonate, as in (7). In principle, sulfates should be at least partially effective as inhibitors through their expected ability to shift the critical potential for stress corrosion cracking to regions that are reasonably removed from normal corrosion potentials, although they are probably less effective in this respect than are nitrates.

[40] S. Parr and F. Straub, *Proc. Amer. Soc. Testing Mat.*, **26**, pt. II, 52 (1926).

GENERAL REFERENCES

"Symposium on the Protection of Motor Vehicles from Corrosion," Soc. Chem. Ind., Monogr. 4, London (1958).

The Corrosion and Oxidation of Metals, U. Evans, Ch. XII, "Boilers and Condensers," pp. 427–480, Ed. Arnold, London, 1960; 1st Suppl. Vol., pp. 169–184 (1968).

Corrosion, Causes and Prevention, F. N. Speller, pp. 414–496, McGraw-Hill, New York, 1951.

Corrosion of Materials by Ethylene Glycol-Water, J. Jackson, P. Miller, F. Fink, and W. Boyd, DMIC Report 216, Battelle Mem. Inst., Columbus, Ohio, 1965.

chapter 18

Alloying for corrosion resistance; stainless steels

Alloying is an effective means for improving the resistance of metals to attack by corrosive environments at either ordinary or elevated temperatures. The beneficial effect of alloyed chromium or aluminum on the oxidation resistance of iron has already been described (p. 200), as well as the beneficial effect of small alloying additions of copper, chromium or nickel on atmospheric resistance (p. 172).

Alloys of gold with copper or silver retain the corrosion resistance of gold above a critical alloy concentration called the reaction limit by Tammann.[1] Below the reaction limit, the alloy corrodes, for example in strong acids, leaving a residue of pure gold either as a porous solid or as a powder. This behavior of noble metal alloys is known as *parting* and is probably related in mechanism to dezincification of copper-zinc alloys (see pp. 327–329).

Pickering and Wagner[2] proposed that the predominant mechanism of parting in a gold-copper alloy containing, for example, 10 at. % Au, occurs by injection of divacancies at the corroding alloy surface which readily diffuse into the alloy at room temperature tending to fill with copper atoms. The latter diffuse in the opposite direction toward the surface, where they enter into aqueous solution. A mechanism involving dissolution of the alloy and subsequent redeposition of gold is not likely because dissolved gold in any amount is not detected. Instead, x-ray examination of the residual depleted porous layer shows that interdiffusion of gold and copper takes place to form solid-solution alloys varying in composition from that of the original alloy to pure

[1] G. Tammann, *Lehrbuch der Metallkunde*, pp. 428–433, Leipzig, 1932.

[2] H. Pickering and C. Wagner, *J. Electrochem. Soc.*, **114**, 698 (1967); H. Pickering, *Ibid.*, **115**, 143 (1968).

gold. In other words, the evidence supports solid-state diffusion of copper from within the alloy to the surface, where it alone undergoes dissolution. Above the critical gold composition for parting, it is possible that conditions for divacancy formation are no longer favorable, or perhaps the vacancies tend to fill with a higher proportion of gold compared to copper atoms and hence copper no longer corrodes preferentially.

Various other noble metal alloys (e.g., Pt-Ni, Pt-Cu, and Pt-Ag) also exhibit reaction limits in HNO_3. Corresponding alloy compositions vary with the corrosive medium, but reaction limits in general, whatever the environment, tend to fall between 25 and 50 at. % of the noble metal component (Table 1).

TABLE 1
Reaction Limits (Tammann)

Corrosive Medium	Cu-Au Alloys		Ag-Au Alloys	
	At. % Au	Wt. % Au	At. % Au	Wt. % Au
$(NH_4)_2S_2$	24.5–25.5	50.2–51.5	32	46.5
H_2CrO_4	50	75.5	49.2	63.9
HNO_3 (Sp. Gr. 1.3)	50	75.5	48.0–49.0	62.8–63.7
H_2SO_4	49–50	74.5–75.5	50	64.7

At higher temperatures the same reaction limits apply with the exception that some attack may initiate above the reaction limit composition for long exposure times. For example at 100°C, exposure of the gold-silver alloys containing more than 50 at. % Au for one or more weeks to nitric acid results in measurable attack.[3]

Alloying becomes an especially effective means for improving corrosion resistance if passivity results from the combination of a metal, otherwise active, with a normally passive metal. The alloying element may either reduce $i_{critical}$, e.g., chromium alloyed with iron (Fig. 11, p. 81) or $i_{passive}$, e.g., nickel alloyed with copper (Fig. 14, p. 84). For >12% Cr-Fe alloys, $i_{critical}$ is reduced to such small values that any small corrosion current exceeds $i_{critical}$ and accounts for initiation of passivity in aerated aqueous solutions. Similarly, if passivity results from a noble metal added in small amount to an active metal or alloy, the corrosion rate

[3] W. Katz, in *Korrosion und Korrosionsschutz,* edited by F. Tödt, pp. 412–414, De Gruyter, Berlin, 1955.

may be reduced by orders of magnitude. The noble metal stimulates the cathodic reaction and in this way increases the anodic current density to the critical value for passivation. Practical examples are palladium or platinum alloyed with stainless steels or with titanium (see p. 68) which resist sulfuric acid at concentrations and temperatures otherwise extremely corrosive. Carbon in steel can act similarly by creating cathodic sites of cementite (Fe_3C) on which reduction of HNO_3 (or HNO_2 in HNO_3) proceeds rapidly, allowing passivity to establish itself in less concentrated acid than is the case for pure iron. In principle, second phases of any kind (e.g., intermetallic compounds) can induce passivity in multicomponent alloys by the same mechanism.

In homogeneous single-phase alloys, passivity usually occurs at and above a composition specific to each alloy which is also dependent on the environment, as explained on page 81. For Ni-Cu alloys, the critical composition comes at 30 to 40% Ni; for Cr-Co, Cr-Ni, and Cr-Fe alloys it comes at 8, 14, and 12% Cr, respectively. The stainless steels are ferrous alloys that contain at least 12% Cr and are passive in many aqueous media, similar to the passivity of pure chromium itself. They represent the most important of the passive alloys.

STAINLESS STEELS

Brief History*

J. Stodart and M. Faraday in England published a report dated 1820[4] on the corrosion resistance of various iron alloys they had prepared; apparently it was this report in which the chromium-iron alloys were first mentioned. The maximum chromium content, however, was below that required for passivity and they narrowly missed discovering the stainless steels. In France in 1821, Berthier,[5] whose attention had been drawn to the work of Stodart and Faraday, found that iron alloyed with considerable chromium was more resistant to acids than was unalloyed iron. His alloys were obtained by direct reduction of the mixed oxides producing what today is called ferrochrome (40–80% Cr). The alloys were brittle, high in carbon, and had no value as structural materials. Berthier prepared some steels using ferrochromium as a component, but the chromium content again was too low to overlap the useful passive properties characteristic of the stainless steels.

Although a variety of chromium-iron alloys was produced in subse-

* Based largely on the more detailed account by Carl Zapffe, in *Stainless Steels*, pp. 5–25, Amer. Soc. Metals, Cleveland, Ohio, 1949.

[4] J. Stodart and M. Faraday, *Quart. J. Sci., Lit., Arts*, 9, 319 (1820); *Phil. Trans. Roy. Soc.*, 112, 253 (1822).

[5] P. Berthier, *Ann. Chimie Phys.*, 17, 55 (1821); *Ann. Mines*, 6, 573 (1821).

quent years by several investigators who took advantage of the high strength and high hardness properties imparted by chromium, the inherent corrosion resistance of the alloys was not observed, largely because the accompanying high carbon content impaired corrosion resistance. Only in 1904 did Guillet[6] of France produce low-carbon chromium alloys overlapping the passive composition range. He studied the metallurgical structure and mechanical properties of the Cr-Fe alloys and also of the Cr-Fe-Ni alloys now called austenitic stainless steels. But recognition of the outstanding property of passivity in such alloys initiating at a minimum of 12% Cr was apparently first described by Monnartz[7] of Germany, who began his researches in 1908 and published a detailed account of the chemical properties of Cr-Fe alloys in 1911. These researches included a description of the beneficial effect of oxidizing compared to reducing environments on corrosion resistance, the necessity of maintaining low carbon contents, and the effects of small quantities of alloying elements (e.g., Ti, V, Mo and W).

The commercial utility of the Cr-Fe hardenable stainless steels as possible cutlery materials was recognized in about 1913 by H. Brearley of Sheffield, England. Looking for a better gun-barrel lining, he noticed that the 12% Cr-Fe alloys did not etch with the usual nitric acid etching reagents and that they did not rust over long periods of exposure to the atmosphere. The austenitic Cr-Fe-Ni stainless steels, on the other hand, were first exploited in Germany in 1912–1914 based on researches of E. Maurer and B. Strauss of the Krupp Steel Works.

The 18% Cr, 8% Ni (18-8) austenitic stainless steel is the most popular of all the stainless steels now produced. In recent years annual production in the United States of all types of stainless steels, including heat-resistant compositions, has reached over 1 million tons.

Classes and Types

There are three main classes of stainless steels designated in accord with their metallurgical structure. Each class is made up of several alloys of somewhat differing composition having related physical, magnetic, and corrosion properties. These are given type numbers by the American Iron and Steel Institute (AISI) by which they are frequently called in practice. A survey of most types produced commercially is given in Table 2. The three main classes are martensitic, ferritic, and austenitic.

1. *Martensitic.* The name of the first class derives from the analogous martensite phase in carbon steels. Martensite is produced by a shear-

---

[6] L. Guillet, *Rev. Met.,* **1,** 155 (1904) ; **2,** 350 (1905) ; **3,** 332 (1906).
[7] P. Monnartz, *Metallurgie,* **8,** 161 (1911).

TABLE 2
Types and Compositions of Wrought Stainless Steels
American Iron and Steel Institute, 150 E. 42nd St., New York, N. Y. 10017, 1963

AISI Type No.	Cr	Ni	C	Mn (max)	P (max)	S (max)	Si (max)	Other Elements	Remarks
Class: *Martensitic* (body-centered cubic, magnetic, heat treatable)									
403	11.5–13.0	—	0.15 max	1.0	0.04	0.03	0.5		Turbine quality
410	11.5–13.5	—	0.15 max	1.0	0.04	0.03	1.0		
414	11.5–13.5	1.25–2.5	0.15 max	1.0	0.04	0.03	1.0		
416	12.0–14.0	—	0.15 max	1.25	0.06	0.15 min	1.0	0.6 Mo (max)	Easy machining, nonseizing
416 Se	12.0–14.0	—	0.15 max	1.25	0.06	0.06	1.0	0.15 Se (min)	Easy machining, nonseizing
420	12.0–14.0	—	Over 0.15	1.0	0.04	0.03	1.0		
431	15.0–17.0	1.25–2.5	0.20 max	1.0	0.04	0.03	1.0		
440 A	16.0–18.0	—	0.6–0.75	1.0	0.04	0.03	1.0	0.75 Mo (max)	
440 B	16.0–18.0	—	0.75–0.95	1.0	0.04	0.03	1.0	0.75 Mo (max)	
440 C	16.0–18.0	—	0.95–1.2	1.0	0.04	0.03	1.0	0.75 Mo (max)	Highest attainable hardness
Class: *Ferritic* (body-centered cubic, magnetic, not heat treatable)									
405	11.5–14.5	—	0.08 max	1.0	0.04	0.03	1.0	0.1–0.3 Al	
430	16.0–18.0	—	0.12 max	1.0	0.04	0.03	1.0		
430 F	16.0–18.0	—	0.12 max	1.25	0.06	0.15 min	1.0	0.6 Mo (max)	Easy machining, nonseizing
430 F, Se	16.0–18.0	—	0.12 max	1.25	0.06	0.06	1.0	0.15 Se (min)	Easy machining, nonseizing
446	23.0–27.0	—	0.20 max	1.5	0.04	0.03	1.0	0.25 N (max)	Resistant to high-temp. oxidation

Class: *Austenitic* (face-centered cubic, nonmagnetic, not heat treatable)

Type	Cr	Ni	Mn	C	P	S	Si	Other	Remarks
201	16.0–18.0	3.5–5.5	5.5–7.5	0.15 max	0.06	0.03	1.0	0.25 N (max)	
202	17.0–19.0	4.0–6.0	7.5–10	0.15 max	0.06	0.03	1.0	0.25 N (max)	
301	16.0–18.0	6.0–8.0	2.0	0.15 max	0.045	0.03	1.0		
302	17.0–19.0	8.0–10.0	2.0	0.15 max	0.045	0.03	1.0		
302 B	17.0–19.0	8.0–10.0	2.0	0.15 max	0.045	0.03	2.0–3.0		Resistant to high-temp. oxidation
303	17.0–19.0	8.0–10.0	2.0	0.15 max	0.20	0.15 min	1.0	0.6 Mo (max)	Easy machining, nonseizing
303 Se	17.0–19.0	8.0–10.0	2.0	0.15 max	0.20	0.06	1.0	0.15 Se (min)	Easy machining, nonseizing
304	18.0–20.0	8.0–10.5	2.0	0.08 max	0.045	0.03	1.0		
304 L	18.0–20.0	8.0–12.0	2.0	0.03 max	0.045	0.03	1.0		Extra low carbon
305	17.0–19.0	10.5–13.0	2.0	0.12 max	0.045	0.03	1.0		Lower rate of work hardening than 302 or 304
308	19.0–21.0	10.0–12.0	2.0	0.08 max	0.045	0.03	1.0		
309	22.0–24.0	12.0–15.0	2.0	0.20 max	0.045	0.03	1.0		
309 S	22.0–24.0	12.0–15.0	2.0	0.08 max	0.045	0.03	1.0		
310	24.0–26.0	19.0–22.0	2.0	0.25 max	0.045	0.03	1.5		
310 S	24.0–26.0	19.0–22.0	2.0	0.08 max	0.045	0.03	1.5		
314	23.0–26.0	19.0–22.0	2.0	0.25 max	0.045	0.03	1.5–3.0		
316	16.0–18.0	10.0–14.0	2.0	0.08 max	0.045	0.03	1.0	2.0–3.0 Mo	
316 L	16.0–18.0	10.0–14.0	2.0	0.03 max	0.045	0.03	1.0	2.0–3.0 Mo	Extra low carbon
317	18.0–20.0	11.0–15.0	2.0	0.08 max	0.045	0.03	1.0	3.0–4.0 Mo	
321	17.0–19.0	9.0–12.0	2.0	0.08 max	0.045	0.03	1.0	Ti: 5 × C (min)	Stabilized grade
347	17.0–19.0	9.0–12.0	2.0	0.08 max	0.045	0.03	1.0	Cb-Ta: 10 × C (min)	Stabilized grade
348	17.0–19.0	9.0–13.0	2.0	0.08 max	0.045	0.03	1.0	Cb-Ta: 10 × C (min)	Stabilized grade 0.1 Ta (max) when radiation conditions require low Ta

type phase transformation on cooling a steel rapidly (quenching) from the austenite region (face-centered cubic structure) of the phase diagram. It is the characteristically hard component of quenched carbon steels as well as of the martensitic stainless steels. In stainless steels, the structure is body-centered cubic and the alloys are magnetic. Typical applications include cutlery, steam turbine blades, and tools.

2. *Ferritic.* Ferritic steels are named after the analogous ferrite phase, or relatively pure iron component of carbon steels cooled slowly from the austenite region. The ferrite or so-called alpha phase for pure iron is the stable phase existing below 910°C. For low-carbon Cr-Fe alloys the high-temperature austenite (or gamma) phase exists only up to 12% Cr; immediately beyond this composition the alloys are ferritic at all temperatures up to the melting point. They can be hardened moderately by cold working but not by heat treatment. Ferritic stainless steels are body-centered cubic in structure and are magnetic. Uses include trim on autos and for constructing synthetic nitric acid plants.

3. *Austenitic.* Austenitic steels are named after the austenite or γ phase, which for pure iron exists as a stable structure between 910 and 1400°C. This phase is face-centered cubic, nonmagnetic, and it is readily deformed. It is the major or only phase of austenitic stainless steels at room temperature, existing as a stable or metastable structure depending on composition. Alloyed nickel is largely responsible for the retention of austenite on quenching the commercial Cr-Fe-Ni alloys from high temperatures, with increasing nickel content accompanying increased stability of the retained austenitie. Alloyed Mn, Co, C and N also contribute to the retention and stability of the austenite phase. Austenitic stainless steels can be hardened by cold working but not by heat treatment. On cold working, but not otherwise, the metastable alloys (e.g., 201, 202, 301, 302, 302 B, 303, 303 Se, 304, 304 L, 316, 316 L, 321, 347, 348; see Table 1) transform in part to the ferrite phase (hence the adjective metastable) having a body-centered cubic structure which is magnetic. This transformation also accounts for a marked rate of work hardening. Alloys 305, 308, 309, 309 S, on the other hand, work harden at a relatively low rate and become only slightly magnetic, if at all. Alloys containing higher chromium and nickel (e.g., 310, 310 S, 314) are essentially stable austenitic alloys and do not transform to ferrite or become magnetic when deformed. Uses of austenitic stainless steels include general purpose applications, architectural and automobile trim, and various structural units for the food and chemical industries.

Mention should also be made of a fourth class called the *precipitation-hardening stainless steels,* which achieve high strength and hardness through low-temperature heat treatment after quenching from high temperatures. These Cr-Fe alloys contain less nickel (or none) than is necessary to stabilize the austenite phase, and in addition they contain alloying elements like aluminum or copper, which produce high hardness values through formation and precipitation of intermetallic compounds along slip planes or grain boundaries. They are applied whenever the improved corrosion resistance imparted by alloyed nickel is desirable, or, more important, when hardening of the alloy is best done after machining operations, using low-temperature heat treatment [e.g., 480°C (900°F)] rather than a high-temperature quench as is required in the case of the martensitic stainless steels. One of the typical precipitation-hardening stainless steels (17-4 PH) contains 16.5% Cr, 4.3% Ni, 0.04% C, 0.25% Cb, and 3.6% Cu.

The highest general corrosion resistance is obtained with the nickel-bearing austenitic types, and in general the highest nickel composition alloys in this class are more resistant than the lowest nickel compositions. For optimum corrosion resistance, austenitic alloys must be quenched (rapidly cooled by water or by an air blast) from about 1050 to 1100°C (1920 to 2000°F). The molybdenum-containing austenitic alloys (316, 316 L, 317) have improved corrosion resistance to chloride-containing environments, dilute nonoxidizing acids, and crevice corrosion. The beneficial effect of molybdenum in this regard does not extend equally to the nickel-free stainless steels. This indicates that the mechanism for improved passivity in specific environments is probably related to electronic interaction between nickel and molybdenum (see p. 89).

The ferritic stainless steels have optimum corrosion resistance when cooled slowly from above 925°C (1700°F) or when annealed at 650 to 815°C (1200–1500°F).*

The martensitic stainless steels, on the other hand, have optimum corrosion resistance as quenched from the austenite region. In this state they are very hard and brittle. Ductility is improved by annealing (650–750°C for 403, 410, 416, 416 Se; 650–730°C for 414; 620–700°C for 431; 680–750°C for 440 A, B, C and 420) but at some sacrifice of corrosion resistance. Resistance to pitting and rusting in 3% NaCl at room temperature reaches a minimum after tempering a 0.2 to 0.3% C, 13% Cr stainless steel at 500°C, and at 650°C for a similar steel

---

* For additional details on heat treatment temperatures and times see *Metals Handbook,* vol. 2, *Heat Treatment, Cleaning and Finishing,* pp. 241–254, Amer. Soc. Metals, Metals Park, Ohio, 1964.

containing 0.06% C.[8] In general, the tempering range 450 to 650°C (840–1200°F) should be avoided if possible. Decrease of corrosion resistance probably results in part from the transformation of martensite containing interstitial carbon into a network of chromium carbides attended by depletion of chromium in the adjoining metallic phase.

Cold working any of the stainless steels usually has only minor effect on corrosion resistance if temperatures are avoided that are high enough to permit appreciable diffusion during or after deformation. Phase changes brought about by cold working the metastable austenitic alloys are not attended by major changes in corrosion resistance. Furthermore, quenched austenitic stainless steel (face-centered cubic) of 18% Cr, 8% Ni composition has approximately the same corrosion resistance as quenched ferritic stainless steel (body-centered cubic), of the same chromium and nickel composition but with lower carbon and nitrogen content.[9] However, if a similar alloy containing a mixture of austenite and ferrite is briefly heated at, for example, 600°C, composition differences are established between the two phases setting up galvanic cells which accelerate corrosion. In other words, composition gradients, whatever their cause, are more important in establishing corrosion behavior than are structural variations of an otherwise homogeneous alloy. This appears to be true of metals and alloys in general.

Intergranular Corrosion

Improper heat treatment of ferritic or austenitic stainless steels causes the alloy separating individual crystals, or what are called the grain boundaries, to become especially susceptible to corrosion. Corrosion of this kind leads to catastrophic reduction in mechanical strength. The specific temperatures and times that induce susceptibility to intergranular corrosion are called *sensitizing* heat treatments. They are very much different for the ferritic and the austenitic stainless steels. In this respect, the transition in sensitizing temperatures for steels containing 18% Cr occurs at about 2.5 to 3% Ni.[10] In other words, stainless steels containing less than this amount of nickel are sensitized in the temperature range typical of nickel-free ferritic steels, whereas those containing more nickel respond to the temperature range typical of the austenitic stainless steels.

[8] J. Truman, R. Perry, and G. Chapman, *J. Iron Steel Inst.*, **202**, 745 (1964).
[9] H. H. Uhlig, *Trans. Amer. Soc. Metals*, **30**, 947 (1942).
[10] J. Upp, F. Beck, and M. Fontana, *Trans. Amer. Soc. Metals*, **50**, 759 (1958).

AUSTENITIC STAINLESS STEELS

For austenitic alloys, the sensitizing temperature range is 400 to 850°C (750–1550°F). The degree of damage to commercial alloys caused by heating in this range depends on time, a few minutes at the higher temperature range of 750°C (1380°F) being equivalent to several hours at a lower (or still higher) temperature range (Fig. 1).[11, 12] Slow cooling

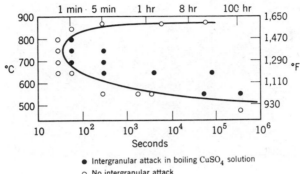

Fig. 1. Effect of time and temperature on sensitization of 18.2% Cr, 11.0% Ni, 0.05% C, 0.05% N stainless steel (Binder, Brown, and Frank).

through the sensitizing temperature range or prolonged welding operations induce susceptibility, but rapid cooling avoids damage. Hence austenitic stainless steels must be and are regularly quenched from high temperatures. Spot welding, in which the metal is rapidly heated by a momentary electric current followed by a naturally rapid cooling, does not cause sensitization. Arc welding, on the other hand, can cause damage, the effect being greater the longer the heating time, especially when heavy-gage material is involved. Sensitizing temperatures are reached some millimeters away from the weld metal itself, the latter being at the melting point or above. Hence on exposure to a corrosive environment, failure of an austenitic stainless steel weld (called weld decay) occurs in zones slightly away from the weld rather than at the weld itself (see Fig. 2).

The extent of sensitization for a given temperature and time depends very much on carbon content. An 18-8 stainless steel containing 0.1% C or more may be severely sensitized after heating for 5 min at 600°C,

[11] W. Binder, C. Brown, and R. Franks, *Trans. Amer. Soc. Metals,* **41,** 1301 (1949).
[12] H. Ebling and M. Scheil, in *Advances in Technology of Stainless Steels and Related Alloys,* p. 275, Amer. Soc. Testing Mat., Philadelphia, Pa., 1965.

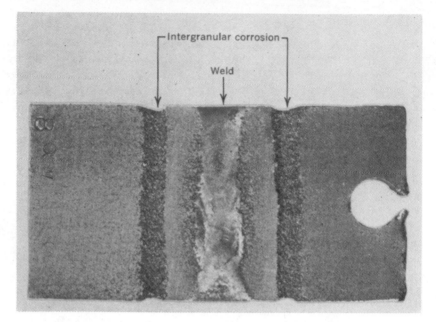

Fig. 2. Example of weld decay, 2X. After sensitization, specimen was exposed to 25% HNO_3.

whereas a similar alloy containing 0.06% C is affected less, and for 0.03% C, the alloy heat treated similarly may suffer no appreciable damage on exposure to a moderately corrosive environment. The higher the nickel content of the alloy, the shorter is the time for sensitization to occur at a given temperature, whereas alloying additions of molybdenum increase the time.[11]

The physical properties of stainless steels after sensitization do not change greatly. Because precipitation of carbides accompanies sensitization, the alloys become slightly stronger and slightly less ductile. Damage occurs only on exposure to a corrosive environment, the alloy corroding along grain boundaries at a rate depending on severity of the environment and the extent of sensitization. In seawater, a sensitized stainless steel sheet may fail within weeks or months; but in a boiling solution containing $CuSO_4 \cdot 5H_2O$ (13 g/liter) and H_2SO_4 (47 ml concentrated acid/liter), which is used as an accelerated test medium, failure occurs within hours.

THEORY AND REMEDIES

Since intergranular corrosion of austenitic stainless steels is associated with carbon content, a low-carbon alloy (<0.02% C) is relatively immune to this type of corrosion.[13] Nitrogen, normally present in commercial alloys to the extent of a few hundredths per cent, is less effective than carbon in causing damage (Fig. 3).[14] At high temperatures, e.g., 1050°C (1920°F), carbon is almost completely dispersed throughout the

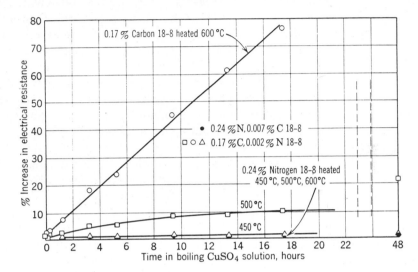

Fig. 3. Intergranular corrosion measured by change in electrical resistance of 18-8 stainless steels containing nitrogen or carbon immersed in 10% $CuSO_4$ + 10% H_2SO_4. All specimens sensitized at stated temperatures for 217 hr.

alloy, but within (or somewhat above) the sensitizing temperature range it rapidly diffuses to the grain boundaries where it combines preferentially with chromium to form chromium carbides (e.g., $M_{23}C_6$, in which M represents presence of some small amount of iron along with chromium). This reaction depletes the adjoining alloy of chromium to the extent that the grain-boundary material may contain less than the 12% Cr necessary for passivity. The affected volume of alloy normally extends some small distance into the grains on either side of the boundary itself, causing apparent grain-boundary broadening of the etched surface. The chromium-depleted alloy sets up passive-active cells of appreciable potential difference, the grains constituting large cathodic areas relative to small anodic areas at the grain boundaries. Electrochemical

[13] E. Bain, R. Aborn, and J. Rutherford, *Trans. Amer. Soc. Metals* (*Steel Treating*) **21**, 481 (1933); B. Strauss, H. Schottky, and J. Hinnüber, *Z. Anorg. Allgm. Chem.*, **188**, 309 (1930).

[14] H. H. Uhlig, *Trans. Electrochem. Soc.*, **87**, 193 (1945).

action results in rapid attack along the grain boundaries and deep penetration of the corrosive medium into the interior of the metal.

If the alloy is rapidly cooled through the sensitizing zone, carbon does not have time either to reach the grain boundaries or to react with chromium if carbon is already concentrated at the grain boundaries. On the other hand if the alloy remains within the sensitizing temperature zone an especially long time (usually several thousand hours) chromium again diffuses into the depleted zones, reestablishing passivity of the grain boundaries and eliminating susceptibility to preferred attack. Although nitrogen forms chromium nitrides, it is less effective than carbon in causing damage, perhaps in part because nitrides precipitate more generally throughout the grains or they form islands along grain boundaries, interrupting a continuous path along which the corrosive agent can proceed.[14] Carbides, on the other hand, form continuous paths of chromium-depleted alloy.

Other mechanisms of intergranular corrosion have been proposed, but the so-called chromium-depletion mechanism appears to fit most of the facts and is probably essentially correct. For example, carbides isolated from grain boundaries of sensitized stainless steels have shown an expected high chromium content. Along the same lines, corrosion products of grain-boundary alloy obtained by choosing corrosive conditions that avoid attack of carbides, show a lower chromium content than corresponds to the alloy. In this respect, Schafmeister,[15] using cold concentrated sulfurous acid acting on sensitized 18% Cr, 8.8% Ni, 0.22% C stainless steel for 10 days, found only 8.7% Cr in the alloy that had corroded from grain-boundary regions. Accompanying analyses for Ni and Fe of 8.4% and 83.0%, respectively, showed no appreciable depletion in nickel and an increase in iron content.

Microprobe scans of sensitized stainless steels have indicated chromium depletion and nickel enrichment at grain boundaries.[16] Radioactive 14C introduced into an austenitic 18% Cr, 12.8% Ni, 0.12% C stainless steel has demonstrated enrichment of carbon at grain boundaries observable immediately after quenching from 1000 to 1350°C. This indicates that sensitization in part consists of carbon reacting to form chromium carbide without the necessity of carbon diffusing to grain boundaries at the sensitizing temperature.

There are at least three effective means for avoiding susceptibility to intergranular corrosion.

1. *Heat Treatment at 1050 to 1100°C (1920–2000°F) Followed by*

---

[15] P. Schafmeister, *Arch. Eisenhüttenw.*, **10**, 405 (1936–1937).

[16] P. LaCombe, in *Proc. 1st Int. Congr. Metallic Corrosion*, pp. 21–35, Butterworths, London, 1962. See also, A. Bäumel et al., *Corros. Sci.*, **4**, 89 (1964).

Quenching. The high-temperature treatment dissolves precipitated carbides and rapid cooling prevents their re-formation. This treatment is recommended, for example, after welding operations. It is not always a possible treatment, however, because of structure size, or because of a tendency of the alloy to warp at high temperatures.

2. *Reduction of Carbon Content.* Carbon content can be reduced in the commercial production of stainless steels, but at extra cost. Alloys of low carbon content (e.g., <0.03% C) are designated by letter L—e.g., types 304 L, 316 L. These alloys can be welded or otherwise heated in the sensitizing temperature range with much less resultant susceptibility to intergranular corrosion. However, they are not immune.

3. *Addition of Titanium or Columbium (Niobium).* By alloying austenitic alloys with a small amount of an element (e.g., Ti or Cb) having higher affinity for carbon than does chromium, carbon is restrained from diffusing to the grain boundaries, and any which is already at the boundary reacts with titanium or columbium instead of with chromium. Alloys of this kind are called *stabilized grades* (e.g., types 321, 347, 348). They can be welded or otherwise heated within the sensitizing zone without marked susceptibility to intergranular corrosion. Optimum resistance to intergranular attack on heating the alloys in the vicinity of 675°C (1250°F) is obtained by a prior so-called stabilizing heat treatment for several hours at about 900°C (1650°F).[12, 17] The latter heat treatment effectively serves to convert available carbon to stable carbides in a temperature range of lower carbon solubility compared to that of the usual higher quench temperatures.

In welding operations, the weld rod usually contains columbium rather than titanium; the latter tends to oxidize at elevated temperatures with the danger of its residual concentration becoming too low to stabilize the weld alloy against corrosion. Columbium, on the other hand, is lost by oxidation to a lesser extent.

INTERGRANULAR CORROSION OF NONSENSITIZED ALLOYS

In strongly oxidizing media (e.g., boiling $5N$ $HNO_3 + Cr^{6+}$) austenitic stainless steels including stabilized grades, quenched from 1050°C, undergo slow intergranular attack.[18] Stress is not necessary. The attack occurs only in the transpassive region, hence oxidizing ions such as Cr^{6+} (0.1–0.5 N $K_2Cr_2O_7$), Mn^{7+} or Ce^{4+} are necessary additions to the boiling

[17] M. Streicher, in *Advances in Technology of Stainless Steels and Related Alloys,* p. 257. *Amer. Soc. Testing Mat.,* Philadelphia, Pa., 1965.

[18] M. Streicher, *J. Electrochem. Soc.,* **106,** 161 (1959).

nitric acid. The rate of attack increases with the amount of alloyed nickel,[19] being more than ten times higher for a 78% Ni, 17% Cr, bal. Fe alloy compared to a similar 10% Ni alloy (70-hr test). This trend is opposite to the beneficial effect of nickel on the resistance of stainless steels to stress corrosion cracking.

Grain boundary attack of 19% Cr, 9% Ni stainless steel is most rapid after quenching from 1100 to 1200°C; it is less pronounced on quenching from either 900 or 1400°C.[20] High-purity alloys are immune. Alloyed carbon, nitrogen, oxygen, or manganese added in small concentrations are without effect, but specifically silicon and phosphorus (>100 ppm) are damaging. Silicon causes increased intergranular attack in the intermediate range of 0.1 to 2% for the 14% Cr, 14% Ni stainless steel; in larger or smaller amounts the alloy is not susceptible.[21, 22]

Similar intergranular attack of 15% Cr, 6% Fe, bal. Ni (Inconel 600) in high-temperature water (350°C) or steam (600–650°C)[19] or of stabilized 18-8 stainless steel in potassium hydroxide solutions (pH 11) at 280°C[23] has been reported. This matter is of prime interest in view of the common use of Inconel 600 and stainless steels in the construction of nuclear reactors for power production. Contaminants in the water such as traces of lead or dissolved oxygen, and the presence of crevices (at which superheating accompanied by solute concentration may occur) are said to cause the observed attack;[24] difference of opinion has been expressed regarding the necessity of any contamination to account for the observed intercrystalline damage.[25] Increased nickel content favors intergranular corrosion[26] of 18 to 20% Cr stainless steels at 200 to 300°C in water containing Cl^- and dissolved oxygen, as it does in boiling HNO_3-Cr^{6+} Silicon (>0.3%) and phosphorus (>0.023%) in 14% Cr, 14% Ni stainless steel exposed to 0.01% $FeCl_3$ at 340°C, analogous to results in $HNO_3 + Cr^{6+}$, cause intergranular corrosion. Unlike the latter medium, the high-temperature 0.01% $FeCl_3$ solution (21-day ex-

[19] H. Coriou et al., *Corrosion,* **22**, 280 (1966).

[20] J. Armijo, *Corros. Sci.,* **7**, 143 (1967).

[21] H. Coriou et al., *Compt. Rend. Acad. Sci. Paris,* **254**, 4467 (1962).

[22] J. Armijo, *Corrosion,* **24**, 24 (1968).

[23] J. Wanklyn and D. Jones, *J. Nuclear Mat.,* **2**, 154 (1959).

[24] H. Copson and G. Economy, *Corrosion,* **24**, 55 (1968).

[25] H. Coriou et al., in *Fundamental Aspects of Stress Corrosion Cracking,* pp. 352–359, Nat. Assoc. Corros. Engrs., Houston, Texas, 1969.

[26] W. Hübner, M. de Pourbaix, and G. Östberg, presented at 4th Int. Congr. Metallic Corrosion, Amsterdam, September, 1969.

posure) caused intergranular corrosion of the nonsensitized stainless steel when it contained $>0.05\%$ C.[27]

The overall results confirm that intergranular attack is the result of specific impurities in the alloy segregating at grain-boundary regions during quenching. The extent of their damaging effect depends on the nature of the chemical environment to which the alloy is exposed, but the mechanism by which they accelerate attack is not known. Applied stress has been stated to increase the observed attack in various media, but stress is apparently not necessary; hence the observed damage is probably better described as intergranular corrosion rather than as stress corrosion cracking.

FERRITIC STAINLESS STEELS

The sensitizing range for ferritic stainless steels lies above 925°C ($>1700°F$) and immunity is restored by heating for a short time (approx. 10–60 min) at 650 to 815°C (1200–1500°F). It should be noted that these temperatures are quite the opposite of those applying to austenitic stainless steels. The same accelerating media (i.e., boiling $CuSO_4$-H_2SO_4 or 65% HNO_3) produce intergranular corrosion in either class, and the extent and rapidity of damage are similar. In welded sections, however, damage to ferritic steels occurs to metal immediately adjacent the weld and to the weld metal itself, whereas in austenitic steels the damage localizes some small distance away from the weld.

Chromium content of ferritic steels, whether high or low (16–28% Cr), has no appreciable influence on their susceptibility to intergranular corrosion.[28] Similar to the situation for austenitic steels, lowering the carbon content is helpful, but the critical carbon content is very much lower. A type 430 ferritic stainless steel containing only 0.009% C was still susceptible.[28] Only on decarburizing low-carbon steels containing 16 or 24% Cr in H_2 at 1300°C for 100 hr was immunity obtained in the $CuSO_4$-H_2SO_4 test solution.[29] Similarly a low-carbon ($\approx 0.002\%$) 25% Cr-Fe alloy was reported to be immune.[30] Introduction of 0.04% C to the latter alloy induced susceptibility again, unlike the addition of 0.2% N, which had no effect. Addition of titanium in the amount of 8 times carbon content or more, provided immunity to the $CuSO_4$ test solution but not to boiling 65% HNO_3.[28] Columbium additions, it was stated, behaved similarly, and only heat treatment as described previously was

[27] R. Duncan, J. Armijo, and A. Pickett, *Materials Prot.*, **8**, 37 (February 1969).
[28] R. Lula, A. Lena, and G. Kiefer, *Trans. Amer. Soc. Metals*, **46**, 197 (1954).
[29] E. Houdremont and W. Tofaute, *Stahl Eisen*, **72**, 539 (1952).
[30] J. Hochmann, *Rev. Met.*, **48**, 734 (1951).

effective in conferring immunity to nitric acid. It was reported, however, that columbium additions ($8 \times C + N$ content) but not titanium additions minimize the observed intergranular corrosion of welds exposed to boiling 65% HNO_3.[31] This behavior may be explained by the observed marked reactivity of titanium carbides, but not columbium carbides, with HNO_3[32] along grain boundaries where such carbides are concentrated.

The theory accounting for susceptibility of ferritic alloys to intergranular corrosion is not entirely clear. It was proposed by Houdremont and Tofaute[29] that carbon-rich austenite forms in equilibrium with ferrite at elevated temperatures which then, on cooling, decomposes into ferrite and easily corroded iron carbides. On heat treatment these carbides supposedly convert into difficult-to-dissolve chromium carbides. This point of view was questioned by Lula et al.[28] who pointed out that alloying additions that favor retention of ferrite and suppress austenite formation (e.g., vanadium, silicon, or increasing additions of chromium) nevertheless have no effect on intergranular susceptibility. It is of interest to the mechanism that the latter investigators found rapid cooling from elevated temperatures (e.g., small specimens quenched into ice-brine) to provide immunity. This experiment proved that the responsible precipitate is not present at the higher temperatures. Accordingly, they proposed that on cooling at other than very fast rates, precipitation of carbides or nitrides from supersaturated ferrite occurs within the grain boundaries. The precipitates produce strain, which supposedly increases dissolution rate of adjoining metal; subsequent heat treatment in turn relieves the strain, thereby restoring corrosion resistance. However, as was discussed on page 127, a strained lattice does not necessarily corrode more rapidly than an annealed lattice unless straining is accompanied by a composition gradient.

It is more likely, therefore, that carbides precipitating along the grain boundaries increases susceptibility to corrosion by upsetting the localized distribution of alloying components (e.g., chromium) and that subsequent heat treatment succeeds in restoring uniform chromium composition. This point of view was also expressed by Bond and Lizlovs.[31] The much more rapid diffusion rates of chromium and carbon in the body-centered cubic compared to the face-centered cubic lattic probably explain the shorter times at a given temperature required in the case of ferritic stainless steels to achieve carbide precipitation in the first place, and subsequently to reestablish a uniform chromium composition after carbides have precipitated.

[31] A. Bond and E. Lizlovs, *J. Electrochem. Soc.,* **116,** 1305 (1969).
[32] A. Bäumel, *Stahl Eisen,* **84,** 798 (1964).

Pitting, Crevice Corrosion

In environments containing appreciable concentrations of Cl^- or Br^-, in which they otherwise remain essentially passive, all the stainless steels tend to corrode at specific areas and to form deep pits. Ions such as thiosulfate ($S_2O_3^{--}$) may also induce pitting. In the absence of passivity, such as in deaerated alkali metal chlorides, nonoxidizing metal chlorides (e.g., $SnCl_2$ or $NiCl_2$), or oxidizing metal chlorides at low pH, pitting does not occur. This holds even though in acid environments general corrosion may be appreciable.

Stainless steels exposed to seawater develop deep pits within a matter of months, the pits initiating usually at crevices or other areas of stagnant electrolyte (*crevice corrosion.*) The tendency is greater in the martensitic and ferritic steels than in the austenitic steels; it decreases in the latter alloys as the nickel content increases. The austenitic 18-8 alloys containing molybdenum (types 316, 316 L, 317) are still more resistant to seawater; however, crevice corrosion and pitting of these alloys eventually develop within a period of 1 to 2.5 years.

Stainless steels exposed at room temperature to chloride solutions containing active depolarizing ions such as Fe^{+3}, Cu^{++}, or Hg^{++} develop visible pits within hours. These solutions have sometimes been used as accelerated test media for pitting susceptibility.

Many extraneous anions, some more effective than others, act as pitting inhibitors when added to chloride solutions. For example, as mentioned in an earlier chaper (see p. 77), addition of 3% $NaNO_3$ to a 10% $FeCl_3$ solution completely inhibited pitting of 18-8 stainless steel as well as avoiding general attack over a period of at least 25 years. Hence, so long as passivity does not break down at crevices because of dissolved oxygen depletion, or for other reasons, localized corrosion does not initiate in inhibited solutions no matter how long the time. Similarly, in neutral chlorides (e.g., NaCl solutions), addition of alkalies inhibits against pitting. In aerated 4% NaCl solution at 90°C, at which temperature the pitting rate of 18-8 is highest, addition of 8 g NaOH/liter was found to eliminate pitting.[33] In refrigerating brines, addition of 1% Na_2CO_3 was effective for at least five years.[34]

Pits develop more readily in a stainless steel that is metallurgically inhomogeneous. Similarly, the pitting tendency of an austenitic steel increases when the alloy is heated briefly in the carbide precipitation (sensitization) range. Pitting resulting from crevice corrosion is also favored whenever a stainless steel is covered by an organic or inorganic

[33] J. Matthews and H. Uhlig, *Corrosion*, **7**, 419 (1951).
[34] P. Schafmeister and W. Tofaute, *Tech. Mitt. Krupp*, **3**, 223 (1935).

film or by marine fouling organisms, which partially shield the surface from access to oxygen. By the same token, crevice corrosion is least in seawater that is moving with some velocity with respect to the surface,[35] movement tending to keep all the surface in contact with aerated water and uniformly passive.

THEORY OF PITTING

The initiation of pits on an otherwise fully passive 18-8 surface, as discussed earlier (p. 76), requires that the corrosion potential exceed the critical potential [0.21 V(s.h.e.) in 3% NaCl]. The oxygen potential in air at pH 7 (0.8 V) or the ferric-ferrous potential ($\phi^0 = 0.77$ V) is sufficiently noble to induce pitting, but the stannic-stannous ($\phi^0 = 0.15$ V) and the chromic-chromous potentials ($\phi^0 = -0.41$ V) are too active, and hence pitting of 18-8 stainless steel is not observed in deaerated stannic or chromic chloride solutions. In sufficient concentration, the nitrate ion shifts the critical potential to a value that is noble to the ferric-ferrous oxidation-reduction potential, and hence pitting in 3% $NaNO_3 + 10\%$ $FeCl_3$ is not observed. Other anions shift the critical potential similarly, their effectiveness in this regard decreasing in the order $OH^- > NO_3^- > SO_4^{--} > ClO_4^-$. Increasing amounts of alloyed chromium, nickel, molybdenum, and rhenium in stainless steels also shift the critical potential in the noble direction accounting for increased resistance to pitting. Interestingly, a reversal in the order of the critical pitting potential at 0°C suggests a greater tendency for initiation of pits in molybdenum-containing 18-8 (type 316) compared to 18-8 (type 304) at this temperature, opposite to the corresponding tendencies at room temperature.[36]

Once a pit initiates, a passive-active cell is set up of 0.5 to 0.6 V potential difference. The resultant high-current density accompanies a high corrosion rate of the anode (pit) and at the same time polarizes the alloy surface immediately surrounding the pit to values below the critical potential. Through flow of current, chloride ions transfer into the pit forming concentrated solutions of Fe^{++}, Ni^{++} and Cr^{+3} chlorides (Fig. 4). The high Cl^- concentration ensures that the pit surface remains active. At the same time, the high specific gravity of such corrosion products causes leakage out of the pit in the direction of gravity, inducing breakdown of passivity wherever the products come into contact with the alloy surface. This accounts for a shape of pits elongated in the direction of gravity, as is often observed in practice. An 18-8 stain-

---

[35] *Corrosion Handbook,* Table 17, p. 414.

[36] J. Horvath and H. Uhlig, *J. Electrochem. Soc.,* **115,** 791 (1968); H. Böhni and H. Uhlig, *Corros. Sci.,* **9,** 353 (1969).

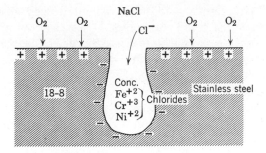

Fig. 4. Passive-active cell responsible for pit growth in stainless steel exposed to chloride solution.

less steel sheet exposed to seawater for one year has been observed to develop an elongated narrow pit reaching 2.5 in. from its point of origin (Fig. 5). The mechanism of growth has been demonstrated in the laboratory[37] by continuously flowing a fine stream of concentrated $FeCl_2$ solution over an 18-8 surface slightly inclined to the vertical and totally

[37] H. H. Uhlig, *Trans. Amer. Inst. Mining Met. Eng.*, **140**, 411 (1940).

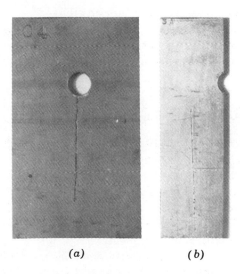

(a)   (b)

Fig. 5. (a) Elongated pit in 18-8 stainless specimen (3 × 5 in.) exposed to Boston Harbor seawater one year (pit began at crevice formed between bakelite rod and interior surface of hole). (b) Artificial elongated pit formed by flowing 50% $FeCl_2$ in a fine stream over 18-8 stainless steel immersed in 10% $FeCl_3$, 4 hr.

immersed in $FeCl_3$, resulting in formation of a deep groove under the $FeCl_2$ stream within a few hours (Fig. 5). A similar groove does not form on an iron surface because a passive-active cell is not established.

A pit stops growing only if the surface within the pit is again passivated, bringing pit and adjacent alloy to the same potential. Extraneous anions such as SO_4^{--} have no effect; on the other hand, dissolved oxygen or passivator ions (e.g., Fe^{3+}) reinitiate passivity on entering a pit. Successful repassivation depends on factors such as pit geometry and stirring rate.

REDUCING OR AVOIDING PITTING CORROSION

There are five principal ways of reducing or avoiding pitting corrosion.

1. Cathodically protect at a potential below the critical pitting potential. An impressed current can be used, or in good conducting media (e.g., seawater) the steel can be coupled to an approximately equal or greater area of zinc, iron, or aluminum.[38] Austenitic stainless steels used to weld mild steel plates, or 18-8 propellers on steel ships, do not pit.

2. Add extraneous anions (e.g., OH^- or NO_3^-) to chloride environments.

3. Reduce oxygen concentration of chloride environments (e.g., NaCl solutions).

4. Operate at the lowest temperature possible.

5. Ensure uniform oxygen or oxidizing concentrations. Avoid crevices, surface films. Agitate, aerate, and circulate solutions. Cleanse alloy surface periodically, using alkaline cleaners with stainless steel wool or the equivalent; or use 10 to 20% HNO_3 at about 60°C (140°F).

Stress Corrosion Cracking, Hydrogen Cracking

In the presence of an applied or residual *tensile* stress, stainless steels may crack transgranularly when exposed to certain environments (Fig. 6). Compressional stresses, to the contrary, are not damaging. The higher the tensional stress the shorter is the time to failure and, although at low stress levels time to cracking may be long, there is in general no practical minimum stress below which cracking will not occur, given sufficient time in a critical environment.

Damaging environments that cause cracking may differ for austenitic compared to martensitic or ferritic stainless steels. For austenitic steels the two major damaging ions are hydroxyl and chloride (OH^- and Cl^-).

[38] *Corrosion Handbook*, p. 416.

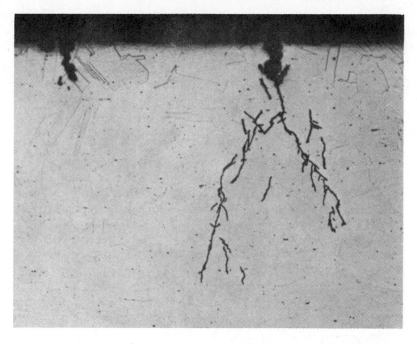

Fig. 6. Stress corrosion cracking of 18-8, type 304 stainless steel; exposed to calcium silicate insulation containing 0.02 to 0.5% chlorides, 100°C, 250X. Note that the cracks for this environment start at a pit. An undulating path accounts for disconnected cracks as viewed in one plane (Dana).

A boiling, relatively concentrated chloride that hydrolyzes to a slightly acid pH, such as $FeCl_2$ or $MgCl_2$, can cause cracking of thick sections of stressed 18-8 within hours. A solution of concentrated $MgCl_2$ boiling at about 154°C, for example, is used as an accelerated test medium. The presence of dissolved oxygen in such solutions is not necessary for cracking to occur, but its availability hastens damage, as does the presence of oxidizing ions such as Fe^{+3}. Pitting is not a preliminary requirement for initiation of cracks. In NaCl and similar neutral solutions, on the other hand, cracking is observed only if dissolved oxygen is present,[39] and the amount of chloride needed to cause damage can be extremely small (Fig. 7). Cracking of 18-8 stainless tubes in heat exchangers has been observed in practice after contact with cooling waters containing 25 ppm Cl^- or less, and cracking has also been induced by small amounts of chlorides contained in magnesia insulation wrapped around

---

[39] W. Williams and J. Eckel, *J. Amer. Soc. Nav. Engrs.*, **68**, 93 (February 1956.)

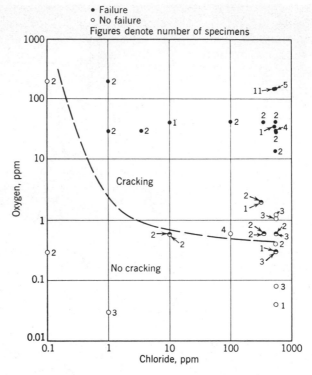

Fig. 7. Relation between chloride and oxygen content of boiler water on stress corrosion cracking of austenitic 18-8 type stainless steels exposed to steam phase with intermittent wetting; pH 10.6, 50 ppm PO_4, 242 to 260°C (467–500°F), 1 to 30 days' exposure (Williams and Eckel).

stainless steel tubes.[40]   In these instances, if small pits form initially at which chlorides concentrate by electrolytic transference (Fig. 4, 6) (see *Theory of Pitting*, p. 310), the cracking tendency is accentuated by concentrated $FeCl_2$ and analogous metal chlorides within the pits.   Hence oxygen may induce stress corrosion cracking in sodium chloride solutions because pitting occurs when it is present.   Another contributing factor is a shift of the corrosion potential in presence of oxygen, but not in its absence, to values that are noble to the critical potential for stress corrosion cracking.   For this situation, stress corrosion cracking can occur regardless of whether corrosion pits develop.

Cracking by alkaline solutions requires relatively high concentrations of OH^-, hence cracking of 18-8 usually occurs not in alkaline boiler

[40] A. Dana, Jr., *Amer. Soc. Testing Mat. Bull.*, **225**, 46 (October 1957).

water, but rather within the splash zone above the water line where dissolved alkalies concentrate by evaporation. Failures can occur in such solutions in the absence of dissolved oxygen.[41] There is no evidence that transgranular stress corrosion cracking occurs in pure water or pure steam.

Cracking of stressed, nickel-free ferritic steels in general does not occur in the chloride media described previouly, these steels being sufficiently resistant to warrant their practical use in preference to austenitic stainless steels in chloride-containing solutions. They (type 430) are also resistant to a 55% $Ca(NO_3)_2$ solution boiling at 117°C or to a 25% NaOH solution boiling at 111°C.[42] When alloyed with more than 1.5% Ni, the 18% Cr-Fe, 0.003% C stainless steels, cold-rolled, become susceptible to transgranular stress corrosion cracking in $MgCl_2$ boiling at 130°C. When annealed at 815°, 1 hr, only the alloy containing 2% Ni fails; higher or lower nickel contents resist cracking up to 200 hr.[43] These results confirm the previous report of Bond and Dundas[44] that small alloying additions of nickel to ferritic stainless steels induce susceptibility in $MgCl_2$. The same authors also report that more than 0.4% Cu in presence of alloyed molybdenum (1–3%) similarly induces susceptibility and that small additions of molybdenum accentuate the damaging effects of alloyed nickel. The effect of nickel is partly explained by an observed shift of potentials such that at and above 1.5% Ni the corrosion potential of the cold-rolled material becomes noble to the critical potential, hence cracking occurs; the reverse order of potentials applies to lower nickel content alloys.[43]

In slightly or moderately acidic solutions, the martensitic steels when heat treated to high hardness values are very sensitive to cracking, particularly in the presence of sulfides, arsenic compounds, or oxidation products of phosphorus or selenium. The specific anions of the acid make little difference so long as hydrogen is evolved, contrary to the situation for austenitic steels, for which only specific anions are damaging. Also, cathodic polarization, rather than protecting against cracking, accelerates failure. All these facts suggest that the martensitic steels under these conditions fail not by stress corrosion cracking but by hydrogen cracking (see p. 142). The more ductile ferritic stainless steels undergo hydrogen blistering instead when they are cathodically protected

[41] C. Edeleanu and P. Snowden, *J. Iron Steel Inst.*, **186**, 406 (1957).
[42] A. Bond, J. Marshall, and H. Dundas in *Stress Corrosion Testing*, Spec. Tech. Publ. No. 425, p. 116, Amer. Soc. Testing Mat., Philadelphia, Pa., 1967.
[43] R. Newberg and H. Uhlig, to be published.
[44] A. Bond and H. Dundas, *Corrosion*, **24**, 344 (1968).

in seawater, especially at high current densities. Austenitic stainless steels are immune to both hydrogen blistering and cracking.

Galvanic coupling of active metals to martensitic stainless steels may also lead to failure because of hydrogen liberated on the stainless (cathodic) surface. Such failures have been demonstrated in laboratory tests.[45] Practical instances have been noted, as mentioned in the chapter on hydrogen cracking, of self-tapping martensitic stainless steel screws cracking spontaneously soon after being attached to an aluminum roof in a sea-coast atmosphere. Similarly, hardened martensitic stainless steel propellers coupled to the steel hull of a ship have failed by cracking soon after being placed in service. Severely cold-worked 18-8 austenitic stainless steels may also fail under conditions that would damage the martensitic types.[46, 47] Here again, sulfides accelerate damage, and since the alloy on cold working undergoes a phase transformation to ferrite, the observed effect is probably another example of hydrogen cracking.

Martensitic stainless steels and also the precipitation-hardening types, heat treated to approximately >180,000 psi yield strength, have been reported to crack spontaneously in the atmosphere, in the salt spray, or immersed in aqueous media, even when not coupled to other metals.[48-51] Martensitic stainless steel blades of an air compressor[52] have failed along the leading edges where residual stresses were high and where condensation of moisture occurred. Stressed to about 75% of the yield strength, and exposed to a marine atmosphere, life of ultra-high-strength 12% Cr martensitic stainless steels was in the order of 10 days or less.[53] Although interpretations differ, convincing evidence has accumulated to show that these steels can fail either by hydrogen cracking or by stress corrosion cracking. At high stress levels, apparently water alone, without the necessary reaction with metal to form interstitial hydrogen, can adsorb and reduce metal bond strength sufficient to initiate failure (stress-sorption cracking).

[45] H. H. Uhlig, *Metal Progr.,* **57,** 486 (1950).

[46] W. Rees, in *Symp. Internal Stresses in Metals and Alloys,* p. 333, Inst. Metals, London (1948).

[47] F. Bloom, *Corrosion,* **11,** 351t (1955).

[48] P. Ffield, in *The Book of Stainless Steels,* Chap. 21C, pp. 679–686, Amer Soc. Metals, 1935.

[49] P. Lillys and A. Nehrenberg, *Trans. Amer. Soc. Metals,* **48,** 327 (1956).

[50] E. Phelps and R. Mears, *Effect of Composition and Structure of Stainless Steels upon Resistance to Stress Corrosion Cracking,* 1st Int. Congr. Metallic Corrosion, p. 319, Butterworths, London, 1962.

[51] J. Truman, R. Perry, and G. Chapman, *J. Iron Steel Inst.,* **202,** 745 (1964).

[52] W. Badger, *Soc. Automot. Eng. Trans.,* **62,** 307 (1954).

[53] E. Phelps, in *Fundamental Aspects of Stress Corrosion Cracking,* p. 398, Nat. Assoc. Corros. Engrs., Houston, Texas, 1969.

METALLURGICAL FACTORS

Austenitic stainless steels containing more than about 45% Ni are immune to stress corrosion cracking in boiling MgCl$_2$ solution and probably in other chloride solutions as well (Fig. 8).[54] Edeleanu and Snow-

[54] H. Copson, in *Physical Metallurgy of Stress Corrosion Fracture*, edited by T. Rhodin, p. 247, Interscience, New York, 1959.

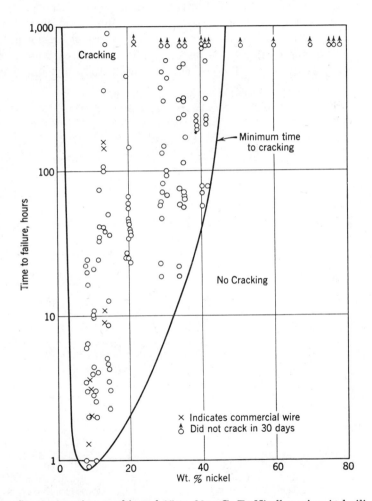

Fig. 8. Stress corrosion cracking of 15 to 26% Cr-Fe-Ni alloy wires in boiling 42% MgCl$_2$. Low-nickel or nickel-free alloys are ferritic and do not crack. Austenitic alloys do not crack above 45% Ni. (From H. R. Copson, in *Physical Metallurgy of Stress Corrosion Fracture*, edited by Thor N. Rhodin, Interscience, New York, 1959).

den[41] noted that high-nickel stainless steels were more resistant to crack-ing in alkalies. Increasing the amount of nickel in austenitic stainless steels shifts the critical potential for stress corrosion cracking in $MgCl_2$ solution in the noble direction more rapidly than it shifts the correspond-ing corrosion potential, hence the alloys become more resistant.[55] At and above approximately 45% Ni, the alloys resist stress corrosion crack-ing regardless of applied potential, indicating that not environmental but metallurgical factors, such as unfavorable dislocation arrays or de-creasing interstitial nitrogen solubility, become more important.

For austenitic steels that are resistant to transformation on cold work-ing (e.g., type 310), nitrogen is the element largely responsible for stress-cracking susceptibility, whereas additions of carbon decrease susceptibility (Fig. 9).[56] The effect is related to alloy imperfection

[55] H. Lee and H. Uhlig, *J. Electrochem. Soc.,* **117**, 18 (1970).

[56] H. Uhlig, R. White, and J. Lincoln, Jr., *Acta Met.,* **5**, 473 (1957); H. Uhlig and R. White, *Trans. Amer. Soc. Metals,* **52**, 830 (1960); H. Uhlig and J. Sava, see *Fracture,* edited by H. Liebowitz, Vol. 3, p. 656, Academic Press, 1971. F. S. Lang reported that, in addition to N, small amounts of P, As, Sb, Bi, Ru, or Al in 18-20 stainless steel are detrimental. *Corrosion,* **18**, 378t (1962).

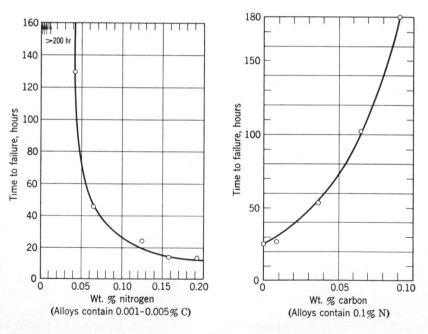

Fig. 9. Stress corrosion cracking of cold-rolled 19% Cr, 20% Ni austenitic stainless steels in boiling $MgCl_2$ (154°C) as affected by carbon or nitrogen content (Uhlig and Sava).

structure rather than to any shift of either critical or corrosion potential.[55] Stabilizing additions effective in preventing intergranular corrosion such as titanium or columbium have no beneficial effect on stress corrosion cracking, nor do alloying additions such as 2 to 3% Mo in type 316 stainless steel. In practice, stress corrosion cracking of austenitic stainless steels in chloride solutions has not been observed below about 80°C,[57] probably, for one reason, because diffusion of interstitial atoms taking part in the cracking mechanism is severely restricted at low temperatures.

TO REDUCE OR ELIMINATE CRACKING

1. Austenitic Stainless Steels

a. Cathodically Protect. The critical potential of 18-8 steel in $MgCl_2$ at 130°C is -0.145 V (s.h.e.). Coupling of stressed 18-8 to a small area of nickel ($\phi_{corros} = -0.18$ V) prevents cracking in this medium, or (as a porous nickel coating) in water containing 50 ppm Cl^- at 300°C.[58]

b. In Acid Environments, Eliminate Cl^-. In neutral or slightly alkaline chloride environments, eliminate dissolved oxygen and other oxidizing ions. Add extraneous ions (e.g., NO_3^-, I^-, acetates).

c. Avoid High Concentrations of OH^-. High concentrations of alkalies occurring initially or incidentally through concentration at crevices or in vapor zones are damaging. Add buffering ions (e.g., PO_4^{-3}).

d. Use alloys containing >45% Ni or reduce the nitrogen content (and other detrimental impurities, if present) to the lowest possible value (<0.04% N in the case of 20% Cr, 20% Ni, 0.001% C stainless steel) (Fig. 9).

e. Substitute Ferritic Alloys (e.g., type 430). However, ferritic alloys may become embrittled by hydrogen or they may blister when galvanically coupled in certain environments.

2. Martensitic, Precipitation Hardening, or Ferritic Stainless Steels

a. Avoid excess current if cathodic protection is applied.

b. Avoid galvanic coupling to more active metals.

[57] W. Rion, Jr., *Ind. Eng. Chem.*, **49**, 73A (1957). However, 18-8 stainless steel is observed to undergo s.c.c. in 10 *N* H_2SO_4 + 0.5 *N* NaCl at room temperature, presumably by a different mechanism. S. Acello and N. Greene, *Corrosion*, **18**, 286t (1962).

[58] P. Neumann and J. Griess, *Corrosion*, **19**, 345t (1963).

c. Temper martensitic or precipitation-hardening steels to lowest possible hardness values. Exposed to the atmosphere, hardness should be below about Rockwell C 40. Maximum susceptibility of types 410 and 420 stainless steels to cracking in the salt-spray or to hydrogen cracking occurs after tempering for 2 hr at 425 to 550°C (800–1000°F); minimum susceptibility to hydrogen cracking occurs after tempering for 2 hr at 260°C (500°F).[49]

Galvanic Coupling, General Corrosion Resistance

Since stainless steels are passive and exhibit a noble potential, they can be coupled successfully to metals that are either passive or inherently noble. This includes metals and alloys like silver; silver solder; copper; nickel; 70% Ni-Cu alloy; 76% Ni, 16% Cr, 7% Fe alloy; and usually, aluminum in environments in which it remains passive.

As discussed under "Pitting, Crevice Corrosion," p. 309, stainless steels are best employed under fully aerated or oxidizing conditions which favor the passive state. Whether used in handling chemicals or exposed to the atmosphere, the alloy surface should always be kept clean and free of surface contamination. Otherwise differential aeration or concentration cells are set up which cause pitting and localized rusting.

In brief, stainless steels are resistant to:

1. Nitric acid, over a wide range of concentrations and temperatures.
2. Very dilute sulfuric acid, room temperature, if aerated; higher concentrations (e.g., 10%) and at boiling temperatures if Fe^{+3}, Cu^{++}, or nitric acid is added as inhibitor;[59] or at lower temperatures if small amounts of Cu, Pt or Pd are alloyed (see p. 69). Also resistant to cold or hot sulfuric acid if anodically protected.
3. Many organic acids, including almost all food acids and acetic acid (but not boiling glacial acetic acid).
4. Sulfurous acid (in the absence of SO_4^{--} or Cl^-).
5. Alkalies. Except under stress in hot concentrated caustic solutions.
6. The atmosphere. Types 302 and 304 have been used successfully as architectural trim of store fronts and buildings (e.g., the Chrysler and Empire State buildings in New York City). These and type 430 are used for auto trim.

Stainless steels are not resistant to:

1. Dilute or concentrated HCl, HBr, and HF. Also not resistant to salts that hydrolyze to these acids.

[59] J. Cobb and H. Uhlig, *J. Electrochem. Soc.,* **99,** 13 (1952).

2. Oxidizing chlorides (e.g., $FeCl_3$, $HgCl_2$, $CuCl_2$, $NaOCl$).

3. Seawater, except for brief periods or when cathodically protected.

4. Photographic solutions, especially fixing solutions containing thiosulfates. (Pitting occurs.)

5. Some organic acids, including oxalic, formic, and lactic acids.

6. Stressed austenitic alloys (e.g., type 304) in waters containing Cl^- plus O_2 at $>80°C$.

GENERAL REFERENCES

Corrosion Handbook, pp. 143–193.

J. Monypenny, *Stainless Iron and Steel,* 2 vols., Chapman and Hall, London, 1951–1954.

Carl Zapffe, *Stainless Steels,* Amer. Soc. Metals, Cleveland, 1949. Contains an elementary explanation of metallurgical changes brought about by heat treatment.

Corrosion Resistance of Metals and Alloys, edited by F. LaQue and H. Copson, pp. 375–445, Reinhold, New York, 1963.

chapter 19

Copper and copper alloys

COPPER

$$Cu \rightarrow Cu^{++} + 2e^- \qquad E° = -0.337 \text{ V}$$

Copper dissolves anodically in most aqueous environments forming the divalent ion Cu^{++}. Equilibrium relations at the metal surface indicate that the reaction $Cu + Cu^{++} \rightleftharpoons 2Cu^+$ is displaced far to the left (see Prob. 1, p. 385). On the other hand, if complexes are formed, as for example between Cu^+ and Cl^- in a chloride solution, the continuous depletion of Cu^+ by conversion to $CuCl_2^-$ favors the univalent ion as the major dissolution product. Comparatively, when the anhydrous ion forms, as in the oxidation of copper in air at elevated temperatures, only Cu^+ forms because Cu_2O has a lower oxygen dissociation pressure than CuO (see *Corrosion Handbook*, Tables 1 and 2, p. 622).

Copper is a metal widely used because of good corrosion resistance combined with mechanical workability, excellent electrical and thermal conductivity, and ease of soldering or brazing. It is noble to hydrogen in the Emf Series, and thermodynamically inert with no tendency to corrode in water and in nonoxidizing acids free of dissolved oxygen. In oxidizing acids or in aerated solutions of ions that form copper complexes (e.g., CN^-, NH_4^+), corrosion can be severe. Copper is also characterized by sensitivity to corrosion by high-velocity water or aqueous solutions, called *impingement attack* (Fig. 1). The rate increases with dissolved oxygen content, whereas in air-free high-velocity water up to at least 25 ft/sec, impingement attack is negligible. In aerated water, attack increases with Cl^- concentration and with decrease of pH.[1] Commer-

---

[1] R. Fish and J. Dankese, S.B. Thesis, Dept. of Chem. Eng., M.I.T., 1954; R. Hadge and D. Revelotis, S.B. Thesis, Dept. of Chem. Eng., M.I.T., 1958.

Fig. 1. Longitudinal cross section of undercut pits associated with impingement attack of condenser alloy by salt water, 7X (flow of water from left to right).

cial Tough Pitch, OFHC (oxygen-free, high conductivity) and electrolytically refined copper are immune to stress-corrosion cracking. Phosphorus-deoxidized copper, however, containing as little as 0.004% P, is susceptible.[2]

Corrosion in Natural Waters

Copper is resistant to seawater, the corrosion rate in temperate climates being about 0.001 to 0.002 ipy (5–10 mdd) in quiet water and somewhat higher in moving water. In tropical climates the rate increases by 1.5 to 2 times. It is one of the very few metals that remains free of fouling organisms, normal corrosion being sufficient to release copper ions in concentrations which poison marine life.

In seawater and in fresh waters, corrosion resistance depends on the presence of a surface oxide film through which oxygen must diffuse in order for corrosion to continue. The film is easily disturbed by high-velocity water or is dissolved by either carbonic acid or the organic acids which are found in some fresh waters or soils, leading to an appreciably high corrosion rate. For example, a hot water in Michigan, zeolite softened with resultant high concentration of $NaHCO_3$, was found to perforate copper water pipe within 6 to 30 months.[3] The same water unsoftened, on the other hand, was not nearly as corrosive because a protective film of $CaCO_3$ containing some silicate was deposited on the metal surface.

Even if the corrosion rate is not excessive and the copper is durable, a water containing carbonic or other acids nevertheless may corrode copper and copper-base alloys sufficiently to cause blue staining of bathroom fixtures and cause an increase in the corrosion rate of iron, gal-

[2] D. Thompson and A. Tracy, *Trans. Amer. Inst. Mining Met. Eng.*, **185**, 100 (1949).

[3] M. Obrecht and L. Quill, *Heating, Piping and Air-Conditioning*, January, pp. 165–169; March, pp. 109–116; April, pp. 131–137; May, pp. 105–113; July, pp. 115–122; September, pp. 125–133 (1960); April, pp. 129–134 (1961). M. Obrecht, *Corrosion*, **18**, 189t (1962).

vanized steel, or aluminum surfaces with which such water comes into contact. Accelerated corrosion in this instance is caused by a replacement reaction in which copper metal is deposited on the base metal, forming numerous small galvanic cells. Treatment of acid waters or waters of negative Saturation Index with lime or with sodium silicate reduces the corrosion rate sufficiently to overcome both staining and accelerated corrosion of other metals, except aluminum. Aluminum is an exception because of its sensitivity to extremely small amounts of Cu^{++} in solution, the usual water treatment being inadequate to reduce Cu^{++} pickup to nondamaging levels.

It is possible to avoid the damaging effects of Cu^{++} contamination of water by using copper piping coated on the inside surface with tin (so-called tinned copper). The tin coating must be pore-free in order to avoid accelerated attack of copper at exposed areas by reason of tin (or copper-tin intermetallic compounds) being cathodic to copper.

Copper piping is usually satisfactory for transporting seawater and soft or hard waters, whether hot or cold. But in addition to the effects just described, a pitting type of corrosion may be induced in waters having relatively good conductivity if dirt or rust from other parts of the system accumulate on the copper surface. Differential aeration cells are formed, supplemented in some cases by turbulent flow, which initiates impingement attack. The resulting corrosion is sometimes called *deposit attack*. Periodic cleaning of the piping or tubing usually prevents corrosion by such deposits.

Even when the copper surface is free of deposits, pitting nevertheless may occur in waters of certain compositions or as a result of pipe fabrication methods. In this respect, Campbell[4] classified fresh waters corrosive to copper into two categories and outlined tests for determining whether a water is of the pitting type. In the first category are *soft waters* containing small amounts of manganese salts. With these he found that pitting usually occurred in the hottest part of the system. Corrosion products within the pit are Cu_2O plus a small amount of chloride, whereas mounds around the pits contain perhaps basic copper sulfate, and the pipe surface is covered by a black scale rich in manganese oxide. In the second category are *hard* or *moderately hard* waters. With these, pitting is usually restricted to the cold portions of the system. Pits contain a higher proportion of cuprous chloride than in soft-water, and mounds consist largely of cupric carbonate. The surrounding surface scale is largely $CaCO_3$ stained green, beneath which is a highly

[4] H. Campbell, *J. Appl. Chem.*, **4**, 633 (1954).

cathodic film of carbon[5] or sometimes a glassy form of Cu_2O. The carbon film results from the residual lubricant used during the drawing of the pipe, or from pitch or resin fillers used for cold bending, which become partially carbonized when melted. The glassy cuprous oxide is caused by abnormal annealing conditions during manufacture of the pipe. In the absence of carbon or glassy Cu_2O, pitting in hard waters is stated to be rare. Mattsson and Fredricksson[6] report that pitting corrosion of copper in hot waters occurs largely at pH values of 7.4 or less, with weight ratios of HCO_3^-/SO_4^{--} less than 1, and that SO_4^{--} is more damaging than Cl^-.

Some waters in which pitting otherwise might be expected because of the factors just described are nevertheless not corrosive. These contain a natural, probably organic, inhibitor found in many surface waters but not in deep wells nor in waters that have been chemically flocculated for clarification. Many properties of the inhibitor have been investigated,[7] but its precise nature has not been established.

In summary, copper is resistant to:

1. Seawater.
2. Fresh waters, hot or cold. Copper is especially suited to convey soft, aerated waters, low in carbonic and other acids.
3. Deaerated, hot or cold, dilute H_2SO_4, H_3PO_4, acetic acid, and other nonoxidizing acids.
4. Atmospheric exposure.

Copper is not resistant to:

1. Oxidizing acids, e.g., HNO_3, hot concentrated H_2SO_4, and aerated nonoxidizing acids (including carbonic acid).
2. NH_4OH (plus O_2). A complex ion forms: $Cu(NH_3)_4^{++}$. Substituted NH_3 compounds (amines) are also corrosive. These compounds are those that cause stress-corrosion cracking of susceptible copper alloys.
3. High velocity aerated waters and aqueous solutions. In corrosive waters (high in O_2 and CO_2, low in Ca^{++} and Mg^{++}) the velocity should be kept[3] below 4 ft/sec.; in less corrosive waters, $<65°C$ ($<150°F$), below 8 ft/sec.
4. Oxidizing heavy metal salts, e.g., $FeCl_3$, $Fe_2(SO_4)_3$.
5. Hydrogen sulfide, sulfur, some sulfur compounds.

---

[5] H. Campbell, *J. Inst. Metals*, **77**, 345 (1950).
[6] E. Mattsson and A. Fredricksson, *Brit. Corros. J.*, **3**, 246 (1968).
[7] H. Campbell, *Proc. Soc. Water Treat. Exam.*, **3**, 100 (1954).

COPPER ALLOYS

Tin bronzes are copper-tin alloys noted for their high strength. Alloys containing more than 5% Sn are especially resistant to impingement attack. The copper-silicon alloys containing 1.5 to 4% Si have better physical properties than copper and similar general corrosion resistance. Resistance to stress-corrosion cracking is poor at about 1% Si, but is relatively high at 4% Si.[2] Seawater immersion tests at Panama showed that 5% Al-Cu was the most resistant of the common copper-base alloys, losing only 20% of the corresponding weight loss of copper after 16 years.[8]

Copper-Zinc Alloys (Brasses)

Copper-zinc alloys have better physical properties than copper alone and they are also more resistant to impingement attack, hence brasses are used in preference to copper for condenser tubes. Corrosion failures of brasses usually occur by dezincification, pitting, or stress corrosion cracking. The tendency for brass to corrode in these ways, except for pitting attack, varies with zinc content as shown in Fig. 2. Pitting

[8] C. Southwell, C. Hummer, Jr., and A. Alexander, *Materials Prot.*, **7**, 41 (January 1968); **7**, *Ibid.*, 61 (March 1968).

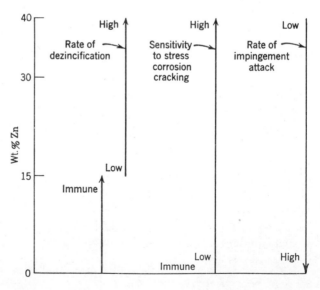

Fig. 2. Trends of dezincification, stress corrosion cracking, and impingement attack with increasing zinc content in copper-zinc alloys (brasses).

is usually caused by differential aeration cells or high-velocity conditions. It can normally be avoided by keeping the brass surface clean at all times and by avoiding velocities and design geometry which lead to impingement attack.

Various names have become attached to brasses of different zinc content. *Muntz metal,* 40% Zn-Cu, is used primarily for condenser systems that employ fresh water as coolant (e.g., Great Lakes water). *Naval brass* is a similar composition containing 1% Sn. *Manganese bronze* is also similar, containing about 1% each of tin, iron, and lead; it is used for ship propellers, among other applications. Dezincification of manganese bronze propellers is avoided to some extent by the cathodic protection afforded by the steel hull.

Yellow brass, 30% Zn-Cu, is used for a variety of applications where easy machining and casting are desirable. The alloy gradually dezincifies in seawater and in soft fresh waters. This tendency is retarded by addition of 1% Sn, the corresponding alloy being called *admiralty metal.* Addition of small amounts of arsenic, antimony, or phosphorus still further retards rate of dezincification; the resulting alloy, called inhibited admiralty metal, finds use in seawater or fresh-water condensers.

Red Brass, 15% Zn-Cu, is relatively immune to dezincification, but is more susceptible to impingement attack than is yellow brass.

Dezincification

Dezincification has already been defined (p. 15). In brasses it takes place either in localized areas on the metal surface, called *plug type* (Fig. 3) or uniformly over the surface, called *layer type* (Fig. 4). Brass

Fig. 3. Plug-type dezincification in brass pipe, 1X.

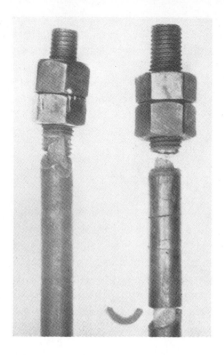

Fig. 4. Layer-type dezincification in brass bolts, 1X.

so corroded retains some strength but has no ductility. Layer-type dezincification in a water pipe may lead to splitting open of the pipe under conditions of sudden pressure increase; and for plug type a plug of dezincified alloy may blow out, leaving a hole. Because dezincified areas are porous, plugs may be covered on the outside surface with corrosion products and residues of evaporated water.

Conditions of the environment that favor dezincification are (1) high temperatures, (2) stagnant solutions, especially if acid, and (3) porous inorganic scale formation. Looking at the situation metallurgically, brasses that contain 15% or less zinc are usually immune. Also, dezincification of so-called α-brasses (up to 40% Zn) can be reduced by alloying additions of about 1% Sn or a few hundredths percent As, Sb, or P with or without Sn.

The detailed mechanism of dezincification is still being discussed. Two points of view are (1) that the alloy corrodes and copper is then redeposited to form a porous outer layer, or (2) that zinc diffuses to the alloy surface, reacting there preferentially, and leaving a copper-rich alloy residue. There is evidence that either mechanism may apply in specific instances of dezincification. The accumulated facts suggest that

(2) is a more common mechanism than (1). Pickering and Wagner[9, 10] proposed that volume diffusion of zinc probably occurs through surface generation of vacancies, in particular divancies. These are produced by anodic dissolution, subsequently diffusing at room temperature into the interior of the alloy (D for divacancies in Cu at 25°C = 1.3×10^{-12} cm^2/sec.)[9], aiding diffusion of zinc in the opposite direction. Interdiffusion of zinc and copper occurs within the depleted alloy as shown by x-ray patterns of dezincified layers of ϵ brass (86 at. % Zn-Cu) or γ brass (65 at. % Zn-Cu). New copper-rich phases appear (e.g., α brass), and composition gradients in such phases are always in the direction of copper enrichment. Similar behavior is noted in the gold-copper alloy system, as mentioned earlier, in which copper corrodes preferentially without detectable corrosion of gold, leaving a porous residue of gold-copper alloy or pure gold. Twin bands in brass are often visible in the complete or incomplete dezincified layer, which constitutes the earlier evidence for a volume diffusion mechanism.[11] Objections have been raised to this evidence,[12] but plausible explanations for the x-ray structure of dezincified layers by a redeposition process have not been forthcoming. Although several possibilities have been described for the mechanism by which arsenic, antimony, or phosphorus inhibits dezincification of α but not of β brass, the mechanism must be considered to be largely unresolved.

Stress Corrosion Cracking (Season Cracking)

When an α brass is subjected to an applied or residual tensional stress in contact with a trace of NH_3 or a substituted ammonia (amine), in the presence of oxygen (or another depolarizer) and moisture, it cracks usually along the grain boundaries (intergranular); see Fig. 5. Cracking through the grains (transgranular) may occur if the alloy is severely deformed plastically.

According to one source, both types of cracking were originally called *season cracking* because of the resemblance of stress corrosion cracks in bar stock to those of seasoned wood. In England, the origin of the term is ascribed to the fact that years ago brass cartridge cases stored in India were observed to crack particularly during the monsoon season.

Traces of nitrogen oxides may also cause stress corrosion cracking,

[9] H. Pickering and C. Wagner, *J. Electrochem. Soc.,* **114,** 698 (1967).

[10] H. Pickering, *Ibid.,* **117,** 8 (1970).

[11] E. Polushkin and H. Shuldener, *Trans. Amer. Inst. Mining Met. Eng.* **161,** 214 (1945).

[12] V. Lucey, *Brit. Corros. J.,* **1,** 9, 53 (1965).

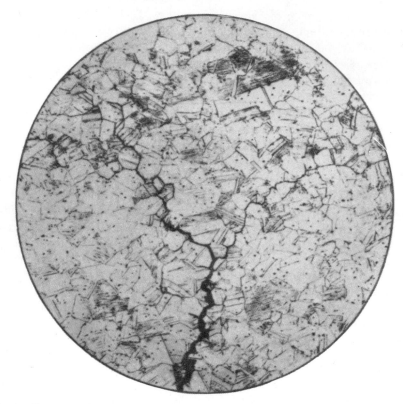

Fig. 5. Intergranular stress corrosion cracking of brass, 75X (specimen stored one year).

probably because such oxides are converted to ammonium salts on the brass surface by chemical reaction with the metal. In one instance of this kind, premature failure of yellow brass brackets in the humidifier chamber of an air-conditioning system was traced to this cause.[13]   The air had passed through an electrostatic dust precipitator, the high voltage field of which generated traces of nitrogen oxides. These, in turn, formed corrosion products on the brass surface which were found by analysis to contain a high proportion of NH_4, causing intergranular cracking of the stressed brackets. Similar cracking of stressed brass could be reproduced in the laboratory over a period of days by using a spark discharge in air of 100% relative humidity.

Stress corrosion cracking of 12% Ni, 23% Zn-Cu alloy (nickel brass)

[13] H. Uhlig and J. Sansone, *Materials Prot.,* **3,** 21 (February 1964).

parts of Central Office Telephone Equipment in Los Angeles occurred within two years for similar reasons.[14] Pollution of Los Angeles air accounts for abnormally high concentrations of nitrogen oxides and suspended nitrates, the latter settling down as dust on the brass parts. Similar failures have not been observed as frequently in New York City, where the air contains less nitrate but also many more sulfate particles than Los Angeles air, indicating that sulfates act as inhibitors.

Mattsson[15] observed that minimum cracking time of 37% Zn-Cu brass in a solution of 1 mol $NH_3 + NH_4^+$, plus 0.05 mol $CuSO_4$/liter occurred at pH 7.3, times to failure being longer at higher pH and considerably longer at lower pH values. Johnson and Leja[16] reported stress corrosion cracking of brass in alkaline cupric citrate or tartrate solutions at pH values in which complexing of Cu^{++} is pronounced.

Annealed brass, if not subject to a high applied stress, does not stress corrosion crack. Whether residual stresses in cold-worked brass may be sufficient to cause stress corrosion cracking in an ammonia atmosphere can be checked by immersing brass in an aqueous solution of 100 g mercurous nitrate ($Hg_2(NO_3)_2$) and 13 ml nitric acid (HNO_3, sp. gr. 1.42) per liter of water. Mercury is released and penetrates the grain boundaries of the stressed alloy. If cracks do not appear within 15 min, the alloy is probably free of damaging stresses.

No alloying additions in small amounts effectively provide immunity to this type of failure in brasses. Low-zinc brasses are more resistant than high-zinc brasses.

High-zinc brasses (e.g., 45 to 50% Zn-Cu) having a so-called β or $\beta + \gamma$ structure, stress corrosion crack through the grains or transgranularly; and, unlike α brasses, only moisture is required to cause failure.[17]

The mechanism of stress corrosion cracking in brasses has been the subject of much study. High-purity alloys or single crystals of α brass continue to crack when stressed in NH_3 atmospheres.[18] In support of an electrochemical mechanism it has been noted that the grain boundaries of polycrystalline brasses are more active in potential than the grains as measured in NH_4OH, but not in $FeCl_3$ solution in which stress-corrosion cracking does not occur.[19] It has been proposed alternatively

[14] N. McKinney and H. Hermance, *Stress Corrosion Testing,* Spec. Tech. Publ. No. 425, p. 274, Amer. Soc. Testing Mat., Philadelphia, Pa. 1967.

[15] E. Mattsson, *Electrochim. Acta,* **3,** 280 (1960–1961).

[16] H. Johnson and J. Leja, *Corrosion,* **22,** 178 (1966).

[17] A. Bailey, *J. Inst. Metals,* **87,** 380 (1959).

[18] G. Edmunds in *Symp. Stress Corrosion Cracking of Metals,* p. 67, ASTM–AIME, Philadelphia, Pa., 1945.

[19] R. Bakish and W. Robertson, *J. Electrochem. Soc.,* **103,** 320 (1956).

that a brittle oxide film forms on brasses which continuously fractures under stress exposing fresh metal underneath to further oxidation.[20, 21] On the other hand, it is possible that the defect structure of stressed, susceptible copper alloys along grain boundaries favors adsorption of Cu complex ions accompanied by reduction of metal bond strength (stress-sorption cracking).

It is likely, in any event, that zinc atoms aided by plastic deformation segregate preferentially at grain boundaries. The resulting composition gradient favors galvanic action between such areas and the grains, accounting for slow intergranular attack in a variety of corrosive media without the necessity of an applied stress (intergranular corrosion). But such areas, subject to plastic deformation, may also favor adsorption of NH_4^+ or complex ammonium ions within a specific potential range, leading to rapid crack formation. Similar effects can occur along slip bands (transgranular cracking). Although segregated zinc may be essential to the observed intergranular corrosion of brasses, it is probable that the defect structure of grain boundaries or of slip bands is more important to stress-corrosion cracking. Hence failure of copper-base alloys by cracking occurs when copper is alloyed not only with zinc but also with a variety of other elements such as silicon, nickel, antimony, arsenic, aluminum, or phosphorus.[2]

Susceptibility to stress corrosion cracking of brasses can be minimized or avoided by four main procedures.

1. *Stress-Relief Heat Treatment.* For 30% Zn-Cu brass, heating at 350°C (660°F) for 1 hr is effective, but recrystallization and some loss of strength of the alloy result. It is reported that heat treatment at 300°C (570°F) improves resistance to cracking without serious loss of physical properties.[22]

2. *Avoiding Contact with NH_3 (or with O_2 and Other Depolarizers in the Presence of NH_3).* It is difficult to guarantee that there will be no contact with NH_3 because of the very small quantities of ammonia that cause cracking. Plastics containing or decomposing to amines have been a source of damage to unannealed brass. Fertilizer washing from farm land, or air over fertilized soil, has similarly caused cracking of brass. On the other hand, brass condenser tubes do not crack in boiler water condensate containing NH_3 because the concentration of oxygen is extremely low.

[20] A. McEvily, Jr., and A. Bond, *J. Electrochem. Soc.,* **112**, 131 (1965).

[21] E. Pugh, J. Craig, and A. Sedriks, in *Proc. Fundamental Aspects of Stress Corrosion Cracking,* p. 118, Nat. Assoc. Corros. Engrs., Houston, Texas, 1969. See also S. Shimodaira and M. Takano, *Ibid.,* p. 202.

[22] C. Bulow, in *Corrosion Handbook,* pp. 78–81.

3. *Cathodic Protection.* Cathodic protection is provided by either impressed current, or by coating brass with a sacrificial metal (e.g., zinc).

4. *Using H₂S as an Inhibitor.*[23] The mechanism may in part involve reaction with available free oxygen.

Condenser Tube Alloys, Copper-Nickel Alloys

For fresh waters, copper, Muntz metal, and admiralty metal (inhibited) are frequently employed. For brackish or seawater, admiralty metal, one of the cupro-nickel alloys (10 to 30% Ni, bal. Cu) and aluminum brass (22% Zn, 76% Cu, 2% Al, 0.04% As) are used. For polluted waters, cupro-nickel alloys are preferred over aluminum brass because the latter is subject to pitting attack. Aluminum brass may also pit readily in unpolluted but stagnant seawater.

Aluminum brass resists high-velocity waters (impingement attack) better than does admiralty metal. Cupro-nickel alloys are especially resistant to high-velocity seawater when they contain small amounts of iron and sometimes manganese as well. For the 10% Ni cupro-nickel alloy, the optimum iron content is about 1.0 to 1.75%, with 0.75% Mn max; for the analogous 30% Ni composition, the amount of alloyed iron is usually less (e.g., 0.40–0.70% Fe accompanied by 1.0% Mn max).[24] The opinion is held by some that especially protective films are formed on condenser tube surfaces when iron is contained in water as a result of corrosion products upstream. In accord with this opinion, the beneficial effect of iron alloyed with copper-nickel alloys is considered to result from similar availability of iron in the formation of protective films. The manner in which iron brings about this improvement, however, is not known.

It is pertinent that the 30% Ni-Cu alloy is relatively resistant to stress corrosion cracking compared to the 10% or 20% Ni-Cu alloys,[2] or compared to any of the 30% Zn-Cu brasses. A detailed general account of the behavior of copper-nickel alloys, especially the 10% Ni-Cu alloy, in seawater is given by Stewart and LaQue.[25]

GENERAL REFERENCES

Corrosion Handbook, "Aqueous Media and the Atmosphere," pp. 61–112; "Elevated Temperatures," pp. 621–630, edited by H. H. Uhlig, Wiley, New York, 1948.
Corrosion Resistance of Metals and Alloys, edited by F. LaQue and H. Copson, pp. 553–599, Reinhold, New York, 1963.

---

[23] E. Camp and C. Phillips, *Corrosion,* **6,** 39 (especially p. 45) (1950).
[24] U. S. Mil. Spec. MIL-C-15726 E (Ships), August 20, 1965.
[25] W. Stewart and F. LaQue, *Corrosion,* **8,** 259 (1952).

chapter 20

Aluminum and magnesium

ALUMINUM

$$Al \rightarrow Al^{+3} + 3e^- \qquad E^\circ = 1.7 \text{ V}$$

There is some evidence[1] that when aluminum dissolves anodically, both Al^{+3} and Al^+ are formed initially, the univalent ion then reducing water to form the trivalent ion in accord with

$$Al^+ + 2H_2O \rightarrow Al^{+3} + H_2 + 2OH^-$$

Hence, when aluminum is anodized (see p. 244) for the purpose of intentionally building a thicker surface oxide film, hydrogen is evolved at the anode as well as at the cathode. Hydrogen evolution at the anode is also explained by some investigators by increased local-action corrosion during anodic dissolution.

Aluminum is a light-weight metal (d. = 2.71 g/cm³) having good corrosion resistance to the atmosphere and to many aqueous media, combined with good electrical and thermal conductivity. It is very active in the Emf Series, but becomes passive on exposure to water. Although oxygen dissolved in water improves the corrosion resistance of aluminum, its presence is not necessary to observed passivity, indicating that the Flade potential of aluminum is active with respect to the hydrogen electrode. It is usually assumed that the passive film is composed of aluminum oxide, which for air-exposed aluminum is variously estimated at about 20 to 100 A in thickness. The observed corrosion behavior of aluminum is sensitive to small amounts of impurities in the metal; all these impurities, with the exception of magnesium, tend to be cathodic to Aluminum. In general, the high-purity metal is much more

[1] E. Raijola and A. Davidson, *J. Amer. Chem. Soc.*, **78**, 556 (1956).

corrosion resistant than the commercially pure aluminum which, in turn, is usually more resistant than aluminum alloys.* Some of the various aluminum alloys produced commercially in the United States are listed in Table 1. For a complete list, the reference given in the table should be consulted.

Clad Alloys

Pure aluminum is soft and weak; it is alloyed therefore largely to obtain increased strength. In order to make use of the good corrosion resistance of pure aluminum, a high-strength alloy may be sandwiched in between pure aluminum or one of the more corrosion-resistant alloys (e.g., 1% Mn-Al) active in the Galvanic Series with respect to the inner alloy. This combination is called alclad or clad. The clad structure, metallurgically bonded at the two interfaces, provides cathodic protection to the inner alloy by sacrificial action of the outer layers, similar to that provided by a zinc coating on steel. In addition to cathodically protecting against pitting corrosion, the less noble coatings also protect against intergranular corrosion and stress corrosion cracking, especially if the high-strength inner alloy should become sensitized through fabrication procedures or by accidental exposure to high temperatures.

Corrosion in Water and Steam

Aluminum tends to pit in waters containing Cl^-, particularly at crevices or at stagnant areas where passivity breaks down through the action of differential aeration cells. The mechanism is analogous to that described for the stainless steels (p. 310) including an observed critical potential below which pitting corrosion does not initiate.[4, 5] Traces of Cu^{++} (as little as 0.1 ppm) or of Fe^{+3} in water react with aluminum, depositing metallic copper or iron at local sites. The copper or iron, being efficient cathodes, shift the corrosion potential in the noble direction to the critical potential, thereby both initiating pitting and, by galvanic action, stimulating pit growth. For this reason, aluminum

* One exception is known with reference to intergranular disintegration of high-purity aluminum in steam or pure water at temperatures above 125°C (260°F). The presence of iron as an impurity in lower grade metal avoids this type of attack or raises the temperature at which it occurs (>200°C for type 1100).[2, 3]

[2] J. Draley and W. Ruther, *Corrosion*, **12**, 441t, 480t (1956).

[3] T. Pearson and H. Phillips, "Corrosion of Pure Aluminum," *Met. Rev. (London)*, **2**, 305 (1957).

[4] H. Kaesche, *Z. Physik. Chem. N.F.*, **34**, 87 (1962).

[5] H. Böhni and H. Uhlig, *J. Electrochem. Soc.*, **116**, 906 (1969).

TABLE 1
Composition Limits for Some Wrought Aluminum Alloys

Maximum Amounts Specified (%)*

AA No.	Si	Fe	Cu	Mn	Mg	Cr	Zn	Ti
1100	1.0 (Si + Fe)		0.05–0.20	0.05			0.10	
2017	0.8	0.7	3.5–4.5	0.40–1.0	0.20–0.8	0.10	0.25	
2024	0.50	0.50	3.8–4.9	0.30–0.9	1.2–1.8	0.10	0.25	
3003	0.6	0.7	0.05–0.20	1.0–1.5			0.10	
3004	0.30	0.7	0.25	1.0–1.5	0.8–1.3		0.25	
5050	0.40	0.7	0.20	0.10	1.1–1.8		0.25	
5052	0.45 (Si + Fe)		0.10	0.10	2.2–2.8	0.15–0.35	0.10	
5056	0.30	0.40	0.10	0.05–0.20	4.5–5.6	0.05–0.20	0.10	
6053	†	0.35	0.10		1.1–1.4	0.15–0.35	0.10	
6061	0.40–0.8	0.7	0.15–0.40	0.15	0.8–1.2	0.04–0.35	0.25	0.15
6063	0.20–0.6	0.35	0.10	0.10	0.45–0.9	0.10	0.10	0.10
7072	0.7 (Si + Fe)		0.10	0.10	0.10		0.8–1.3	
7075	0.40	0.50	1.2–2.0	0.30	2.1–2.9	0.18–0.35	5.1–6.1	0.20
7079	0.30	0.40	0.40–0.8	0.10–0.30	2.9–3.7	0.10–0.25	3.8–4.8	0.10

Code for major alloying element:

1XXX, > 99.00% Al	4XXX, Si	7XXX, Zn
2XXX, Cu	5XXX, Mg	
3XXX, Mn	6XXX, Si + Mg	

* Except when composition ranges are given.

† 45–65% of the Mg content.

Aluminum Standards and Data, 1970–1971, The Aluminum Association, 750 Third Ave., New York, N. Y. 10017.

is not a satisfactory material for piping to handle potable or industrial waters, which all contain traces of heavy metals ions. On the other hand, for distilled water or for a water from which heavy metal ions have been removed, aluminum is satisfactory. Aluminum piping of high-purity metal or of type 1100 has been used satisfactorily for distilled water lines over many years.

Extraneous anions in dilute chloride solutions act as pitting inhibitors by shifting the critical pitting potential to more noble values.[5] The following ions are effective in decreasing order: nitrate > chromate > acetate > benzoate > sulfate.

In uncontaminated seawater at Panama (tropical environment) commercially pure aluminum (1100) or 0.6% Si, 0.8% Mg, 0.2% Cu-Al alloy (6061-T) corroded at decreasing rates with time. After 16 years' exposure, the total weight loss was 670 and 630 mg/dm^2 and the corresponding observed deepest pits were 0.033 and 0.079 in. (0.084 and 0.20 cm), respectively, for small-size test panels.[6] At the same location in fresh water, probably contaminated with heavy metals, the corresponding weight losses for 16 years were higher—3470 and 1030 mg/dm^2—and the deepest pits measured 0.11 in. (0.28 cm) each.

Various commercial aluminum alloys immersed in seawater at Key West, Fla. for 368 days pitted or not according to their corrosion potentials.[7] Alloys ranging from −0.4 to −0.6 V (s.h.e.) (most of which contained some alloyed copper) showed a mean depth of pitting equal to 0.006 to 0.039 in. (0.015–0.099 cm), whereas those alloys of more active potential (−0.7 to −1.0 V) were essentially not pitted. These potential ranges can be compared with the critical pitting potential of pure aluminum measured in 3% NaCl equal to −0.45 V (see p. 77). Coupling of susceptible alloy panels to a smaller area panel of an active aluminum alloy (see p. 217), which polarized the couple to approximately −0.85 V (s.h.e.), succeeded in largely preventing pitting over the same period of time. These field tests support the concept of critical potentials below which neither aluminum nor its alloys undergo pitting corrosion.

As the temperature increases, the corrosion rate of type 1100 in de-aerated distilled water increases, but the value at 70°C is still very low, being about 0.07 mdd.[2] This rate is reduced to 0.04 mdd in presence of 10^{-3} N H_2O_2 and still lower by 2 ppm $K_2Cr_2O_7$. At 150°C the rate in distilled water is about 0.4 mdd, and at 200°C it is 2.5 mdd, the

[6] C. Southwell, A. Alexander, and C. Hummer, Jr., *Materials Prot.*, **4**, 30 (December 1965).

[7] R. Groover, T. Lennox, Jr., and M. Peterson, *Materials Prot.*, **8**, 25 (November 1969).

attack in all these instances being uniform and without marked pitting. At higher temperatures (e.g., 275°C), but usually only after several days of exposure, the rate becomes more rapid by a factor of 10 to 20 times, and corrosion proceeds mostly along grain boundaries, causing rapid failure. At 315°C complete disintegration to Al_2O_3 may occur within a few hours. It is important to note that this catastrophic corrosion does not occur at the cited high temperatures if the aluminum is coupled to a sufficient area of stainless steel or to zirconium, which are both cathodic to aluminum. The same beneficial effect is obtained by small alloying additions of nickel and iron. For example, the corrosion rate at 350°C for type 1100 aluminum alloyed with 1% Ni is about 18 mdd.[2]

Draley and Ruther[2, 8] explain their findings by formation of hydrogen gas at the metal-oxide interface which supposedly ruptures the protective oxide film. When aluminum is coupled to more cathodic metals or alloyed with nickel and iron, it is assumed that H^+ ions discharge on the cathodic surfaces instead and that consequently the oxide film remains unimpaired. The beneficial effect of cathodic areas, however, can also be explained[9] as one of anodic passivation or anodic protection of aluminum in the same manner as alloying additions or coupling of platinum or palladium to stainless steels or to titanium passivates such metals in acids (p. 69).

At 500 to 540°C the oxidation rate of pure aluminum in steam is appreciably lower than at 300 to 450°C,[10] presumably because a more protective oxide film forms within the higher temperature range.

Effect of pH

Aluminum corrodes more rapidly both in acids and in alkalies compared with distilled water, the rates in acids depending on the nature of the anion. Figure 1 provides data at 70 to 95°C,[2] showing that the minimum rate, when using sulfuric acid for adjustment of pH in the acid region, occurs between pH 4.5 and 7. At room temperature, the minimum rate occurs in the pH range approximating 4 to 8.5. Corrosion rates of aluminum in the alkaline region greatly increase with pH, unlike iron and steel which remain corrosion resistant. The reason for this difference is that Al^{+3} is readily complexed by OH^-, forming

[8] J. Draley and W. Ruther, *J. Electrochem. Soc.*, **104**, 329 (1957).

[9] *The Corrosion of Aluminum and Aluminum Alloys in High Temperature Water*, D. MacLennan, A. McMillan, and J. Greenblatt, 1st Int. Congr. Metallic Corrosion, Butterworths, London, 1961.

[10] R. Hart and J. Maurice, *Corrosion*, **21**, 222 (1965).

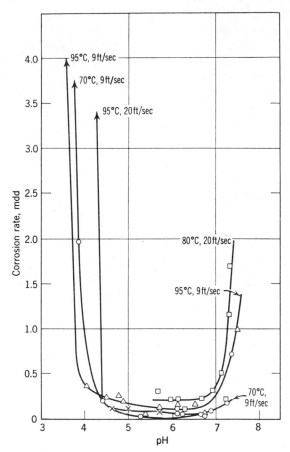

Fig. 1. Effect of pH on corrosion of commercially pure aluminum (no. 1100), aerated solutions; pH, measured at room temperature, was adjusted with H_2SO_4 or NaOH. Most solutions contained $1-10 \times 10^{-5}$ N H_2O_2, 68 ppm $CaSO_4$, 30 ppm $MgSO_4$, 1–2 ppm NaCl (Draley and Ruther).

AlO_2^- in accord with

$$Al + NaOH + H_2O \rightarrow NaAlO_2 + \tfrac{3}{2}H_2 \qquad (1)$$

This reaction proceeds rapidly at room temperature, whereas for iron a similar reaction forming $NaFeO_2$ and Na_2FeO_2 requires concentrated alkali and high temperatures.

Corrosion Characteristics

As mentioned previously, aluminum is characterized by (1) sensitivity to corrosion by alkalies and (2) pronounced attack by traces of Cu ions in aqueous media. In addition, (3) aluminum is subject to rapid attack by mercury metal and mercury ions and (4) by anhydrous chlorinated solvents (e.g., CCl_4, ethylene dichloride, propylene dichloride).[11]

The rate of attack can be appreciable in either dilute or concentrated alkalies. For this reason when aluminum is cathodically protected, overprotection must be avoided in order to ensure against damage to the metal by accumulation of alkalies at the cathode surface. Lime, $Ca(OH)_2$ and some of the strongly alkaline organic amines (but not NH_4OH) are corrosive. Fresh Portland cement contains lime and is also corrosive, hence aluminum surfaces in contact with wet concrete may evolve hydrogen visibly. The corrosion rate is reduced when the cement sets, but continues if the concrete is kept moist or contains deliquescent salts (e.g., $CaCl_2$).

A drop of mercury in contact with an aluminum surface rapidly breaks down passivity accompanied by amalgamation. In the presence of moisture, the amalgamated metal quickly converts to aluminum oxide, causing perforation of piping or sheet. Mercury ions present in solution in only trace amounts similarly accelerate corrosion, producing intolerably high rates of attack.

REACTION WITH CARBON TETRACHLORIDE

Severe accidents have been caused by the rapid reaction of aluminum with anhydrous chlorinated solvents such as in the degreasing of castings, ball milling of aluminum flake with carbon tetrachloride (CCl_4), or even in the case of aluminum used as a container at room temperature for mixed chlorinated solvents. The reaction with CCl_4, for example, has been shown to follow

$$2Al + 6CCl_4 \rightarrow 3C_2Cl_6 + 2AlCl_3 \qquad (2)$$

forming hexachlorethane, C_2Cl_6, with evolution of heat. The corrosion rate for 99.99% Al in boiling anhydrous CCl_4 is very high, e.g., 37,500 mdd (20 ipy).[12] Should the temperature reach the melting point of aluminum, the reaction may proceed explosively. An induction time in the order of 55 min, during which corrosion is negligible, precedes the rapid rate (Fig. 2). This induction time is lengthened either by the presence of water in the CCl_4 (480 min) or by some alloying additions, e.g., Mn or Mg (30 hr for type 5052). On the other hand, addition of $AlCl_3$ or $FeCl_3$ to CCl_4 decreases the induction time to zero without

[11] A. Hamstead, G. Elder, and J. Canterbury, *Corrosion*, **14,** 189t (1958).
[12] M. Stern and H. Uhlig, *J. Electrochem. Soc.*, **99,** 381, 389 (1952).

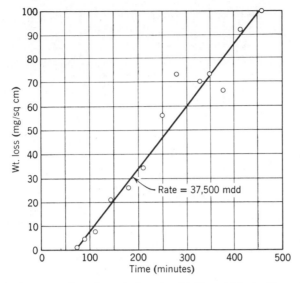

Fig. 2. Weight loss as a function of time for 99.99% Al in boiling, distilled CCl₄ containing 0.0011% H₂O (Stern and Uhlig).

an appreciable effect on the corrosion rate. The corrosion rate of aluminum in water-saturated CCl₄, following a prolonged induction time, is about twice that in the anhydrous solvent (Fig. 3).

In practice, volatile inhibitors are often added to chlorinated solvents in order to suppress the fast reaction. Examples include various ketones, quinones, and amines, which presumably react preferentially with and destroy free radicals (e.g., ·CCl₃ and ·Cl). The latter are considered to play an important part in the reaction mechanism.[13] A proposed sequence of intermediate reactions, for example, is the following:

$$CCl_4 \rightarrow {}^{\cdot}CCl_3 + {}^{\cdot}Cl$$
$$Al + {}^{\cdot}Cl \rightarrow AlCl$$
Initiation

$$AlCl + CCl_4 \rightarrow AlCl_2 + {}^{\cdot}CCl_3$$
$$AlCl_2 + CCl_4 \rightarrow AlCl_3 + {}^{\cdot}CCl_3$$
$$AlCl_3 + CCl_4 \rightarrow CCl_3{}^{+}[AlCl_4]^{-}$$
$$CCl_3{}^{+}[AlCl_4]^{-} \rightarrow AlCl_3 + {}^{\cdot}CCl_3 + {}^{\cdot}Cl$$
Propagation

$$2{}^{\cdot}CCl_3 \rightarrow C_2Cl_6$$
$${}^{\cdot}Cl + {}^{\cdot}CCl_3 \rightarrow CCl_4$$
Termination

[13] M. Stern and H. Uhlig, *J. Electrochem. Soc.*, **100**, 543 (1953).

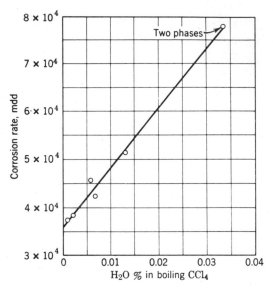

Fig. 3. Effect of water on corrosion rate of 99.99% Al in boiling CCl_4 (Stern and Uhlig).

Water itself may react with and destroy initiating species until the water is consumed by chemical reaction or until sufficient $AlCl_3$ forms. Hence the induction time is extended when water is present in the solvent. By the same token in anhydrous CCl_4, vacuum treatment of aluminum specimens results in a shorter induction time of 5 min compared to 55 min not treated, probably by reducing the H_2O content of the oxide film. Substances like $AlCl_3$ decrease the induction time by forming a complex with CCl_4 which later dissociates into free radicals taking part in the chain reaction. Similarly, alloying elements like magnesium are considered to react preferentially with free radicals, accounting for the observed improved corrosion resistance of magnesium-aluminum alloys to CCl_4 compared to pure aluminum. The behavior of alloying constituents in general (often opposite to their effect in aqueous environments), the observed lack of effect of galvanic coupling or of an impressed nominal voltage on the rate, and the same observed corrosion rate in the vapor phase as in the boiling liquid, are strong arguments that the reaction does not follow an electrochemical mechanism.[14] The red color of CCl_4 on reacting with aluminum (probably

[14] For a different point of view, the reader should consult R. Brown, E. Cook, M. Brown, and S. Minford, *J. Electrochem. Soc.*, **106**, 192 (1959).

˙CCl₃) and the ease with which many organic substances suppress the fast reaction (free radicals are very reactive) support the free-radical mechanism.

In summary, aluminum (type 1100) is resistant to:

1. Hot or cold NH_4OH.
2. Hot or cold acetic acid. Highest rates of attack are for the boiling 1 to 2% acid. Ninety-nine percent acetic acid is not appreciably corrosive, but removal of the last 0.5% H_2O increases attack over a hundredfold. Formic acid and Cl^- contamination also increase attack. Aluminum is resistant to citric, tartaric, and malic acids.
3. Fatty acids. Aluminum equipment is used for distillation of fatty acids.
4. Nitric acid, >80% up to about 50°C (120°F).[15] See Fig. 4.

[15] E. Cook, R. Horst, and W. Binger, *Corrosion*, **17**, 25t (1961).

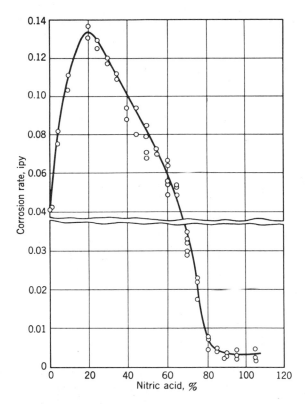

Fig. 4. Corrosion rates of commercially pure aluminum (no. 1100) in nitric acid, room temperature (E. Cook, R. Horst, and W. Binger).

5. Distilled water.

6. Atmospheric exposure. Excellent resistance to rural, urban, and industrial atmospheres; lesser resistance to marine atmospheres.

7. Sulfur, sulfur atmospheres, H_2S. Aluminum is used for mining of sulfur.

8. Fluorinated refrigerant gases, (e.g., Freon), but *not* to methyl chloride or bromide.

Aluminum is not resistant to:

1. Strong acids, such as HCl and HBr (dilute or concentrated), H_2SO_4 (satisfactory for special applications at room temperature below 10%), HF, $HClO_4$, H_3PO_4, and formic, oxalic, and trichloroacetic acids.

> High-purity Aluminum corrodes intergranularly in concentrated hydrochloric acid at a rate that depends on rate of cooling from about 600°C (1100°F) and on the amount of iron present as an impurity. Perryman[16] reported that 0.009% Fe-Al was more susceptible to intergranular corrosion when furnace cooled instead of water quenched, but that the reverse situation applied to aluminum containing 0.02 to 0.055% Fe.

2. Alkalies. Lime and fresh concrete are corrosive, as well as strong alkalies, e.g., NaOH and the very alkaline organic amines. Corrosion by soap solutions is inhibited by the addition of a few tenths per cent of sodium silicate (not effective for strong alkalies).

3. Mercury and mercury salts.

4. Seawater. Pitting occurs at crevices and surface deposits, especially when trace amounts of heavy metal ions are present.

5. Waters containing heavy metal ions, (e.g., mine waters or waters previously passing through copper, brass or ferrous piping).

6. Chlorinated solvents.

7. Anhydrous ethyl, propyl, or butyl alcohols at elevated temperatures. A trace of water acts as an inhibitor.[15]

8. Contact with wet woods, in particular beech wood.[17] Any wood impregnated with copper preservatives is especially damaging.

For behavior in soils, see p. 180.

Galvanic Coupling

Cadmium has the nearest potential to that of aluminum in many environments, hence cadmium-plated steel screws, bolts, trim etc., can be used in direct contact with aluminum. Tin coatings are also said to be satisfactory. Zinc is somewhat further removed in potential, but

[16] E. Perryman, *Trans. Amer. Inst. Min. Met. Eng., J. Metals,* **5,** 911 (1953).
[17] R. Farmer, *Metallurgia,* **68,** 161 (1963).

it is also usually satisfactory. Zinc is anodic to and hence cathodically protects aluminum in neutral or acid media (see p. 224). In alkalies, however, the polarity reverses and zinc accelerates the corrosion of aluminum. Magnesium is anodic to aluminum, but the potential difference and the resultant current flow when the metals are coupled (e.g., in seawater) is so high that aluminum may be overprotected cathodically with resultant damage to aluminum. Aluminum is damaged in this respect to a lesser extent when alloyed with magnesium. High-purity aluminum, it is said, can be coupled to magnesium without damage to either metal,[18] because galvanic currents are reduced in absence of iron, copper, and nickel impurities which act as efficient cathodes.

It is imperative that aluminum never be coupled to copper or copper alloys because of the resulting damage to aluminum. In this connection it is also important to avoid contact of aluminum with rain water that has washed over copper flashing or gutters because of the damage caused by small amounts of Cu^{++} dissolved in such water.

In rural areas, coupling of steel to aluminum is usually satisfactory, but in sea-coast areas attack of aluminum is accelerated. In fresh waters, aluminum may be anodic or cathodic to steel, depending on small composition differences of the water.

Aluminum Alloys

The usual alloying additions to aluminum in order to improve physical properties include Cu, Si, Mg, Zn, and Mn. Of these, manganese may actually improve the corrosion resistance of wrought and cast alloys. One reason is that the compound $MnAl_6$ forms and takes iron into solid solution. The compound $(MnFe)Al_6$ settles to the bottom of the melt, in this way reducing the harmful influence on corrosion of small quantities of alloyed iron present as an impurity.[19] No such incorporation occurs in the case of cobalt, copper, and nickel, so that manganese additions would not be expected to counteract the harmful effects of these elements on corrosion behavior.

The Duralumin alloys (e.g., types 2017, 2024) contain several percent copper, deriving their improved strength from the precipitation of $CuAl_2$ along slip planes and grain boundaries. Copper is in solid solution above the homogenizing temperature of about 480°C (900°F), which on quenching remains in solution. Precipitation takes place slowly at room temperature with progressive strengthening of the alloy. Should the

[18] M. Bothwell, *J. Electrochem. Soc.,* **106,** 1014 (1959).
[19] M. Whitaker, *Metal Ind.,* **80,** 1 (May 9, 1952).

alloy be quenched from solid-solution temperatures into boiling water, or, after quenching should it be heated (artificially aged) above 120°C (250°F), the compound $CuAl_2$ forms preferentially along the grain boundaries. This results in a depletion of copper in the alloy adjacent to the intermetallic compound, accounting for grain boundaries anodic to the grains, and a marked susceptibility to *intergranular corrosion*. Prolonged heating (overaging) restores uniform composition alloy at grains and grain interfaces, thereby eliminating susceptibility to this type of corrosion, but at some sacrifice of physical properties. In practice the alloy is quenched from about 490°C (920°F), followed by room-temperature aging.

STRESS CORROSION CRACKING

Pure aluminum is immune to stress corrosion cracking (s.c.c.). Should a Duralumin alloy, on the other hand, be stressed in tension in the presence of moisture, it may crack along the grain boundaries. Sensitizing the alloy by heat treatment, as described previously, makes it more susceptible to this type of failure. In aging tests at 160 to 205°C (320 to 400°F), maximum susceptibility was observed at times somewhat short of maximum tensile strength.[20] Hence in heat-treatment procedures, it is better practice to aim at a slightly overaged rather than an underaged alloy.

Magnesium alloyed with aluminum increases susceptibility to s.c.c., especially when added in amounts >4.5%. To avoid such failures, slow cooling (50°C/hr) from the homogenizing temperature is necessary supposedly in order to coagulate the β phase (Al_3Mg_2), a process which is promoted by addition of 0.2% Cr to the alloy.[21] Edeleanu[22] showed that cathodic protection stopped the growth of cracks that had already progressed into the alloy immersed in 3% NaCl solution. On aging the alloy at low temperatures, maximum susceptibility to s.c.c. occurred before maximum hardness values were reached. This behavior paralleled that of a Duralumin alloy cited previously. Accordingly, Edeleanu proposed that the susceptible material along the grain boundary which caused cracking was not the equilibrium β phase responsible for hardness, but instead was made up of Mg atoms segregating at the grain boundary before the formation of the intermetallic compound. On this basis, susceptibility to stress corrosion cracking decreased on

[20] W. Robertson, *Trans. Amer. Inst. Mining Met. Eng.*, 166, 216 (1946); see also U. Evans, *The Corrosion and Oxidation of Metals*, p. 674, Ed. Arnold, London, 1960.

[21] P. Brenner, *Metal Progr.*, 65, 112 (1954).

[22] C. Edeleanu, *J. Inst. Metals*, 80, 187 (1951).

continued aging of the alloy because the separated β phase consumed the original segregated grain-boundary material responsible for susceptibility. A similar mechanism probably applies to the copper-aluminum alloy series.

High concentrations of zinc in aluminum (4–20%) also induce susceptibility to cracking of the stressed alloys in presence of moisture. Traces of H_2O, for example, contained in the surface oxide film are sufficient to cause cracking; carefully baked-out specimens in dry air do not fail.[23] Oxygen is not necessary, nor is a liquid aqueous phase required. These conditions, plus the susceptibility of a high-strength aluminum alloy (7075) to failure in organic solvents,[24] suggest that the mechanism of failure is one of stress-sorption cracking caused by adsorbed water or organic molecules on appropriate defect sites of the strained alloy.

Many high-strength aluminum alloys are available (some are listed in Table 1); specific composition ranges and heat treatments for these alloys are usually chosen with the intent of minimizing susceptibility to s.c.c. Service temperatures—especially those above room temperature—which can cause artificial aging, sometimes induce susceptibility followed by premature failure in the presence of moisture or sodium chloride solutions. Susceptibility of any of the wrought alloys is greatest when stressed at right angles to the rolling direction (in the short transverse direction), probably because more grain-boundary area of elongated grains along which cracks propagate comes into play.

As mentioned earlier, cladding of alloys can serve to cathodically protect them from either intergranular corrosion or s.c.c. Compressive surface stresses are effective for avoiding s.c.c., hence practical structures are sometimes shot peened.

MAGNESIUM

$$Mg \rightarrow Mg^{++} + 2e^- \qquad E^\circ = 2.34 \ V$$

Evidence has been obtained by several investigators that magnesium dissolves anodically to a large extent as Mg^+ and that this ion in solution reduces water to form Mg^{++} plus H_2 (see p. 222).

Magnesium is the most active metal in the Emf Series used for structural purposes. Its low density (1.7 g/cm^3) makes it useful wherever weight is a factor. It becomes passive on exposure to water in the presence or absence of oxygen. The corrosion rate is only slightly

[23] G. Wassermann, *Z. Metallk.*, **34**, 297 (1942).

[24] H. Paxton and R. Procter, in *Fundamental Aspects of Stress Corrosion Cracking*, p. 509, edited by R. Staehle et al., Nat. Assoc. Corros. Engrs., Houston, Texas (1969).

affected by dissolved oxygen; instead, corrosion is attended, for the most part, by hydrogen evolution.

Corrosion resistance depends on purity of the metal even more than in the case of aluminum. Distilled magnesium corrodes in seawater, for example, at the rate of 0.01 ipy, which is about twice the rate for

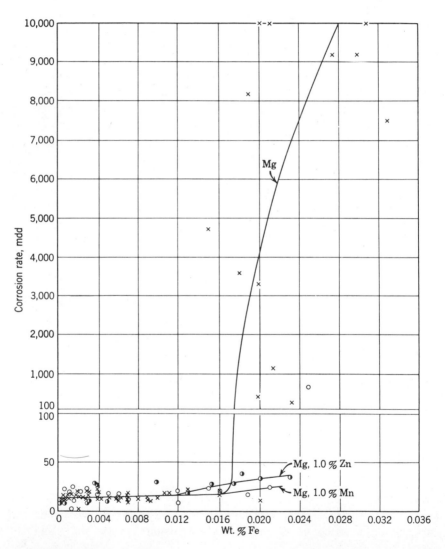

Fig. 5.   Corrosion of magnesium in 3% NaCl, alternate immersion, 16 weeks, showing tolerance limit for iron and beneficial effect of alloyed zinc and manganese (Hanawalt, Nelson, and Peloubet).

iron, but commercial magnesium corrodes at a rate 100 to 500 times higher, with visible hydrogen evolution. The impurities in commercial magnesium largely responsible for the higher rate are iron and, to a lesser extent, nickel and copper. Their effect becomes marked above a critical concentration called the tolerance limit.[25] For iron the tolerance limit is 0.017%; similarly for nickel and copper it is 0.0005% and 0.1%, respectively (Fig. 5). Additions of manganese or zinc to the metal raise these limits to higher values. Manganese is said to surround iron particles and to render them inactive as local cathodes.[26] An additional mechanism appears to involve formation of an insoluble intermetallic compound which includes iron in its structure. This is heavier than magnesium and hence, parallel to the situation for aluminum (p. 345), it settles to the bottom of the melt, in this way reducing the actual Fe content of the cast metal.

Major limitations in the use of magnesium alloys for structural parts are:

1. Sensitivity to stress corrosion cracking in moist air. The 1.5% Mn and the 3% Zn, 0.7% Zr alloys are stated to be among the magnesium alloys that are relatively resistant.[27] Unlike aluminum alloys, crack paths are typically transgranular.

2. Tendency of magnesium and its alloys to corrode when galvanically coupled to other metals. Aluminum alloys containing some magnesium (e.g., types 5050, 5052, 5056) are least affected by alkalies generated by a magnesium-aluminum couple, and hence can be applied in contact with magnesium. Pure aluminum is also satisfactory. In general, however, magnesium must be insulated from all dissimilar metals, or large insulating washers must be used under bolt or screw heads in order to extend the path of electrolyte resistance and thereby minimize galvanic action.

Various protective anodized coatings are available as produced in electrolytes composed mainly of fluorides, phosphates, or chromates.[28]

Magnesium, in brief, is resistant to:

1. Atmospheric exposure. The controlled purity 3% Al, 1.5% Zn-Mg alloy exposed for eight years to the tropical marine atmosphere at Panama lost 0.094 in. (0.24 cm) per year, compared to 0.004 in. (0.01

[25] J. Hanawalt, C. Nelson, and J. Peloubet, *Trans. Amer. Inst. Mining Met. Eng., Inst. Metals Div.,* **147**, 273 (1942).

[26] H. Robinson and P. George, *Corrosion,* **10**, 182 (1954).

[27] E. Perryman, *J. Inst. Metals,* **78**, 621 (1951).

[28] E. Emley, *Principles of Magnesium Technology,* pp. 692–705, Pergamon Press, New York, 1966.

cm) per year at the temperate marine atmosphere of Kure Beach, N.C.[6] Most alloys must be stress relieved, avoiding subsequent high applied stress in order to avoid stress corrosion cracking in moist atmospheres.

2. Distilled water. Same precautions as for 1.

3. >2% HF. A protective film of magnesium fluoride forms. Pitting may occur at the water-air interface. Corrosion rate of 8% Al, 0.2% Mn alloy in 5% HF is 23 mdd; in 48% HF it is 0.5 mdd.

4. Alkalies. Unlike aluminum, magnesium is resistant to alkalies. The rate in 48% NaOH–4% NaCl is 2 mdd (high-purity Mg). Above 60°C (140°F) the rate increases appreciably.

Magnesium is not resistant to:

1. Waters containing traces of heavy metal ions.
2. Seawater.
3. Inorganic or organic acids and acid salts (e.g., NH_4 salts).
4. Methanol (anhydrous); magnesium methylate forms. The reaction can be inhibited by $(NH_4)_2S$, H_2CrO_4, turpentine, or dimethyglyoxime. The higher alcohols are resisted satisfactorily.
5. Leaded gasoline mixtures.
6. Freon (CCl_2F_2) plus water. The anhydrous gas is not corrosive.

GENERAL REFERENCES

ALUMINUM

R. Mears, in *Corrosion Handbook,* edited by H. H. Uhlig, pp. 39–56, Wiley, New York, 1948.

M. Whitaker, *Metal Ind.,* **80,** 1 (1952).

F. Haynie and W. Boyd, *Stress Corrosion Cracking of Al Alloys,* DMIC Rept., Battelle Mem. Inst., Columbus, Ohio, 1966.

MAGNESIUM

W. Loose, in *Corrosion Handbook,* edited by H. H. Uhlig, pp. 218–252, Wiley, New York, 1948.

C. Sheldon Roberts, *Magnesium and its Alloys,* Wiley, New York, 1960.

H. Godard, W. Jepson, M. Bothwell, and R. Kane, *The Corrosion of Light Metals,* Wiley, New York, 1967.

chapter 21
Lead

$$Pb \rightarrow Pb^{++} + 2e^- \qquad E^\circ = 0.126 \text{ V}$$

Lead is a relatively active metal in the Emf Series, becoming passive in many corrosive media that form insoluble lead compounds (e.g., H_2SO_4, HF, H_3PO_4, H_2CrO_4) by reason of thick diffusion-barrier coatings (Def. 2, p. 61). In these acids, corrosion resistance is good provided relative velocity of metal and acid is below the value that causes erosion of protective films. Lead finds varied uses, therefore, in the chemical industry as lining and piping.

The metal is corroded by dilute nitric acid and by several dilute aerated organic acids (e.g., acetic and formic). In contact with certain green woods, (e.g., Douglas fir or oak), which slowly exude volatile acids, corrosion can be serious. Woods not causing damage in this respect are seasoned cedar or hemlock.[1]

Being amphoteric, lead is corroded by alkalies at moderate or high rates, depending on aeration, temperature, and concentration. It is attacked, for example, by calcium hydroxide solutions at room temperature, including waters that have been in contact with fresh Portland cement.

An alloy of 2% Ag-Pb is used as a corrosion-resistant anode in the cathodic protection of marine structures (see p. 221). Alloying with 6 to 12% Sb increases strength [only at temperatures $<120°C$ ($<250°F$)] of the otherwise weak metal, but corrosion resistance of the alloy in some media is below that of pure lead.

Lead is resistant to seawater. It is also durable for use in contact with fresh waters; however, the toxic properties of trace amounts of

[1] G. Hiers, in *Corrosion Handbook*, p. 216.

lead salts make it mandatory to exclude its use, and the use of its alloys, for soft potable waters, carbonated beverages, and all food products. The rate of corrosion in aerated distilled water is high (approximately 90 mdd; see Fig. 2, p. 210, *Corrosion Handbook*) and the rate increases with concentration of dissolved oxygen. In the absence of dissolved oxygen, the corrosion rate in waters or dilute acids is either negligible or very low.

Lead is resistant to atmospheric exposures, particularly to industrial atmospheres in which a protective film of lead sulfate forms. Buried underground, the corrosion rate may exceed that of steel in some soils (e.g., those containing organic acids), but in soils high in sulfates the rate is low. Soluble silicates, which are components of many soils and natural waters, also act as effective corrosion inhibitors.

In applications where thermal cycling occurs, the high coefficient of expansion (30×10^{-6}/°C) of lead may cause intergranular cracking due to fatigue or corrosion fatigue.

In summary, lead is resistant to:

1. Many strong acids.
   a. H_2SO_4, $<96\%$, room temperature; the rate for $<80\%$ H_2SO_4, boiling, is <0.08 ipy; for 20% H_2SO_4, boiling, the rate is 0.003 ipy. (See Fig. 4, p. 212, *Corrosion Handbook*.)
   b. Commercial H_3PO_4 (hot or cold).
   c. H_2CrO_4 (as used in chromium plating).
   d. HF <60 to 65% (room temperature).
   e. H_2SO_3.
2. Atmospheric exposures, particularly industrial.
3. Seawater (0.0005 ipy).
4. Chlorine, wet or dry, <100°C (but only dry Br_2 and at lower temperatures), SO_2, SO_3, H_2S.

Lead is not resistant to:

1. HNO_3, $<70\%$.
2. HCl.
3. Concentrated H_2SO_4 ($>96\%$, room temperature).
4. Alkalies. For some chemical applications, the rate in caustic alkalies is considered tolerable.
5. HF, gaseous.
6. Many aerated organic acids.

chapter 22

Nickel and nickel alloys

$$\text{Ni} \rightarrow \text{Ni}^{++} + 2e^- \qquad E^\circ = 0.250 \text{ V}$$

Nickel is active in the Emf Series with respect to hydrogen, but noble with respect to iron. It does not react rapidly with dilute nonoxidizing acids (e.g., H_2SO_4, HCl) unless dissolved oxygen is present. It is thermodynamically inert to (will not corrode in) deaerated water at room temperature in which Ni(OH)_2 forms as corrosion product. It is passive in contact with many aerated aqueous solutions, but the passive film is not so stable as, for example, that on chromium (for nickel, Flade potential $\phi_F^\circ = 0.2$ V).[1] It corrodes by pitting when exposed to seawater (passive-active cells).

Nickel is outstandingly resistant to hot or cold alkalies. Only silver and possibly zirconium are more resistant. Nickel exposed to boiling 50% NaOH corrodes at the rate of 0.6 mdd (0.0001 ipy). It also resists fused NaOH, low-carbon nickel being preferred for this application in order to avoid intergranular attack of the stressed metal; stress relief anneal, 5 min at 875°C (1100°F) is advisable. Nickel is attacked by aerated aqueous ammonia solutions, a soluble complex $\text{Ni(NH}_3)_6^{++}$ forming as a corrosion product. It is also attacked by strong hypochlorite solutions with formation of pits. A small amount of sodium silicate acts as inhibitor.[2]

Nickel is not subject to stress corrosion cracking except in high concentrations of alkali or in fused alkali, as mentioned previously.

Nickel has excellent resistance to oxidation in air up to 800 to 875°C (1500–1600°F) and is often used in service at still higher temperatures.

---

[1] J. Osterwald and H. Uhlig, *J. Electrochem. Soc.*, **106**, 515 (1961).

[2] *Corrosion Handbook*, p. 263.

When subject to alternate oxidizing and reducing atmospheres at elevated temperatures, it tends to oxidize along grain boundaries. It also fails along grain boundaries in presence of sulfur-containing environments above about 315°C (600°F). Fused salts contaminated with sulfur, or with sulfates in the presence or absence of organic or other chemically reducing impurities, may damage nickel or high nickel alloys in this manner.

In summary, nickel is resistant to:

1. Alkalies, hot or cold. Including fused alkalies.
2. Dilute nonoxidizing inorganic and organic acids. Resistance is improved if acids are deaerated.
3. The atmosphere. In industrial atmospheres, a nonprotective film forms, composed of basic nickel sulfate (fogging). Fogging is minimized by a thin chromium electroplate over nickel. Good resistance to oxidation in air at elevated temperatures.

Nickel is not resistant to:

1. Oxidizing acids (e.g., HNO_3).
2. Oxidizing salts (e.g., $FeCl_3$, $CuCl_2$, $K_2Cr_2O_7$).
3. Aerated ammonium hydroxide.
4. Alkaline hypochlorites.
5. Seawater.
6. Sulfur or sulfur-containing reducing environments, $>315°C$ ($>600°F$).

NICKEL ALLOYS

General Behavior

Nickel alloyed with *copper* improves, to a moderate degree, resistance under reducing conditions (e.g., nonoxidizing acids). Also, in line with the resistance of copper to pitting, the tendency of nickel-copper alloys to form pits in seawater is less pronounced than for nickel, and pits tend to be shallow rather than deep. Above about 60 to 70 at. % Cu (62–72 wt. %), the alloys lose the passive characteristics of nickel and tend to behave more like copper (see p. 83), retaining, however, appreciably improved resistance to impingement attack. Hence 10 to 30% Ni, bal. Cu alloys (cupro-nickels) do not pit in stagnant seawater and stand up well in fast-moving seawater. Such alloys containing a few tenths to 1.75% Fe, which still further improves resistance to impingement attack, find application for seawater condenser tubes. The 70%

Ni-Cu alloy (Monel), on the other hand, pits in stagnant seawater and is best used only in rapidly moving, aerated seawater to ensure uniform passivity. It does not pit under conditions that provide cathodic protection, such as when the alloy is coupled to a more active metal (e.g., iron).

Additions of *chromium* to nickel impart resistance to oxidizing conditions (e.g., HNO_3, H_2CrO_4). The critical minimum chromium content obtained from critical current densities for anodic passivation in sulfuric acid is 14 wt. % Cr.[3] At the same time, the alloys become more sensitive than nickel to attack by Cl⁻ and by HCl, and deep pits form when the alloys are exposed to stagnant seawater. Chromium also imparts to nickel improved resistance to oxidation at elevated temperatures. One important commercial alloy contains 20% Cr, 80% Ni (see p. 204).

Alloying nickel with molybdenum provides, to a marked degree, improved resistance under reducing conditions. The corrosion potentials of such alloys in acids, whether aerated or deaerated, tend to be more active than their Flade potentials,[4, 5] hence the alloys are not passive in the sense of Def. 1 (p. 61). For example, the corrosion potentials of nickel alloys containing 3 to 22.8% Mo in 5% H_2SO_4, hydrogen-saturated, all lie within 2 mV of a platinized platinum electrode in the same solution.[4] Notwithstanding an active corrosion potential, the alloy containing 15% Mo, for example, corrodes at $\frac{1}{12}$ the rate of nickel in deaerated 10% HCl, 70°C, and the rate decreases still further with increasing molybdenum content ($\frac{1}{100}$ the rate for nickel at 25% Mo). Polarization measurements show that modybdenum alloyed with nickel has little effect on hydrogen overvoltage but increases anodic polarization instead; hence the corrosion rate of the alloy is anodically controlled. The mechanism of increased anodic polarization is related most likely to a sluggish hydration of metal ions imparted by molybdenum, or alternatively to a porous diffusion-barrier film of molybdenum oxide, rather than to formation of a passive film typical of chromium or the passive chromium-nickel alloys.

Because the binary molybdenum-nickel alloys have poor physical properties (low ductility, poor workability), other elements, e.g., iron, are combined to form ternary or multicomponent alloys. These are still difficult to work, but they mark an improvement over the binary alloys. Resistance of such alloys to hydrochloric and sulfuric acids is better than that of nickel, but it is not improved with respect to

[3] P. Bond and H. Uhlig, *J. Electrochem. Soc.*, **107**, 488 (1960).
[4] H. Uhlig, P. Bond, and H. Feller, *J. Electrochem. Soc.*, **110**, 650 (1963).
[5] G. Masing and G. Roth, *Werkst. Korros.*, **3**, 176, 253 (1952).

oxidizing media (e.g., HNO_3). Since the Ni-Mo-Fe alloys have active corrosion potentials and are not able therefore to establish passive-active cells, they do not pit in the strong acid media to which they are usually exposed in practice.

By alloying nickel with both molybdenum and chromium, an alloy is obtained resistant to oxidizing media imparted by alloyed chromium, as well as to reducing media imparted by molybdenum. One such alloy, which also contains a few per cent iron and tungsten (Hastelloy C), is immune to pitting and crevice corrosion in seawater (10-year exposure) and does not tarnish appreciably when exposed to marine atmospheres. Alloys of this kind, however, despite improved resistance to Cl⁻, corrode more rapidly in hydrochloric acid than do the nickel-molybdenum alloys free of chromium.

Nominal compositions of some commercial nickel-base alloys containing copper, molybdenum, or chromium are given in Table 1. The Ni-Cu alloys are readily rolled and fabricated whereas the Ni-Cr alloys are less readily, and the Ni-Mo-Fe and Ni-Mo-Cr alloys are difficult to work or fabricate.

Some commercial Cr-Ni-Fe-Mo alloys corresponding in composition to high-nickel stainless steels also contain a few per cent copper. They are designed to resist, among other media, sulfuric acid over a wide range of concentrations. The function of alloyed copper is the same as that of alloyed palladium in titanium mentioned on p. 69, namely, to hasten the cathodic reaction (H⁺ or O_2 reduction) to the point where the anodic current density reaches or exceeds the critical value for anodic passivation.

70% Ni, 30% Cu Alloy (Monel)

Since Monel is resistant to high-velocity seawater, it is often used for valve trim and pump shafts. It is also used for industrial hot fresh-water tanks and for varied equipment employed in the chemical industry. It resists boiling sulfuric acid in concentrations less than 20%, the corrosion rate being <0.008 ipy (23-hr test).[6] It is outstandingly resistant to unaerated HF at all concentrations and temperatures up to boiling. [Rate in N_2-saturated 35% HF, 120°C (248°F) is 0.001 ipy; when air saturated, 0.15 ipy.][7] Resistance to alkalies is good, *except* in hot concentrated caustic solutions or aerated NH_4OH.

The alloy is not resistant to oxidizing media (e.g., HNO_3, $FeCl_3$, $CuSO_4$, H_2CrO_4) nor to wet Cl_2, Br_2, SO_2, NH_3. Stress corrosion crack-

[6] *Corrosion Handbook*, p. 271.
[7] *Corrosion Handbook*, p. 273.

TABLE 1
Nominal Composition (%) of Commercial Wrought Nickel-Base Alloys

	Ni	Mo	Co	Cr	W	Fe	Si	Mn	C	
Monel, Alloy 400*	66					1.4	0.2	0.9	0.12	31.5% Cu
Inconel, Alloy 600*	76			16		7.2	0.2	0.2	0.04	
Hastelloy B†	61 approx.	26–30	2.5 max	1.0 max		4.0–7.0	1.0 max	1.0 max	0.05 max	
Hastelloy C†	54 approx.	15–17	2.5 max	14.5–16.5	3–4.5	4–7	1.0 max	1.0 max	0.08 max	
Hastelloy F†	46 approx.	5.5–7.5	2.5	21–23	1.0 max	20	1.0 max	1–2	0.05 max	Cb + Ta, 1.75–2.5%

* Trademark, International Nickel Co.
† Trademark, Haynes Stellite Co.

ing has been known to occur in fluosilicic acid and in concentrated hot NaOH.[8]

76% Ni, 16% Cr, 7% Fe Alloy (Inconel)

The resistance of Inconel is good to oxidizing aqueous media [e.g., mine waters, $Fe_2(SO_4)_3$, $CuSO_4$, HNO_3]. In nitric acid, resistance is best above 20% HNO_3, including red fuming acid, but is not so good in general as the stainless steels. Resistance to alkalies is good, except in concentrated hot caustic solutions. It is also resistant to all concentrations of NH_4OH at room temperature. Like the stainless steels, this alloy tends to pit in seawater and also in oxidizing metal chlorides (e.g., $FeCl_3$). The alloy does not stress corrosion crack in boiling magnesium chloride nor in many other chemical media commonly encountered. Intergranular corrosion has been known to occur in boiling 25% HNO_3[9] and in hot caustic solutions. It has also been observed in contaminated high-temperature water; it is still not certain whether such attack also occurs in pure water (see p. 306). Susceptibility to intergranular corrosion in nitric acid, but not necessarily in other media, can be minimized by heat treatment of the cold-worked alloy at 900°C (1650°F).[9]

Inconel finds extensive use where oxidation resistance at elevated temperatures is required (see p. 204).

60% Ni, 30% Mo, 5% Fe Alloy (Hastelloy B)

Hastelloy B and commercial alloys of similar composition are usefully resistant to hydrochloric acid of all concentrations and temperatures up to the boiling point (see Hast. B, Fig. 1). [Rate = 0.009 ipy in boiling 10% HCl; 0.02 ipy in boiling 20% HCl; 0.002 ipy in 37% HCl at 65°C (150°F).][10] Resistance to boiling sulfuric acid is good up to 60% H_2SO_4 (<0.007 ipy). In phosphoric acid, rate of attack is low for all concentrations and temperatures, the highest rate for the pure acid applying to the boiling 86% H_3PO_4 (0.03 ipy). In various organic acids, hot or cold, resistance is also good.

The alloy is not resistant to oxidizing conditions, e.g., to HNO_3, or to oxidizing metal chlorides, e.g., $FeCl_3$.

---

[8] O. Fraser, in *Symp. Stress Corrosion Cracking of Metals,* p. 458, ASTM-AIME, Philadelphia, Pa., 1945.

[9] *Effect of Nickel Content on Resistance to Stress Corrosion Cracking of Fe-Ni-Cr Alloys in Chloride Environments,* H. Copson, p. 328, 1st Int. Congr. Metallic Corrosion, London, Butterworths, 1961.

[10] Manufacturer's literature.

It is subject to intergranular attack when heated in the range 500 to 700°C (930–1300°F). Proper heat treatment to avoid such attack consists of annealing at 1150 to 1175°C (2100–2150°F), followed by rapid cooling in air or water.

54% Ni, 15% Cr, 16% Mo, 4% W, 5% Fe Alloy (Hastelloy C)

Because of its chromium content, Hastelloy C is resistant to such oxidizing media as HNO_3, HNO_3-H_2SO_4 mixed acids, H_2CrO_4, $Fe_2(SO_4)_3$, $FeCl_3$, and $CuCl_2$. Corrosion rates fall below 0.02 ipy in <50% HNO_3 at 65°C (150°F), but are higher above this concentration. In boiling >15% HNO_3 (see Hast. C, Fig. 1), rates are high. Resistance is excellent to wet or dry Cl_2 at room temperature; pitting in wet Cl_2 may occur at higher temperatures. It also resists wet or dry SO_2 up to about 70°C (155°F).[10] The outstanding resistance of Hastelloy C to pitting or crevice attack in seawater has already been mentioned. Resistance to acetic acid is excellent. In boiling 40% formic acid, the rate is 0.01 ipy.

It has good resistance to hydrochloric acid at room temperature (0.001 ipy for 37% HCl), but not at higher temperatures [0.2 ipy for 20% HCl, 65°C (150°F)]. It is recommended for boiling sulfuric acid up to 10% acid (0.01 ipy) and for boiling phosphoric acid up to at least 50% acid (0.01 ipy).

When heated in the temperature range 500 to 700°C (930–1300°F) the alloy tends to corrode intergranularly. This is avoided by a final heat treatment at 1210 to 1240°C (2210–2260°F), followed by rapid cooling in air or water.[11]

A different alloy containing less molybdenum, more chromium and iron, and about 2% Cb + Ta is also resistant under oxidizing and reducing conditions (see Hast. F, Fig. 1 and Table 1). It is used for sulfite pulp digesters because of good resistance to H_2SO_3 and SO_2. It is resistant to stress-corrosion cracking, consistent with a high nickel content. Heat treatment of the alloy should follow the same anneal-quench procedure outlined in the preceding paragraph.

GENERAL REFERENCES

Corrosion Handbook, edited by H. H. Uhlig, pp. 253–298, Wiley, New York, 1948.
Corrosion Resistance of Metals and Alloys, edited by F. LaQue and H. Copson, pp. 467–551, Reinhold, New York, 1963.

---

[11] See also C. Samans, A. Meyer, and G. Tisinai, *Corrosion,* **22,** 336 (1966).

chapter 23

Titanium, zirconium, tantalum

TITANIUM

Titanium is a relatively high melting point metal [m.p. = 1668°C (3035°F) d. = 4.5 g/cm³] with high strength-to-weight ratio. It is active in the Emf Series, the calculated standard oxidation potential for $Ti \rightarrow Ti^{++} + 2e^{-}$ being 1.63 V. The metal is readily passivated in aerated aqueous solutions, including dilute acids and alkalies. Its galvanic potential in seawater is near the noble value for stainless steels. The Flade potential region is relatively active,[1, 2] ($\phi_F^{\circ} \approx -0.05$ V) indicating stable passivity. Only in strong acids or alkalies does passivity break down, accompanied by appreciable corrosion.

Titanium is markedly resistant to oxidizing media in presence of Cl^{-}, e.g., heavy metal chlorides ($FeCl_3$), aqua regia at room temperature, or wet chlorine. It is better resistant to nitric acid at elevated temperatures than are the stainless steels. It resists alkalies at room temperature, but is attacked by hot concentrated or fused sodium hydroxide.

On exposure of the metal for one or more hours, stressed or unstressed, to fuming HNO_3 containing 2.5 to 28% NO_2 and not more than 1.25% H_2O, a dark substance forms over the surface (97.5% Ti metal by x-ray analysis).[3] In the dry state, this surface film is pyrophoric when scratched, and it explodes when abraded in contact with concentrated HNO_3. The 8% Mn-Ti alloy is especially sensitive in this regard.

Titanium, presumably as sponge, in contact with liquid oxygen is reported to be sensitive to detonation by impact.[4]

---

[1] M. Stern and H. Wissenberg, *J. Electrochem. Soc.*, **106**, 755 (1959).

[2] N. Thomas and K. Nobe, *J. Electrochem. Soc.*, **116**, 1748 (1969).

[3] J. Rittenhouse, *Trans. Amer. Soc. Metals*, **51**, 871 (1951).

[4] T.M.L. Report No. 88, Battelle Mem. Inst., Columbus, Ohio, November, 1957; *Metals Handbook*, 8th ed., vol. 1, p. 1152, Amer. Soc. Metals, Cleveland, Ohio, 1961.

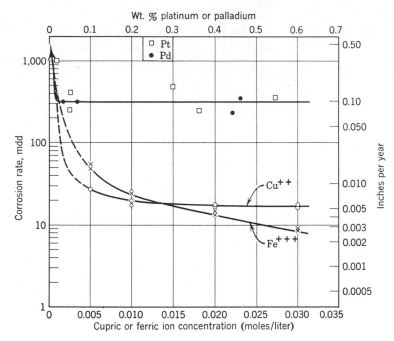

Fig. 1. Corrosion of titanium in boiling 10% HCl as a function of Fe^{+3} and Cu^{++} concentration and alloyed palladium or platinum.

Although corrosion resistance is poor in boiling HCl or H$_2$SO$_4$ (4.5 ipy in boiling 10% HCl), resistance is improved by orders of magnitude in the presence of small amounts of Cu^{++} or Fe^{+3} (0.005 ipy in 10% HCl, boiling, containing 0.02 mole Cu^{++} or Fe^{+3}/liter),[5] or by small alloying additions of palladium or platinum (0.1 ipy in 10% HCl, boiling, for 0.1% Pd-Ti alloy)[6] (Fig. 1). The latter substances accelerate the cathodic reaction (Cu^{++} and Fe^{+3} reduction, or H$^+$ discharge) to a level at which the corresponding anodic current density reaches or exceeds the critical value for anodic passivation (see p. 69). Anodic passivation (or anodic protection) is also practical in HCl, H$_2$SO$_4$, and some other acids using a small impressed current.[7]

Titanium is characterized by marked resistance to pitting and crevice corrosion in seawater. Only very slight uniform weight loss occurs over

[5] J. Cobb and H. Uhlig, *J. Electrochem. Soc.,* **99,** 13 (1952).
[6] M. Stern and H. Wissenberg, *J. Electrochem. Soc.,* **106,** 759 (1959); M. Stern and C. Bishop, *Trans. Amer. Soc. Metals,* **52,** 239 (1960).
[7] J. Cotton, *Chem. Ind. (London),* **1958,** 68.

periods of exposure amounting to several years (<0.0001 ipy). No pitting or appreciable weight loss occurred when titanium was buried in a variety of soils for eight years.[8] Resistance to pitting is accounted for by a very noble critical pitting potential of about 12 to 14 V in dilute chloride solutions at room temperature.[9, 10] At high Cl⁻ concentrations and elevated temperatures, the critical potential becomes much more active (Fig. 2), hence pitting corrosion is observed in hot concentrated $CaCl_2$ and similar solutions.[11] Addition of 1% Mo to titanium is effective in shifting the critical potential to more noble values above 125°C (260°F), with accompanying greater resistance to pitting in this temperature range.

Titanium does not undergo crevice corrosion in seawater at room tem-

[8] B. Sanderson and M. Romanoff, *Materials Prot.*, **8**, 29 (April 1969).

[9] C. Hall, Jr., and N. Hackerman, *J. Phys. Chem.*, **57**, 262 (1953).

[10] I. Dugdale and J. Cotton, *Corros. Sci.*, **4**, 397 (1964).

[11] F. Posey and E. Bohlmann, *2nd Symp. Fresh Water From the Sea*, Athens, Greece, May 1967.

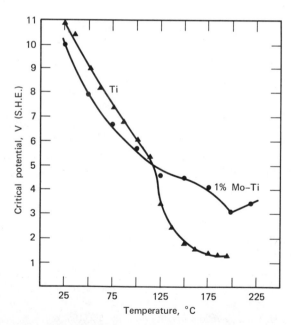

Fig. 2. Critical pitting potentials of commercial titanium and 1% Mo-Ti alloy in 1*M* NaCl as a function of temperature (Posey and Bohlmann).

perature, but instances have been recorded in hot seawater such as under an asbestos gasket at 95 to 120°C (200–250°F).[12] Attack of this kind, it is reported, does not occur at temperatures below about 95°C (200°F) ; also such attack is more commonly observed in acid and neutral rather than in alkaline chloride solutions. In laboratory tests, Griess[13] showed that titanium undergoes crevice attack in $1M$ NaCl containing dissolved oxygen at 150°C (300°F). For this type of corrosion, Cl⁻ was not necessary, contrary to the situation for pitting corrosion; crevice attack was also observed in hot solutions of I⁻, Br⁻, or SO_4^{--}. Susceptibility was related to acid anodic corrosion products accumulating in the crevice which reached a pH as low as 1.0. Hence the 0.1% Pd-Ti alloy, which resists breakdown of passivity in acids, was found to resist crevice corrosion better than did titanium.,[13, 14] In deaerated NaCl in which differential aeration cells are not established, crevice corrosion is suppressed. Various case histories of crevice and pitting corrosion observed in practice have been described.[15]

Intergranular Corrosion and Stress Corrosion Cracking

Intergranular corrosion of titanium (and various of its alloys) occurs in fuming nitric acid at room temperature (3–16 hr tests). Addition of 1% NaBr acts as inhibitor.[16] Similar corrosion of commercially pure titanium occurs in methanol solutions containing Br_2, Cl_2, or I_2; or Br⁻, Cl⁻, or I⁻.[17] Addition of water acts as an inhibitor.

The 6% Al, 4% V-Ti alloy has failed by stress corrosion cracking (s.c.c.) in liquid N_2O_4 within 40 hr at 40°C (105°F).[18] A slight excess of NO (or the presence of H_2O) inhibits such failure.

Various titanium alloys, including 8% Al, 1% Mo, 1% V-Ti (8-1-1) heated in air in contact with moist sodium chloride (e.g., from fingerprints) at 260°C (500°F) or higher, undergo s.c.c. (or intergranular

[12] J. Jackson and W. Boyd, in *Applications Related Phenomena in Titanium Alloys,* Spec. Tech. Publ. No. 432, pp. 218–226, Amer. Soc. Testing Mat., Philadelphia, Pa., 1968.

[13] J. Griess, Jr., *Corrosion,* **24,** 96 (1968).

[14] A. Takamura, *Corrosion,* **23,** 306 (1967).

[15] *Materials Prot.,* **6,** 222 (October 1967).

[16] G. Kiefer and W. Harple, *Metals Progr.,* **63,** 2, 74 (1953).

[17] A. Sedricks and J. Green, presented at Amer. Inst. Chem. Eng. Meeting, Atlanta, Ga., February 1970; *Corrosion,* **25,** 324 (1969).

[18] J. Jackson and W. Boyd, *Corrosion of Ti,* DMIC Memo 218, Battelle Mem. Inst., Columbus, Ohio, September 1966.

corrosion?) usually along grain boundaries.[19-21] Pure titanium is resistant to this type of failure.

In aqueous media at room temperature, Cl^-, Br^-, and I^- are unique in causing transgranular s.c.c. of various titanium alloys, including the 8-1-1 alloy. On the other hand, F^-, SO_4^{--}, OH^-, S^{--}, NO_3^-, and ClO_4^- do not cause failure, and, to the contrary, may inhibit crack propagation that otherwise occurs in distilled water (e.g., about 100 ppm KNO_3 is effective).[22, 23] Various anions of the group just listed also act as inhibitors of s.c.c. in presence of halide ions; in this respect their behavior simulates the effect of extraneous anions in the case of austenitic stainless steels (see p. 319). Beck[23] found that cathodic polarization of the 8-1-1 alloy to -0.76 V (s.h.e.) prevents failure in the presence of Cl^-, Br^-, or I^-. Leckie[24] reported protection of 7% Al, 2% Nb, 1% Ta-Ti alloy in 3% NaCl at -1.1 V (s.h.e.) and also at -1.3 V, at which hydrogen evolution was copious. Successful anodic protection against s.c.c. of the 8-1-1 alloy was reported in the presence of Cl^- but not in the presence of Br^- or I^-.[23]

Failure of the 8-1-1 alloy also occurs in pure methanol (CH_3OH). Interestingly, adding a small amount of Cl^- to distilled water or to methanol did not increase velocity of crack propagation, but less potassium nitrate (10 ppm) was needed to inhibit cracking in methanol as compared to water.[22] The stressed alloy was also found to be sensitive to cracking in the pure nonaqueous solvents CCl_4 and CH_2Cl_2.

The 8-1-1 alloy is a mixture of mostly α (hexagonal close-packed) and some β phase (body-centered cubic), observed cracks proceeding across grains of the α alloy but with the β phase failing by ductile fracture. Heat-treatment procedures and composition changes (e.g., lowering the aluminum content) that favor β phase increase resistance to s.c.c. However, the composition of the phase is also a determining factor, because the β phase of several other titanium alloys is found to be susceptible to s.c.c.[25] Although the mechanism of cracking is not

[19] "Stress Corrosion Cracking of Titanium," Spec. Tech. Publ. No. 397, Amer. Soc. Testing Mat., Philadelphia, Pa., 1966. Also S. Rideout, et al. in Ref. 25, p. 650.

[20] R. Newcomber, H. Tourkakis, and H. Turner, *Corrosion*, **21**, 307 (1965).

[21] H. Gray, *Corrosion*, **25**, 337 (1969); H. Gray and J. Johnston, *Met. Trans.*, **1**, 3101 (1970).

[22] T. Beck and M. Blackburn, *Amer. Inst. Aeronaut. Astron. J.*, **6**, 326 (1968).

[23] T. Beck, *J. Electrochem. Soc.*, **114**, 551 (1967).

[24] H. Leckie, *Corrosion*, **23**, 187 (1967).

[25] R. Adams and E. Von Tiesenhausen, in *Proc. Fundamental Aspects of Stress Corrosion Cracking*, edited by R. Staehle et al., p. 691, Nat. Assoc. Corros. Engrs., Houston, Texas, 1969.

yet resolved, the sensitive effect of alloy structure, the specificity of the environment, and the effects of extraneous anions and applied potential are largely similar to effects observed in the stainless steels (see p. 312), suggesting that the mechanisms governing s.c.c. of titanium and of stainless steels probably run parallel.

In brief, titanium is resistant to:

1. Seawater, 0.00003 ipy, 4.5-year exposure, including high-velocity conditions (0.0002 ipy, 138 ft/sec).[26] Known to resist pitting and crevice attack for 5 years and longer.
2. Wet chlorine (>0.9% H_2O, less in flowing gas).[27] In dry Cl_2, titanium may ignite.
3. Nitric acid. All concentrations and temperatures up to boiling (see Fig. 1, p. 358). But not to fuming HNO_3.
4. Oxidizing solutions, hot or cold (e.g., $CuCl_2$, $FeCl_3$, $CuSO_4$, $K_2Cr_2O_7$).
5. Hypochlorites.

Titanium is not resistant to:

1. Aqueous hydrogen fluoride.
2. Fluorine.
3. Hydrochloric and sulfuric acids, except when dilute, or in moderate concentrations when inhibited by oxidizing metal ions (e.g., Fe^{+3}, Cu^{++}) or other oxidizing substances (e.g., $K_2Cr_2O_7$, $NaNO_3$) or when alloyed with platinum or palladium.
4. Oxalic, >10% formic, anhydrous acetic[28] acids.
5. Boiling $CaCl_2$, >55%.[29]
6. Concentrated hot alkalies.
7. Molten salts (e.g., NaCl, LiCl, fluorides).
8. High-temperature exposure to air, nitrogen, hydrogen. Oxidation in air occurs above 450°C (840°F) with formation of titanium oxides and nitrides. Titanium hydride forms rapidly above 250°C (480°F). Absorption of oxygen, nitrogen, or hydrogen at elevated temperatures leads to embrittlement.

ZIRCONIUM

Zirconium is an active metal in the Emf Series, normally exhibiting very stable passivity. The metal [m.p. = 1852°C (3366°F), d. = 6.45

[26] T. May, International Nickel Co., private communication, 1962.
[27] E. Mellway and M. Kleinman, *Corrosion*, **23**, 88 (1967).
[28] K. Risch, *Chem.-Ing. Tech.*, **39**, 385 (1967).
[29] P. Gegner and W. Wilson, *Corrosion*, **15**, 341t (1959).

g/cm³] reacts readily at elevated temperatures with O_2, N_2, and H_2. An unusual property is the high solid solubility of the metal for oxygen, the metal dissolving up to 29 at. % (6.7 wt. %) according to the oxygen-zirconium phase diagram.[30] Similarly, α zirconium absorbs N in solid solution up to 25 at. % (4.8 wt. %). It reacts with air to form both zirconium oxides and nitrides. The reaction is slow enough, however, to permit hot rolling at 600 to 750°C (1100–1400°F).[31]

Commercial zirconium, as used primarily for corrosion resistance in the chemical industry,[32] contains up to 2.5% hafnium which is difficult to separate because of the similar chemical properties of zirconium and hafnium. This amount of alloyed hafnium has no marked effect on corrosion properties. The pure metal low in hafnium (0.02% max) has a low thermal neutron capture, making it specifically useful for nuclear power applications.

The outstanding corrosion property of zirconium is its resistance to alkalies at all concentrations up to the boiling point. It also resists fused sodium hydroxide. In this respect, it is distinguished from tantalum and, to a lesser extent, titanium, which are attacked by hot alkalies. In acids, zirconium is resistant to hydrochloric and nitric acids at all concentrations and to <70% H_2SO_4 up to boiling temperatures. In HCl and similar media the metal must be low in carbon (<0.06%) for optimum resistance. In boiling 20% HCl, a transition or "breakaway" point is observed in the corrosion rate (see below) after a specific time of exposure. The final rate, which is higher than the initial rate, is usually less than 0.0045 ipy.[33] Zirconium is not resistant to oxidizing metal chlorides (e.g., $FeCl_3$; pitting occurs), nor to HF or fluosilicic acid.

Critical pitting potentials of 0.38 V (s.h.e.) in $1N$ NaCl and 0.45 V in $0.1N$ NaCl[34] indicate that the metal is vulnerable to pitting in seawater. It undergoes intergranular s.c.c. in anhydrous methyl or ethyl alcohol containing HCl, but not when a small amount of water is added.[35] This behavior, parallel to that of commercial titanium, suggests that stress may not be necessary and that the failure is perhaps an example of intergranular corrosion.

[30] M. Hansen, *Constitution of Binary Alloys*, p. 1079, McGraw-Hill, New York, 1958.

[31] W. Hurford, in *The Metallurgy of Zirconium*, edited by B. Lustman and F. Kerze, Jr., p. 263, McGraw-Hill, New York, 1955.

[32] W. Brady, *Materials Prot.*, **7**, 40 (July 1968).

[33] W. Kuhn, *Corrosion*, **15**, 103t (1959).

[34] Ja. Kolotyrkin, *Corrosion*, **19**, 261t (1963).

[35] K. Mori, A. Takamura, and T. Shimose, *Corrosion*, **22**, 29 (1966).

Behavior in Hot Water and Steam

The good resistance of zirconium to deaerated hot water and steam is of special interest to nuclear power applications. The metal or its alloys can be exposed in general for prolonged periods without pronounced attack at temperatures below about 425°C (800°F). The rate of attack is characteristically low at first; but after a certain time of exposure, ranging from minutes to years depending on temperature, the rate suddenly increases. This "transition" phenomenon is reported to occur for pure or impure zirconium after a weight gain in the order of 35 to 50 mg/dm² and similar additional accelerated oxidation may occur at still higher weight gains.[36] It occurs at lower temperatures if the zirconium is contaminated with nitrogen (>0.005 %) or with carbon (>0.04 %).[37] The damaging effect of nitrogen in this respect is offset by alloying additions of 1.5 to 2.5% tin in combination with lesser amounts of iron, nickel, and chromium. Such alloys are called Zircaloys.

Marker experiments indicate that oxidation proceeds by diffusion of oxygen ions toward the metal-oxide interface (anion defect lattice). It has been suggested on this basis that trivalent nitrogen ions in the ZrO_2 lattice increase anion-defect concentration, thereby increasing the diffusion rate of oxygen ions. But were this the mechanism, oxidation in oxygen would also be affected, which is not the case. Adding to the complexity is the observation that alloyed tin appreciably shortens corrosion life of zirconium in water, but tin in the presence of small amounts of alloyed iron, nickel, or to a lesser extent, chromium, again increases corrosion resistance, the combination overcoming the detrimental effect of alloyed nitrogen.

The mechanism accounting for the transition phenomenon is not well understood. It has been explained on the basis of cracks forming in the oxide because of stresses accumulating as the oxide thickens. However, an increased corrosion rate does not occur when the metal oxidizes in oxygen except for much longer times and much thicker oxide films. Hydrogen formed by the decomposition of H_2O during reaction appears to exert a dominant role, especially that portion which dissolves in the metal, causing higher oxidation rates.[36] X-ray data for the oxides that form in H_2O show a monoclinic modification of ZrO_2 either before or after the breakaway time, but with some evidence that the initial oxide is a tetragonal modification.[37]

[36] B. Cox, *Corrosion,* **18,** 33t (1962).
[37] D. Thomas, Ref. 31, pp. 608–640.

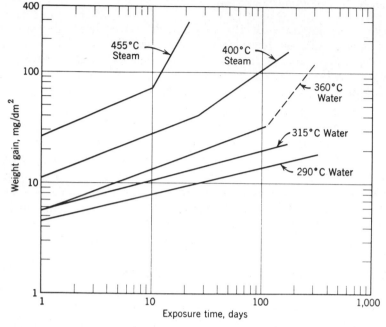

Fig. 3. Corrosion of Zircaloy 2 in high-temperature water and steam, showing transition points (Thomas).

The oxidation behavior of Zircaloy 2 [1.5% Sn, 0.12% Fe, 0.10% Cr, 0.05% Ni, 60 ppm N (max.) ; 50 ppm Al (max.) ; 50 ppm Ti (max.)] in water and steam is shown in Fig 3.

In brief, zirconium is resistant to:

1. Alkalies. All concentrations up to boiling point, including fused caustic.

2. Hydrochloric acid. All concentrations up to boiling point. Embrittlement of metal and higher corrosion rates occur above boiling temperatures under pressure. (See Fig. 1, p. 359).

3. Nitric acid. All concentrations up to boiling point, including red fuming acid.

4. Sulfuric acid. $<70\%$, boiling. (See Fig. 1, p. 359.)

5. Phosphoric acid. $<55\%$ H_3PO_4, boiling. (See Fig. 1, p. 358.)

6. Boiling formic, acetic, lactic, or citric acids.

Zirconium is not resistant to:

1. Oxidizing metal chlorides (e.g., $FeCl_3$, $CuCl_2$).

2. Hydrofluoric acid, fluosilicic acid.

3. Wet chlorine.
4. Oxygen, nitrogen, and hydrogen at elevated temperatures.
5. Aqua regia.
6. Trichloroacetic or oxalic acids, boiling.
7. Boiling $CaCl_2$, $>55\%$.[29]

TANTALUM

Tantalum [m.p. = 3000°C (5430°F), d. = 16.6 g/cm^3] exhibits the most stable passivity among known metals. It retains passivity in boiling acids (e.g., HCl, HNO$_3$, or H$_2$SO$_4$) and in moist chlorine or FeCl$_3$ solutions at above room temperature. Corrosion resistance of this order suggests a Flade potential much more active than the hydrogen electrode potential, and a low passive current insensitive to Cl$^-$. High corrosion resistance to acids makes tantalum useful for special applications in the chemical industry (e.g., H$_2$SO$_4$ concentrators or HCl absorption systems). Liners of tantalum sheet may average only 0.013 in. (0.03 cm) thick, permitting varied applications of the metal despite high cost.

Tantalum is attacked by alkalies and by hydrofluoric acid. It is readily embrittled by hydrogen at room temperature when the metal is cathodically polarized, or when coupled in an electrolyte to a metal more active in the Galvanic Series. The damage by cathodic hydrogen can be avoided by coupling tantalum to a very small area of a low-overvoltage metal like platinum.[36] Hydrogen ions then discharge on the platinum instead of entering the tantalum lattice. The benefit of coupled platinum extends to avoiding embrittlement caused by hydrogen generated through a corrosion reaction. For example, even though tantalum is not embrittled by concentrated hydrochloric acid at boiling temperatures, it becomes embrittled at 190°C (375°F) under pressure. This damage is not experienced if an area ratio of platinum to tantalum is provided in the order of at least 1:10,000. Platinum can be attached to tantalum by riveting, welding, or electrodeposition. Metal embrittled by cathodic hydrogen or by that entering at elevated temperatures can be restored to normal properties only by heating in vacuum.

In brief, tantalum is resistant to:

1. Hydrochloric acid. All concentrations up to boiling point. (See Fig. 1, p. 359.)
2. Nitric acid. All concentrations up to and above boiling point. (See Fig. 1, p. 358.)
3. Sulfuric acid. All concentrations (except fuming) <175°C (<350°F). Fuming acid attacks tantalum at room temperature. (See Fig. 1, p. 359.)

---

[36] C. Bishop and M. Stern, *Corrosion*, **17**, 379t (1961).

4. Chromic acid, hot or cold.

5. Phosphoric acid. Resistant to all concentrations up to and, in some cases, above boiling temperatures (see Fig. 1, p. 358). For 85% H_3PO_4 at 225°C, rate = 0.0035 ipy. Attack occurs at lower temperatures when the acid is contaminated with HF (>4 ppm).

6. Halogen gases. Wet or dry Cl_2 up to 150°C (300°F); Br_2 up to 175°C (350°F).

7. Aqua regia.

8. Oxidizing metal chlorides, hot or cold (e.g., $FeCl_3$, $CuCl_2$).

9. Organic acids: lactic, oxalic, acetic.

Tantalum is attacked by:

1. Alkalies. Embrittlement occurs, e.g., with 5% NaOH, 100°C.

2. Hydrofluoric acid and fluorides. Trace amounts.

3. Fuming sulfuric acid.

4. Oxygen, nitrogen, and hydrogen at elevated temperatures. Oxidation in air becomes appreciable above 250°C (500°F).

GENERAL REFERENCES

TITANIUM

D. Stough, F. Fink, and R. Peoples, *The Corrosion of Titanium,* Battelle Mem. Inst., Columbus, Ohio, T.M.L. Report 57, Oct. 29, 1956.
J. Cotton and H. Bradley, "The Corrosion Resistance of Titanium," *Chem. Ind.* (*London*), **1958**, 640.
Stress-Corrosion Cracking of Titanium, Spec. Tech. Publ. No. 397, Amer. Soc. Testing Mat., Philadelphia, Pa., 1966.

ZIRCONIUM

The Metallurgy of Zirconium, edited by B. Lustman and F. Kerze, Jr., McGraw-Hill, New York, 1955.
Zirconium, G. Miller, Academic Press, New York, 1957.
Corrosion of Zirconium Alloys, Spec. Tech. Publ. No. 368, Amer. Soc. Testing Mat., Philadelphia, Pa., 1964.

TANTALUM

C. Balke, in *Corrosion Handbook,* edited by H. H. Uhlig, pp. 320–323, Wiley, New York, 1948.
C. Hampel, "Corrosion Resistance of Ti, Zr, Ta Used for Chemical Equipment," *Corrosion,* **17,** 9 (October 1961).

chapter 24

Silicon-iron and silicon-nickel alloys

Silicon alloyed with iron or nickel imparts corrosion resistance to a variety of chemical media, in particular to strong nonoxidizing acids. The alloys are brittle and are therefore sensitive to fracture by thermal shock or by impact. The silicon-nickel alloy is less sensitive in this regard than the silicon-iron alloy. These alloys are available only as castings, and usually any subsequent forming must be done by grinding. The silicon-nickel alloy can be machined with some difficulty. The hardness of the latter alloy is greater the more rapidly it is cooled from about 1025°C (1875°F).

Corrosion rates of the silicon-iron alloys in 10% H_2SO_4 at 80°C (175°F) as a function of silicon content are shown in Fig. 1. Optimum resistance for minimum silicon content comes at 14.5% Si, and this is the composition of the commercial alloy. The nickel alloy contains 8.5 to 10% Si, which is below the optimum for corrosion resistance, but this composition range has improved mechanical properties compared to higher silicon compositions. Nominal compositions of both alloys are given in Table 1.

The silicon-iron alloy resists strong acids such as H_2SO_4, HNO_3, H_3PO_4 (pure), acetic, formic, and lactic, at all concentrations and temperatures up to boiling. It is also used for corrosion-resistant anodes in electrowinning of copper and in cathodic protection systems. It is not satisfactorily resistant to halogens, fused alkalies, HCl, HF, H_3PO_4 contaminated with HF, H_2SO_3, $FeCl_3$, hypochlorites, and aqua regia. The alloy is generally anodic to nickel, copper, or carbon, and cathodic to zinc and aluminum, with reversals occurring in some media.[1] Fair

[1] W. Richardson, *Trans. Electrochem. Soc.,* **39,** 61 (1921).

TABLE 1
Nominal Compositions of Commercial Si-Fe and Si-Ni Alloys

		Si	C	Mn	Fe	Cu	Cr	Co
			%					
Duriron* Regular	Si-Fe	14.5	0.85	0.65				
Hastelloy D†	Si-Ni	8.5–10	0.12 max.	0.5–1.25	2.0 max.	2–4	1.0 max.	1.5 max.

* Trademark, The Duriron Co., Inc.
† Trademark, Haynes Stellite Co.

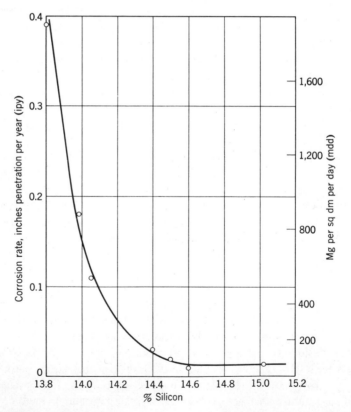

Fig. 1. Effect of alloyed silicon on corrosion of iron in 10% H_2SO_4 at 80°C (175°F) (W. Bryan, in *Corrosion Handbook*, p. 202).

resistance to HCl is obtained by alloying 14.5% Si-Fe with about 3% Mo [0.06 ipy, 30% HCl, 80°C (175°F)].[2]

The silicon-nickel alloy is resistant to H_2SO_4, H_2SO_3, H_3PO_4 (tech. or pure) (see Hast. D, Fig. 1, p. 358), organic acids, and acid salts under a wide variety of temperatures and concentrations. It is not usefully resistant to HNO_3, to HCl at elevated temperatures, or to $FeCl_3$.

Initial corrosion rates for both silicon alloys tend to be much higher than final rates. This behavior is commonly ascribed to slow build-up of a protective silica (SiO_2) surface film. Corrosion rates cited here and in the literature are only *final* rates, obtained after initial exposure of the alloys to a corrosive medium ranging from hours to weeks. Final rates in media for which these alloys are considered satisfactorily resistant usually fall within <0.001 to 0.05 ipy.

[2] *Corrosion Handbook*, p. 206.

chapter 25

Problems*

CHAPTER 2, ELECTROCHEMICAL MECHANISMS

1. Derive the general relation between ipy and mdd.

2. Magnesium corrodes in seawater at a rate of 14.5 mdd. What is the rate in ipy? If this corrosion rate applies to lead, what is the corresponding rate in ipy?

3. Laboratory corrosion tests on three alloys in an industrial waste solution show the following results:

Material	Density of Material (g/cm³)	Wt. Loss (mdd)	Pitting Factor
A	2.7	400	1
B	9.0	620	2
C	7.8	56	9.2

Calculate maximum penetration in inches for each material at the end of 1 year.

CHAPTER 3, CORROSION TENDENCY AND ELECTRODE POTENTIALS

(Temperature = 25°C unless otherwise stated)

1. Calculate the value for 2.303 RT/F at 50°C.   *Ans:* 0.0643 V.

2. Calculate the exact half-cell potential for the Ag-AgCl electrode in $1M$ NaCl.   *Ans:* 0.233 V.

* Answers when given are usually to slide rule accuracy.

3. Calculate the exact half-cell potential of zinc in $0.01M$ $ZnCl_2$.

Ans: -0.827 V.

4. Calculate the half-cell potential of the hydrogen electrode in a solution of pH $=$ 7 and partial pressure of $H_2 = 0.5$ atm at 40°C.

Ans: -0.426 V.

5. The emf of a cell made up of zinc (anode) and hydrogen electrode (cathode) immersed in $0.5M\,ZnCl_2$ is $+0.590$ V. What is the pH of the solution? *Ans:* 3.28.

6. Calculate the theoretical tendency of zinc to corrode (in volts) with evolution of hydrogen when immersed in $0.01M$ $ZnCl_2$ acidified to pH $=$ 2. *Ans:* 0.709 V.

7. Calculate the theoretical tendency of nickel to corrode (in volts) in deaerated water of pH $=$ 7. Assume corrosion products $= H_2$, and $Ni(OH)_2$ the solubility product of which is 1.6×10^{-16}. *Ans:* -0.110 V.

8. (*a*) What is the emf of a cell made up of copper electrode and hydrogen electrode ($p_{H_2} = 2$ atm) in copper sulfate (activity $Cu^{++} = 1$) of pH $=$ 1. (*b*) What is the polarity of the cell and which electrode is anode? *Ans:* -0.405 V.

9. (*a*) Calculate whether copper will corrode in deaerated $CuSO_4$, pH $=$ 0, to form Cu^{++} (activity $= 0.1$) and H_2 (1 atm). What is the corrosion tendency in volts? *Ans.* -0.307 V.

(*b*) Similarly, calculate whether copper will corrode in deaerated KCN (activity $CN^- = 0.5$) of pH $=$ 10, assuming $Cu(CN)_2^-$ is formed whose activity $= 10^{-4}$.

$(2CN^- + Cu \rightarrow Cu(CN)_2^- + e^-$ $E° = +0.446$ V)   *Ans:* 0.056 V.

10. (*a*) Calculate the emf of the following cell:

Pt; Fe^{+3} (activity $= 0.1$), Fe^{++} (activity $= 0.001$),

Ag^+ (activity $= 0.01$); Ag

(*b*) Write the spontaneous reaction for the cell. What is the polarity and which electrode is anode? *Ans:* Ag is negative and the anode. Emf $= -0.207$ V.

11. (*a*) Calculate the emf of a concentration cell made up of copper electrodes in $0.1M$ $CuSO_4$ and $0.5M$ $CuSO_4$, neglecting liquid junction potential. (*b*) Write the spontaneous reaction of the cell and indicate which electrode is anode.

12. (*a*) Calculate the emf of the following cell at 40°C:

O_2 (760 mm Hg pressure), Pt; H_2O; O_2 (76 mm Hg pressure), Pt

Ans: -0.0155 V.

(*b*) What is the polarity and which electrode is anode?

13. (*a*) Calculate the emf of a cell made up of hydrogen electrode ($p_{H_2} = 1$ atm) and oxygen electrode ($p_{O_2} = 0.5$ atm) in $0.5M$ NaOH. (*b*) What is the polarity and which electrode is anode? (Assume that the oxygen electrode is reversible.) *Ans:* 1.224 V.

14. Calculate whether silver will corrode if immersed in $0.1M$ $CuCl_2$ to form solid AgCl. What is the corrosion tendency in volts?
Ans: 0.019 V.

15. Calculate whether silver will corrode with hydrogen evolution in deaerated KCN solution, pH = 9, when CN^- activity = 1.0 and $Ag(CN)_2^-$ activity is 0.001. *Ans:* No. Emf = -0.04 V.

16. Zinc is immersed in a solution of $CuCl_2$ with activity of $Cu^{++} = 0.1$. What is the reaction and at what concentration of Cu^{++} will the reaction stop?

17. Calculate the emf of a cell made up of iron and lead electrodes in a solution of Fe^{++} and Pb^{++} of equal activity. Which electrode tends to corrode when the cell is short-circuited?
Ans: 0.314 V; Pb is cathode.

18. Calculate the emf as in Prob. 17 in an air-saturated alkaline solution of pH = 10. Which electrode corrodes on short-circuiting the cell? (Assume $HPbO_2^-$ forms as corrosion product and its activity = 0.1; also that iron is passive and its potential approximates the oxygen electrode.

$$Pb + 3OH^- \rightarrow HPbO_2^- + H_2O + 2e^- \qquad E^\circ = 0.54 \text{ V}$$

Ans: -0.84 V; Pb is anode.

19. Given
$$Fe \rightarrow Fe^{++} + 2e^- \qquad E^\circ = 0.440 \text{ V}$$
and
$$Fe^{++} \rightarrow Fe^{+3} + e^- \qquad E^\circ = -0.771 \text{ V}$$
calculate E° for $Fe \rightarrow Fe^{+3} + 3e^-$. *Ans:* 0.036 V.

20. Given $4OH^- \rightarrow O_2 + 2H_2O + 4e^- \qquad E^\circ = -0.401$ V, calculate E° for $2H_2O \rightarrow O_2 + 4H^+ + 4e^-$. *Ans:* -1.23 V.

21. Calculate the pressure (fugacity) of hydrogen required to stop corrosion of iron immersed in $0.1M$ $FeCl_2$, pH = 3.

22. Calculate the pressure of hydrogen, as in Prob. 21, in deaerated water with $Fe(OH)_2$ as corrosion product. (Solubility product $Fe(OH)_2$ = 1.8×10^{-15}.) *Ans:* 42 atm.

23. Calculate the pressure of hydrogen required to stop corrosion of cadmium at 25°C in deaerated water, with $Cd(OH)_2$ as corrosion product. (Solubility product $Cd(OH)_2 = 2.0 \times 10^{-14}$.) *Ans:* 0.2 atm.

24. A copper storage tank containing dilute H_2SO_4 at pH = 0.1 is

blanketed with hydrogen at 1 atm. Calculate the maximum Cu^{++} contamination of the acid in moles Cu^{++}/liter. What is the corresponding contamination if the hydrogen partial pressure is reduced to 10^{-4} atm?

Ans: 2.5×10^{-12}; 2.5×10^{-8} mole/liter.

CHAPTER 4, POLARIZATION AND CORROSION RATES

1. The potential of an iron electrode when polarized as cathode at 0.001 A/cm^2 is -0.916 V vs. $1N$ calomel half-cell. The pH of the electrolyte is 4.0. What is the value of the hydrogen overvoltage?

Ans: -0.40 V.

2. The potential of a cathode at which H^+ discharges at 0.001 A/cm^2 is -0.92 V versus Ag-AgCl in $0.01N$ KCl at 25°C. (a) What is the cathode potential on the standard hydrogen scale? (b) If the pH of the electrolyte is 1, what is the value of hydrogen overvoltage?

Ans: (a) -0.58; (b) -0.52 V.

3. The potential of a platinum anode at which oxygen evolves in an electrolyte of pH 10 is 1.30 V with respect to the saturated calomel electrode. What is the value of oxygen overvoltage? *Ans:* 0.90 V.

4. The potential of a copper electrode on which Cu^{++} deposits from $0.2M$ CuSO$_4$ versus $1N$ calomel electrode is -0.180 V. What is the polarization of the electrode in volts? Is the electrode polarized in the active or noble direction? *Ans:* -0.188 V, active direction.

5. Closely separated short-circuited zinc and mercury electrodes are immersed in deaerated HCl of pH = 3.5. What is the current through the cell if total exposed area of each electrode is 10 cm^2? What is the corresponding corrosion rate of zinc in mdd? (Corrosion potential of zinc versus $1N$ calomel half-cell = -1.03 V.)

Ans: 2.2×10^{-7} A; 0.064 mdd.

6. The corrosion potential of mild steel in a deaerated solution of pH = 2 is -0.64 V versus saturated Cu-CuSO$_4$ half-cell. The hydrogen overvoltage (V) for the same steel follows the relation $0.7 + 0.1 \log i$ where $i = $ A/cm^2. Assuming that approximately all the steel surface acts as cathode, calculate the corrosion rate in ipy. *Ans:* 0.0053 ipy.

7. Derive an expression for the slope of corrosion rate versus pH for a dilute cadmium amalgam in a deaerated cadmium ion solution. Neglect concentration polarization and assume that the amalgam is approximately all cathode.

Ans: $di_{corros}/d\mathrm{pH} = -0.059 i_{corros}/\beta$.

8. The potential of platinum versus saturated calomel electrode when polarized cathodically in deaerated H$_2$SO$_4$, pH = 1.0 at 0.01 A/cm^2 is

−0.334 V and at 0.1 A/cm^2 it is −0.364 V. Calculate β and i_0 for discharge of H$^+$ on platinum in this solution.

Ans: $\beta = 0.03$ V; $i_0 = 8 \times 10^{-4}$ A/cm^2.

9. The limiting current density, i_L (A/cm^2) for the discharge of H$^+$ is equal to nFD(H$^+$)$10^{-3}/\delta$ where D is the diffusion constant for hydrogen ions (7.39 $\times$ 10^{-5} cm^2/sec), δ is the thickness of the stagnant layer of electrolyte at the surface of the electrode (in unstirred solutions, δ is approximately 0.05 cm), and (H$^+$) is the activity of hydrogen ions (moles H$^+$/1000 g H$_2$O $\times$ γ). Derive an expression for the relation of limiting current density to pH of the electrolyte.

10. The corrosion rate of iron in deaerated HCl of pH = 3 is 30 mdd. Calculate the corrosion potential of iron in this acid versus 0.1N calomel electrode.

Ans: −0.719 V.

11. The linear polarization slope $d\phi/dI$ at low current densities for iron in a corrosive solution equals 2 mV/μA-cm^2. Using equation (8), calculate the corrosion rate in mdd. Assume $\beta_a = \beta_c = 0.1$ V.

Ans: 27.3 mdd.

12. It is observed that anodic activation overvoltage, η, for small applied current density, i, follows the relation $\eta = ki$. Derive the value of k in terms of the exchange current density, i_0, assuming $\beta_a = \beta_c = 0.1$ V.

Ans: $k = 0.0217/i_0$.

CHAPTER 5, PASSIVITY

1. Calculate the minimum concentration of oxygen in milliliters per liter necessary to passivate iron in 3% Na$_2$SO$_4$. Also 12% Cr-Fe alloy. (D for O$_2$, 25°C = 2 $\times$ 10^{-5} cm^2/sec.) Hint: Equate limiting diffusion current for reduction of oxygen to critical current density for passivity.

Ans: 31,800; 0.32 ml/liter.

2. Calculate the standard Flade potential at 25°C for iron, assuming that the passive film substance represented schematically as O·M in equation (2), is (a) Fe$_2$O$_3$, (b) Fe$_3$O$_4$, (c) chemisorbed oxygen. Data: $\Delta G°$ of formation for Fe$_2$O$_3$ = −177.1 kcal/mole Fe$_2$O$_3$; for Fe$_3$O$_4$ = −242.4 kcal/mole Fe$_3$O$_4$; for H$_2$O(l) = −56.69 kcal/mole H$_2$O; for H$^+$ = 0. For chemisorption of oxygen on Fe, $\Delta H° = -75$ kcal/mole O$_2$; $\Delta S° = -46.2$ cal/mole O$_2$-°C—estimated from value for chemisorbed N$_2$.

Ans: (a) −0.051 V; (b) −0.086 V; (c) 0.56V.

3. Calculate the temperature coefficient of Flade potential for iron at 25°C and compare with observed value of about 3 mV/°C. Assume passive film substance is (a) Fe$_2$O$_3$, (b) Fe$_3$O$_4$, (c) chemisorbed oxygen. Data: Molal entropy for Fe = 6.49; for H$_2$O(l) = 16.72; for Fe$_2$O$_3$ = 21.5; for Fe$_3$O$_4$ = 35.0 cal/°C.

4. Calculate the maximum emf of the cell:

stainless steel; O_2 (0.2 atm), $0.001M$ $ZnSO_4$ (pH = 3.0); Zn

assuming that passive stainless steel acts as a reversible oxygen electrode. What is the change of emf per 1 atm O_2 pressure?

$Ans:$ -1.89 V; -0.032 V/atm at 0.2 atm O_2 pressure.

5. From the standard Flade potential for iron, calculate the free energy of formation of the passive film per gram atom of oxygen. Similarly for nickel and chromium. $Ans:$ for iron = -27.6 kcal/g-at. O.

6. A pit in 18-8 stainless steel exposed to seawater grows to a depth of 6.5 cm in one year. To what average current density at the base of the pit does this rate correspond?

7. The steady-state passive current density for iron in $1N$ H_2SO_4 is 7 $\mu A/cm^2$. How many atom layers of iron are removed from a smooth electrode surface every minute? $Ans:$ 0.67.

8. Derive the relation $\log(Cl^-) = k \log(\text{anion}) + \text{const}$, where (anion) is the minimum activity of anion necessary to inhibit pitting of a passive metal in a chloride solution of activity (Cl^-). Assume that the amount of ion a adsorbed per unit area follows the Freundlich adsorption isotherm, $a = k_1 (\text{anion})^{1/n_1}$ where k_1 and n_1 are constants, and that at a critical ratio of adsorbed Cl^- to adsorbed anion the passive film is displaced by Cl^-, allowing a pit to initiate.

CHAPTER 6, IRON AND STEEL

1. Five iron rivets, 0.5 in.^{2} total exposed area each, are inserted in a copper sheet of 8 ft^2 exposed area. The sheet is immersed in an aerated, stirred, conducting solution in which iron corrodes at 0.0065 ipy. (a) What is the corrosion rate of the rivets in ipy? (b) What is the corrosion rate of an iron sheet in which five copper rivets are placed, with the same dimensions? $Ans:$ (a) 3.0 ipy.

2. Water entering a steel pipeline at the rate of 40 liters/min contains 5.50 ml O_2/liter (25°C, 1 atm). Water leaving the pipe contains 0.15 ml O_2/liter. Assuming that all corrosion is concentrated at a heated section 3000 dm^2 in area forming Fe_2O_3, what is the corrosion rate in mdd?

$Ans:$ 313 mdd.

3. Two steels immersed in dilute deaerated sulfuric acid corrode at different rates. When they are arranged as a cell, a small potential difference is observed between them. Which is anode? Illustrate by a polarization diagram. $Ans:$ slower corroding steel is anode.

4. Calculate the Saturation Index of each of the following waters at (a) 25°C (77°F) using equation (10); (b) using chart; (c) at 75°C (167°F) using charts. (Note that the calcium concentration is listed in the chart as ppm $CaCO_3$).

Water	Alkalinity (ppm CaCO$_3$)	Ca (ppm)	$pK_2' - pK_s'$	pH	Total Solids (ppm)
Boston, Mass.	13	5	2.2	6.90	43
Chicago, Ill.	120	34	2.3	8.0	157
Detroit, Mich.	73	26	2.2	7.7	153

Ans:

Sat. Index (using charts)

Water	25°C	75°C
Boston	−2.7	−2.2
Chicago	0.2	0.6
Detroit	−0.5	−0.1

CHAPTER 7, EFFECT OF STRESS

1. (*a*) Engell and Bäumel (*Physical Metallurgy of Stress Corrosion Fracture*, p. 354, Interscience, New York, 1959) report that cracks in iron stressed in boiling calcium nitrate solution propagate at the rate of 0.2 mm/sec. To what current density does this rate correspond? If this rate is typical, what does your answer indicate with regard to the electrochemical mechanism of crack growth?   *Ans:* 545 A/cm^2.

(*b*) If instead cracking is assumed to be continuous, with an average rate of 0.00018 mm/sec [Uhlig and Sava, *Trans. Amer. Soc. Metals*, **56**, 361 (1963)], how would this affect your conclusion?

2. The residual energy of a cold-worked steel as determined calorimetrically is 5 cal/g. Calculate the increased tendency in volts for iron to corrode when cold worked. (Assume that the entropy change is negligible.)   *Ans:* 0.006 V.

3. The critical potential for stress corrosion cracking of 18-8 stainless steel in deaerated MgCl$_2$ solution at 130°C is −0.145 V (s.h.e.). Above what predicted maximum value of pH of MgCl$_2$ at 130°C will stress corrosion cracking not initiate?   *Ans:* 1.8.

CHAPTER 10, OXIDATION AND TARNISH

1. Calculate Md/nmD for aluminum forming Al$_2$O$_3$ and for sodium forming Na$_2$O. Indicate whether the oxides are protective.
   Ans: 1.28; 0.58.

2. Estimate the free energy of formation of NiO at 780°C from the cell

described on p. 195. Assume that NiO in borax is saturated. Compare with reported values as follows:

$\Delta G°$ (cal/mole)

1000°K	−35,080
1100°K	−32,900

3. Copper oxidizes within the parabolic range at a higher rate the higher the oxygen pressure. (*a*) What is the quantitative relation? (Use equation (3), p. 192.) (*b*) Make the same calculation for nickel.

Ans: (*a*) rate is proportional to $p_{O_2}^{1/6}$.

CHAPTER 11, STRAY-CURRENT CORROSION

1. (*a*) A direct current of 10 A enters and leaves a steel water pipe 2 in. outside diameter, 0.25 in. wall thickness, containing water whose resistivity measures 10^4 Ω-cm. Calculate the current carried by the steel and that carried by the water. Assume that the resistivity of the pipe equals 10^{-5} Ω-cm. (*b*) Similarly, if the pipe conveys seawater of resistivity 20 Ω-cm.

2. (*a*) A direct current of 10 A passes through external clamps 2 meters apart attached to a copper pipe of 5 cm outside diameter containing water whose conductivity is 10^{-4} Ω^{-1} cm^{-1}. If the pipe wall thickness is 0.35 cm, calculate the total weight of the copper in grams per year which corrodes internally at the positive clamp because of this current. (*b*) Similarly if the pipe contains seawater of conductivity equal to 0.05 Ω^{-1} cm^{-1}. (*c*) What weight of copper would in theory corrode in 1 year if an insulating joint were inserted between the clamps?

Ans: (*a*) 5.0×10^{-5}; (*b*) 0.025; (*c*) 1.04×10^5 g.

3. A long pipeline 8 in. in diameter is buried 6 ft underground. The potential difference between two Cu-CuSO$_4$ reference electrodes located on the soil surface over the pipe and a point at right angles 60 ft distant, is 1.25 V. The electrode over the pipe is negative to the other, and soil resistivity measures 3000 Ω-cm. (*a*) Does current flow from or to the pipe? How many amperes per linear foot of pipe? *Ans:* 0.0173 A/ft.

(*b*) What is the per cent error in calculated current if it is found later that the pipe is buried 7 ft instead of 6 ft below the surface?

Ans: 6.4%.

4. Assuming that a pipe is buried h meters underground in soil with a resistivity of 3000 Ω-cm, plot ΔE versus y for a constant current, j, entering the pipe per linear length of pipe in meters.

5. In measuring resistivity by the four-electrode method, what dis-

tance between electrodes must be specified in order to apply the simplified formula:

$$\rho = \frac{1000 \, \Delta\phi}{I}$$

Ans: 159 cm (5.22 ft).

6. A stray current of 0.7 A leaves a section of steel pipe 2 in. in diameter and 2 ft long. What is the initial corrosion rate in ipy caused by this current?

7. (*a*) If a spherical electrode of radius *a* is half buried in earth of resistivity *σ*, what is the resistance, *R*, to earth as a function of distance, *D*, from the electrode center? (*b*) At what distance is the resistance within 99% of maximum?

$$\text{Ans: (a)} \ R = \frac{\sigma}{2\pi} \left(\frac{1}{a} - \frac{1}{D} \right); \text{(b) } 100a.$$

CHAPTER 12, CATHODIC PROTECTION

1. Sketch the polarization diagram of a buried steel pipe connected to a sacrificial magnesium anode. Indicate on the diagram (*a*) the potential near the pipe surface with respect to the open-circuit potentials of anodic and cathodic areas of the pipe, (*b*) the potential of the pipe with reference electrode located far from and at right angles to the pipe, (*c*) the change in potential of the pipe proceeding toward the vicinity of the magnesium anode.

2. Calculate the minimum potential versus $Cu\text{-}CuSO_4$ reference electrode to which cadmium must be polarized for complete cathodic protection, assuming $Cd(OH)_2$ as corrosion product on the surface. (Solubility product of $Cd(OH)_2 = 2 \times 10^{-14}$.) *Ans:* −0.86 V.

3. From sketches of polarization diagrams for a corroding metal, compare applied current, as required for complete cathodic protection, to the normal corrosion current when (*a*) the corrosion rate is anodically controlled, (*b*) cathodically controlled.

4. Current between a magnesium anode and a 50-gal steel tank filled with air-saturated hot water is 100 mA. Disregarding local-action currents, what interval of time is required between filling and emptying the tank to ensure minimum corrosion of outlet steel water piping? (Solubility of oxygen in inlet water, 25°C, = 6 ml/liter). *Ans:* 49.8 hr.

5. Iron corrodes in seawater at a rate of 25 mdd. Assuming that all corrosion is by oxygen depolarization, calculate the minimum initial current density (A/dm^2) necessary for complete cathodic protection.

Ans: 0.0010 A/dm².

6. Calculate the minimum potential versus saturated calomel electrode to which copper must be polarized in $0.1M$ $CuSO_4$ for complete cathodic protection. *Ans:* 0.042 V.

7. A copper bar 3 dm² total exposed area coupled to an iron bar 0.5 dm² area is immersed in seawater. What minimum current must be applied to the couple in order to avoid corrosion of both iron and copper. (Corrosion rate of iron in seawater is 0.005 ipy.) *Ans:* 3.8×10^{-3} A.

8. To what minimum potential versus saturated calomel electrode must indium be polarized in $0.01M$ $In_2(SO_4)_3$ solution for complete cathodic protection? *Data:* $In \rightarrow In^{+3} + 3e^-$, $E^\circ = 0.342$ V; $\gamma\, 0.01M$ $In_2(SO_4)_3 = 0.142$. *Ans:* -0.634 V.

9. (*a*) How many milliliters H_2 (N.T.P.) are evolved per square centimeter per day from a steel surface maintained at the critical potential for cathodic protection: $\phi_H = -0.59$ V?

(*b*) Similarly, if cathodic protection is increased to a potential 0.1 V more active than the critical value? *Data:* Assume hydrogen overvoltage (V) on steel $= 0.105 \log i/(1 \times 10^{-7})$ where $i = $ A/cm², pH of steel surface $= 9.5$. *Ans:* (*a*) 0.0019 ml/cm²-day (*b*) 0.018 ml/cm²-day.

CHAPTER 19, COPPER AND COPPER ALLOYS

1. Calculate the equilibrium constant at 25°C for the reaction $Cu + Cu^{++} \rightarrow 2Cu^+$. Since equilibrium tends to be maintained between ions and metal, which valence ion predominates when Cu is dissolved anodically?

Data:
$$Cu \rightarrow Cu^{++} + 2e^- \qquad E^\circ = -0.337 \text{ V}$$
$$Cu \rightarrow Cu^+ + e^- \qquad E^\circ = -0.521 \text{ V}$$
Ans: 5.9×10^{-7}.

2. As in Prob. 1, estimate, from the analogous equilibrium constant, which valence ion predominates when Cr in the active state is dissolved anodically.

Data:
$$Cr \rightarrow Cr^{++} + 2e^- \qquad E^\circ = 0.91 \text{ V}$$
$$Cr \rightarrow Cr^{+3} + 3e^- \qquad E^\circ = 0.74 \text{ V}$$
Ans: Cr^{++} predominates.

chapter 26

Appendix

ACTIVITY AND ACTIVITY COEFFICIENTS OF STRONG ELECTROLYTES

Let μ be the partial molal free energy, μ° the partial molal free energy in the standard state $(a = 1)$, a the activity, M the molality, and γ the activity coefficient.

By definition.

$$\mu = \mu^\circ + RT \ln a$$
$$a = \gamma M$$

where $\gamma \rightarrow 1$ as $M \rightarrow 0$.

For a binary electrolyte of molality M, assuming total dissociation

$$\mu = \mu^\circ + RT \ln a_- + RT \ln a_+$$
$$= \mu^\circ + RT \ln \gamma_- M_- + RT \ln \gamma_+ M_+$$
$$= \mu^\circ + 2RT \ln \gamma_\pm M$$

where $M_- = M_+ = M$ and $\gamma_\pm$ is the mean ion activity coefficient (it is impossible to measure the activities of individual ions). For an electrolyte like lanthanum sulfate of molality M

$$La_2(SO_4)_3 \rightarrow 2La^{+3} + 3SO_4^{--}$$
$$\mu = \mu^\circ + 2RT \ln \gamma_\pm M_{La^{+3}} + 3RT \ln \gamma_\pm M_{SO_4^{--}}$$

Since $M_{La^{+3}} = 2M$ and $M_{SO_4^{--}} = 3M$

$$\mu = \mu^\circ + RT \ln \gamma_\pm^5 (2M)^2 (3M)^3$$
$$= \mu^\circ + 5RT \ln \gamma_\pm M (2^2 \times 3^3)^{1/5}$$

For the general case in which one mole of electrolyte dissociates into ν_1 moles of cation and ν_2 moles of anion, the mean activity is given by

$$a_\pm = \gamma_\pm M (\nu_1^{\nu_1} \nu_2^{\nu_2})^{1/(\nu_1 + \nu_2)}$$

See also, C. W. Davies, *Electrochemistry*, George Newnes, Ltd., London, 1967. Activity coefficients are listed in Table 1

Example. Calculate the emf of the following cell:

$$Zn; ZnCl_2 (0.01M); Pt (Cl_2, 0.5 atm)$$

$$\gamma_\pm \text{ for } 0.01M \text{ ZnCl}_2 = 0.71 \text{ (p. 390)}$$

$$Zn \rightarrow Zn^{++} + 2e^- \qquad E° = 0.763 \text{ V} \qquad (1)$$

$$2Cl^- \rightarrow Cl_2 + 2e^- \qquad E° = -1.360 \text{ V} \qquad (2)$$

Tentative reaction: $(1) - (2)$

$$Zn + Cl_2 \rightarrow Zn^{++} + 2Cl^- \qquad (3)$$

The corresponding Nernst equation is

$$E = 0.763 + 1.360 - \frac{0.0592}{2} \log \frac{(Zn^{++})(Cl^-)^2}{p_{Cl_2}}$$

$$= 2.123 - \frac{0.0592}{2} \log \frac{(0.01)(0.71)[(0.02)(0.71)]^2}{0.5}$$

$$= 2.30 \text{ V}$$

[Reaction (3) is spontaneous (ΔG is negative); Zinc is anode ($-$) and Platinum is cathode ($+$)].

DERIVATION OF STERN-GEARY EQUATION FOR CALCULATING CORROSION RATES FROM POLARIZATION DATA OBTAINED AT LOW CURRENT DENSITIES

Assume that the corrosion current, I_{corros}, occurs at a value within the Tafel region for both anodic and cathodic reactions. Also assume that concentration polarization and IR drop are negligible. If the corroding metal is polarized as cathode by means of an external current to a potential, ϕ', differing only slightly from ϕ_{corros}, polarization follows line $\phi_c OA$. Accompanying larger cathodic current, I_c, automatically decreases anodic current I_a by reason of the relation (see Fig. 1)

$$I_c = I_a + I_{applied} \qquad (1)$$

Similarly, for anodic polarization, for which $I_{applied}$ changes sign

$$I_a = I_c - I_{applied} \qquad (2)$$

$$\phi_{corros} - \phi' = \Delta\phi = -\beta_c \log \frac{I_{corros}}{i_{0c} A_c} + \beta_c \log \frac{I_c}{i_{0c} A_c} = \beta_c \log \frac{I_c}{I_{corros}} \qquad (3)$$

TABLE 1
Activity Coefficients of Strong Electrolytes (M = molality)

$M \rightarrow$	0.001	0.002	0.005	0.01	0.02	0.05	0.1	0.2	0.5	1.0	2.0	3.0	4.0
HCl	0.966	0.952	0.928	0.904	0.875	0.830	0.796	0.767	0.758	0.809	1.01	1.32	1.76
HBr	0.966	—	0.929	0.906	0.879	0.838	0.805	0.782	0.790	0.871	1.17	1.67	—
HNO_3	0.965	0.951	0.927	0.902	0.871	0.823	0.785	0.748	0.715	0.720	0.783	0.876	0.982
$HClO_4$	—	—	—	—	—	—	—	—	—	0.81	1.04	1.42	2.02
HIO_3	0.96	0.94	0.91	0.86	0.80	0.69	0.58	0.46	0.29	0.19	0.10	0.073	0.060
H_2SO_4	0.830	0.757	0.639	0.544	0.453	0.340	0.265	0.209	0.154	0.130	0.124	0.141	0.171
NaOH	—	—	—	—	—	—	—	—	0.69	0.68	0.70	0.77	0.89
KOH	—	—	0.92	0.90	0.86	0.82	0.80	0.73	0.73	0.76	0.89	1.08	1.35
CsOH	—	—	—	0.92	0.88	0.83	0.80	0.76	0.74	0.78	—	—	—
$Ba(OH)_2$	—	0.853	0.773	0.712	0.627	0.526	0.443	0.370	—	—	—	—	—
$AgNO_3$	—	—	0.92	0.90	0.86	0.79	0.72	0.64	0.51	0.40	0.28	—	—
$Al(NO_3)_3$	—	—	—	—	—	—	0.20	0.16	0.14	0.19	0.45	1.0	—
$BaCl_2$	0.88	—	0.77	0.72	—	0.56	0.49	0.44	0.39	0.39	0.44 (1.8M)	—	1.2
$Ba(NO_3)_2$	0.88	0.84	0.77	0.71	0.63	0.52	0.43	0.34	—	—	—	—	—
$Ba(IO_3)_2$	0.83	0.79	0.71	0.64	0.55	—	—	—	—	—	—	—	—
$CaCl_2$	0.89	0.85	0.785	0.725	0.66	0.57	0.515	0.48	0.52	0.71	0.35	0.37	0.42
$Ca(NO_3)_2$	0.88	0.84	0.77	0.71	0.64	0.54	0.48	0.42	0.38	0.35	—	—	—
$CdCl_2$	0.76	0.68	0.57	0.47	0.38	0.28	0.21	0.15	0.09	0.06	—	—	—
CdI_2	0.76	0.65	0.49	0.38	0.28	0.17	0.11	0.068	0.038	0.025	0.018	—	—
$CdSO_4$	0.73	0.64	0.50	0.40	0.31	0.21	0.17	0.11	0.067	0.045	0.035	0.036	—
CsF	0.98	0.97	0.96	0.95	0.94	0.91	0.89	0.87	0.85	0.87	—	—	—
CsCl	—	—	0.92	0.90	0.86	0.79	0.75	0.69	0.60	0.54	0.49	0.48	0.47
CsBr	—	—	0.93	0.90	0.86	0.79	0.75	0.69	0.60	0.53	0.48	0.46	0.46
CsI	—	—	—	—	—	—	0.75	0.69	0.60	0.53	0.47	0.43	—
$CsNO_3$	—	—	—	—	—	—	0.73	0.65	0.52	0.42	—	—	—

TABLE 1 (continued)
Activity Coefficients of Strong Electrolytes

M→	0.001	0.002	0.005	0.01	0.02	0.05	0.1	0.2	0.5	1.0	2.0	3.0	4.0
CsAc	...	...	...	...	...	...	0.79	0.77	0.76	0.80	0.95	1.15	...
$CuCl_2$	0.89	0.85	0.78	0.72	0.66	0.58	0.52	0.47	0.42	0.43	0.51	0.59	...
$CuSO_4$	0.74	...	0.53	0.41	0.31	0.21	0.16	0.11	0.068	0.047	...	...	...
$FeCl_2$	0.89	0.86	0.80	0.75	0.70	0.62	0.58	0.55	0.59	0.67	...	...	...
$In_2(SO_4)_3$	...	...	...	0.142	0.092	0.054	0.035	0.022	...	...	...	...	...
KF	...	0.96	0.95	0.93	0.92	0.88	0.85	0.81	0.74	0.71	0.70	...	...
KCl	0.965	0.952	0.927	0.901	0.872	0.815	0.769	0.719	0.651	0.606	0.576	0.571	0.579
KBr	0.965	0.952	0.927	0.903	0.88	0.822	0.777	0.728	0.665	0.625	0.602	0.603	0.622
KI	0.965	0.951	0.927	0.905	...	0.84	0.80	0.76	0.71	0.68	0.69	0.72	0.75
$K_4Fe(CN)_6$	...	...	...	...	...	0.19	0.14	0.11	0.067	...	...	...	...
$KClO_3$	0.967	0.955	0.932	0.907	0.875	0.813	0.755	...	...	...	...	...	...
K_2CO_3	0.89	0.86	0.81	0.74	0.68	0.58	0.50	0.43	0.36	0.33	0.33	0.39	...
$KClO_4$	0.965	0.951	0.924	0.895	0.857	0.788	...	...	...	...	...	...	...
K_2SO_4	0.89	...	0.78	0.71	0.64	0.52	0.43	0.36	...	...	...	...	...
$LaCl_3$	...	...	...	...	...	0.38	0.33	0.28	0.27	0.36	0.33	0.39	0.49
$La(NO_3)_3$	...	...	...	...	...	0.39	0.32	0.27	...	...	...	...	...
LiCl	0.963	0.948	0.921	0.89	0.86	0.82	0.78	0.75	0.73	0.76	0.91	1.18	1.46
LiBr	0.966	0.954	0.932	0.909	0.882	0.842	0.810	0.784	0.783	0.848	1.0	1.35	...
LiI	...	...	...	...	...	...	0.81	0.80	0.81	0.89	1.19	1.70	...
$LiNO_3$	0.966	0.953	0.930	0.904	0.878	0.834	0.798	0.765	0.743	0.76	0.84	0.97	...
$LiClO_3$	0.967	0.955	0.933	0.911	0.884	0.842	0.810	0.782	0.77	0.81	...	...	...
$LiClO_4$	0.967	0.956	0.935	0.915	0.890	0.853	0.825	0.805	0.82	0.91	...	...	...
$MgCl_2$	0.88	0.84	0.77	0.71	0.64	0.55	0.56	0.53	0.52	0.62	1.05	2.1	...
$Mg(NO_3)_2$	...	...	...	...	...	...	0.51	0.46	0.44	0.50	0.69	0.93	...
$MgSO_4$	...	...	...	0.40	0.32	0.22	0.18	0.13	0.088	0.064	0.055	0.064	...
$MnSO_4$	...	...	...	...	...	...	0.25	0.17	0.11	0.073	0.058	0.062	0.079

TABLE 1 (*continued*)
Activity Coefficients of Strong Electrolytes

M→	0.001	0.002	0.005	0.01	0.02	0.05	0.1	0.2	0.5	1.0	2.0	3.0	4.0
NiSO₄	…	…	…	…	…	…	0.18	0.13	0.075	0.051	0.041	…	…
NH₄Cl	0.961	0.944	0.911	0.88	0.84	0.79	0.74	0.69	0.62	0.57	…	…	…
NH₄Br	0.964	0.949	0.901	0.87	0.83	0.78	0.73	0.68	0.62	0.57	…	…	…
NH₄I	0.962	0.946	0.917	0.89	0.86	0.80	0.76	0.71	0.65	0.60	…	…	…
NH₄NO₃	0.959	0.942	0.912	0.88	0.84	0.78	0.73	0.66	0.56	0.47	…	…	…
(NH₄)₂SO₄	0.874	0.821	0.726	0.67	0.59	0.48	0.40	0.32	0.22	0.16	…	…	…
NaF	…	…	0.93	0.90	0.87	0.81	0.75	0.69	0.62	…	…	…	…
NaCl	0.966	0.953	0.929	0.904	0.875	0.823	0.780	0.730	0.68	0.66	0.67	0.71	0.78
NaBr	0.966	0.955	0.934	0.914	0.887	0.844	0.800	0.740	0.695	0.686	0.734	0.826	0.934
NaI	0.97	0.96	0.94	0.91	0.89	0.86	0.83	0.81	0.78	0.80	0.95	0.44	0.41
NaNO₃	0.966	0.953	0.93	0.90	0.87	0.82	0.77	0.70	0.62	0.55	…	…	…
Na₂SO₄	0.887	0.847	0.778	0.714	0.641	0.53	0.45	0.36	0.27	0.20	…	…	…
NaClO₄	0.97	0.95	0.93	0.90	0.87	0.82	0.77	0.72	0.64	0.58	…	…	…
PbCl₂	0.86	0.80	0.70	0.61	0.50	…	…	…	…	…	…	…	…
Pb(NO₃)₂	0.88	0.84	0.76	0.69	0.60	0.46	0.37	0.27	0.17	0.11	…	…	…
RbCl	…	…	0.93	0.90	…	…	0.76	0.71	0.63	0.58	0.54	0.54	0.54
RbBr	…	…	…	…	…	…	0.76	0.70	0.63	0.58	0.53	0.52	0.51
RbI	…	…	…	…	…	…	0.76	0.70	0.63	0.57	0.53	0.52	0.51
RbNO₃	…	…	…	…	…	…	0.73	0.65	0.53	0.43	0.32	0.25	0.21
RbAc	…	…	…	…	…	…	0.73	0.65	0.52	0.42	…	…	…
TlCl	0.96	0.95	0.93	0.90	…	0.77	0.70	0.60	…	…	…	…	…
TlNO₃	…	…	…	…	…	0.79	0.73	0.65	0.59	…	…	…	…
TlClO₄	…	…	…	…	…	0.80	0.74	0.68	0.53	…	…	…	…
TlAc	…	…	…	…	0.64	0.56	0.50	0.45	0.38	0.33	…	…	…
ZnCl₂	0.88	0.84	0.77	0.71	…	…	0.51	0.45	…	0.51	0.44	0.40	0.38
ZnSO₄	0.70	0.61	0.48	0.39	…	…	0.15	0.11	0.065	0.045	0.036	0.04	…

Oxidation Potentials, W. Latimer, Prentice-Hall, Englewood Cliffs, N. J., 1952 (with permission).

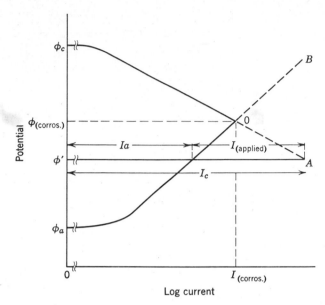

Fig. 1. Polarization diagram for corroding metal polarized cathodically from ϕ_{corros} to ϕ'.

where A_c is the fraction of area that is cathode, and i_{0c} is the exchange current density for the cathode reaction. Note that I_c and i_{0c} in the Tafel equation refer to current per unit local area (not total area), hence I_{corros}/A_c is the current density to which i_{0c} applies. Likewise if the metal is polarized an equal amount in the anodic direction

$$\Delta\phi = -\beta_a \log \frac{I_a}{I_{\text{corros}}} \tag{4}$$

or

$$I_c = I_{\text{corros}} 10^{\Delta\phi/\beta_c} \quad \text{and} \quad I_a = I_{\text{corros}} 10^{-(\Delta\phi/\beta_a)}$$

Then

$$I_{\text{applied}} = I_{\text{corros}} [10^{\Delta\phi/\beta_c} - 10^{-(\Delta\phi/\beta_a)}] \tag{5}$$

Expressed as a series

$$10^x = 1 + 2.3x + \frac{(2.3x)^2}{2!} + \text{etc.}$$

Hence if $\Delta\phi/\beta_c$ and $\Delta\phi/\beta_a$ are small, higher terms can be neglected and

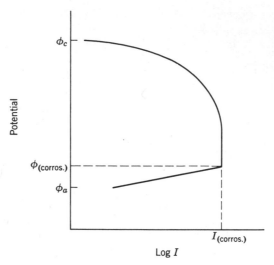

Fig. 2. Polarization diagram for metal corroding under control by oxygen depolarization.

(5) is approximated by

$$I_{\text{applied}} = 2.3 I_{\text{corros}} \Delta\phi \left(\frac{1}{\beta_c} + \frac{1}{\beta_a} \right) \qquad (6)$$

or

$$I_{\text{corros}} = \frac{1}{2.3} \frac{I_{\text{applied}}}{\Delta\phi} \left(\frac{\beta_c \beta_a}{\beta_a + \beta_c} \right) \qquad (7)$$

Equation 7 is the Stern–Geary equation. Should the cathodic reaction be controlled by concentration polarization, as occurs in corrosion reactions controlled by oxygen depolarization, the corrosion current equals the limiting diffusion current (Fig. 2). This situation is equivalent to a large or infinite value of β_c in (7). Hence when concentration polarization of this kind controls, (7) becomes

$$I_{\text{corros}} = \frac{\beta_a}{2.3} \frac{I_{\text{applied}}}{\Delta\phi} \qquad (8)$$

If a noncorroding metal that polarizes only slightly (high value of i_{0a}) is polarized anodically at moderate current densities, then i_{0a} can be substituted for I_{corros} in Fig. 1 and

$$\Delta\phi = \frac{I_{\text{applied}}}{2.3 i_{0a}} \frac{\beta_c \beta_a}{\beta_a + \beta_c} \qquad (9)$$

indicating, as observed, that anodic polarization of many metals at low current densities is a linear function of applied current.

POURBAIX DIAGRAM FOR IRON

Each line of Fig. 3 represents conditions of thermodynamic equilibrium for some reaction. A horizontal line is for a reaction that does not involve H^+ or OH^-. A vertical line involves H^+ or OH^- but not electrons. A

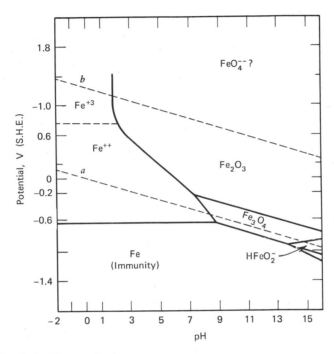

Fig. 3. Pourbaix diagram for iron.

sloping line involves H^+, OH^-, and electrons. For example, the sloping line separating Fe^{++} from Fe_2O_3 represents the reaction $2Fe^{++} + 3H_2O \rightarrow Fe_2O_3 + 6H^+ + 6e^-$. It should be noted that the pH referred to in the figure is that of solution in immediate contact with the metal surface. This value, in some instances, differs from that of the solution itself.

Oxygen is evolved above but not below line b, in accord with the reaction $H_2O \rightarrow \frac{1}{2}O_2 + 2H^+ + 2e^-$. Hydrogen is evolved below but not above line a, in accord with the reaction $H^+ \rightarrow \frac{1}{2}H_2 - e^-$. Soluble hypoferrites ($HFeO_2^-$) can form in very alkaline solutions within a restricted active potential range. Soluble ferrates (FeO_4^{--}) can form in alkaline solutions at very noble potentials, but the stable field is not well defined.

When any reaction involves ions other than H^+ or OH^- it is assumed, by and large, that their activity equals 10^{-6}. Thus the horizontal line at -0.62 V means that iron will not corrode below this value to form a solution of $>10^{-6}M$ Fe^{++} in accord with $Fe \rightarrow Fe^{++} + 2e^-$, $\phi = -0.44 + (0.059/2) \log (10^{-6}) = -0.62$ V.

The fields marked Fe_2O_3 and Fe_3O_4 are sometimes labeled "passivation" on the assumption that iron reacts in these regions to form protective oxide films. This is correct insofar as passivity is accounted for by a diffusion barrier oxide layer (Def. 2, p. 61). Actually the Flade potential above which, but not below, passivity of iron is observed in media like sulfuric or nitric acid, parallels line a and b intersecting 0.6 V at pH $= 0$. For this reason, the passive film (Def. 1) is probably not any of the stoichiometric iron oxides, as is further discussed on pp 70-74.

Other than the general atlas cited on p. 27, the following papers discuss Pourbaix diagrams in a general way: E. Verinck, Jr., *Corrosion*, **23,** 371 (1967); A. Guy and F. Rhines, *Metal Progr*, **85,** 117 (1964).

DERIVATION OF EQUATION EXPRESSING THE SATURATION INDEX OF A NATURAL WATER

$$K_s' = (Ca^{++})(CO_3^{--}) \tag{1}$$

$$K_2' = \frac{(H^+)(CO_3^{--})}{(HCO_3^-)} \tag{2}$$

(alk) = alkalinity, or titratable equivalents of base per liter (by titrating with acid) obtained by using methyl orange as indicator.

Assume that salts of weak acids other than carbonic acid are absent. Then when a water is titrated, equivalents of added acid equal equivalents of carbonate and bicarbonate, plus OH^- or minus H^+, depending on pH of the water:

$$(alk) + (H^+) \rightarrow 2(CO_3^{--}) + (HCO_3^-) + (OH^-) \tag{3}$$

Concentrations of (H^+) and (OH^-) are small between pH values 4.5 and 10.3 and may be neglected.

From equation (2),

$$(CO_3^{--}) = \frac{K_2'(HCO_3^-)}{(H^+)} \tag{4}$$

From equation (3),

$$(CO_3^{--}) = \frac{(alk) - (HCO_3^-)}{2} \tag{5}$$

Therefore

$$(HCO_3^-) = \frac{(alk)}{1 + \dfrac{2K_2'}{(H^+)}} \tag{6}$$

Equation (6) was derived by Langelier[1] on the assumption that K_s' and K_2' are based on concentrations (moles/liter) rather than on activities. For example, referring to (1), if K_s is the true activity product, then $K_s = K_s'\gamma_\pm^2$ where $\gamma_\pm^2$ refers to the mean ion activity coefficient for $CaCO_3$. The activity coefficient was approximated by Langelier employing the Debye-Hückel theory: $-\log \gamma = 0.5z^2\mu^{\frac{1}{2}}$, where μ is the ionic strength and z is the valence. Hence concentrations of CO_3^{--} and HCO_3^- obtained by titration can be equated to corresponding concentrations of these species in K_s' and K_2'. Accordingly, K_s' and K_2' vary not only with temperature but also with total dissolved solids because of the effect of ionic strength of a solution on activities of specific ions.

Substituting (6) into (4),

$$(CO_3^{--}) = \frac{K_2'}{H^+} \frac{(alk)}{1 + \dfrac{2K_2'}{(H^+)}} \tag{7}$$

and (7) into (1)

$$(Ca^{++}) \frac{K_2'}{(H^+)} \frac{(alk)}{1 + \dfrac{2K_2'}{(H^+)}} = K_s' \tag{8}$$

Taking logarithms of both sides and using the notation $\log \dfrac{1}{\alpha} = p\alpha$,

$$pH_s = (pK_2' - pK)_s' + p(Ca^{++}) + p(alk) + \log\left[1 + \frac{2K_2'}{(H^+)_s}\right] \tag{9}$$

where pH_s is the pH of a given water at which solid $CaCO_3$ is in equilibrium with its saturated solution.

The last term is ordinarily small and can be omitted when pH_s falls

[1] W. Langelier, *J. Amer. Water Works Assoc.*, **28**, 1500 (1936).

within 6.5 and 9.5.* Based on the value $K_2' = 4.8 \times 10^{-11}$, typical values as a function of pH_s in the alkaline range at 25°C are given in Table 2.

TABLE 2

pH_s	10.3	10.0	9.7	9.4	9.1
$\log\left[1 + \dfrac{2K_2'}{(H^+)_s}\right]$	0.47	0.29	0.17	0.09	0.05

Values of $(pK_2' - pK_s')$ decrease with temperature as follows: 0°C, 2.48; 20°C, 2.04; 25°C, 1.96; 50°C, 1.54. In the presence of other salts (e.g., NaCl, Na_2SO_4, or $MgSO_4$), the increasing ionic strength of the solution depresses the activity of other ions in solution. This effect increases values of $(pK_2' - pK_s')$. For example, at 25°C, at a total dissolved solids content of 100 ppm, the value is 2.13, and for 500 ppm it is 2.35.

A nomogram for obtaining pH_s of a water at various temperatures and dissolved solids content was constructed by C. Hoover.[2] A chart for the same purpose as prepared by Powell, Bacon, and Lill[3] is reproduced in Fig. 4. To use the chart, we must know the alkalinity of a water and calcium ion concentration calculated as ppm $CaCO_3$, total dissolved solids in ppm, and the temperature.

The Saturation Index is then the algebraic difference between the measured pH of a water and the computed pH_s

$$\text{Saturation Index} = pH_{\text{measured}} - pH_s \qquad (10)$$

To calculate the Saturation Index at above-room temperature, the actual pH of water at the higher temperature should be used. This can be estimated from the room temperature value by using Fig. 5,[3] which gives values for two waters of differing alkalinity.

DERIVATION OF POTENTIAL CHANGE ALONG A CATHODICALLY PROTECTED PIPELINE

Assume that current enters pipe from the soil side through a porous insulating coating, returning to the anode by way of the pipe (Fig. 6).

---

* Actually any $pH_s < 9.5$. Langelier probably meant that pH_s seldom falls below pH 6.5.

[2] C. Hoover, *J. Amer. Water Works Assoc.*, **30**, 1802 (1938).

[3] S. Powell, H. Bacon, and J. Lill, *Ind. Eng. Chem.*, **37**, 842 (1945).

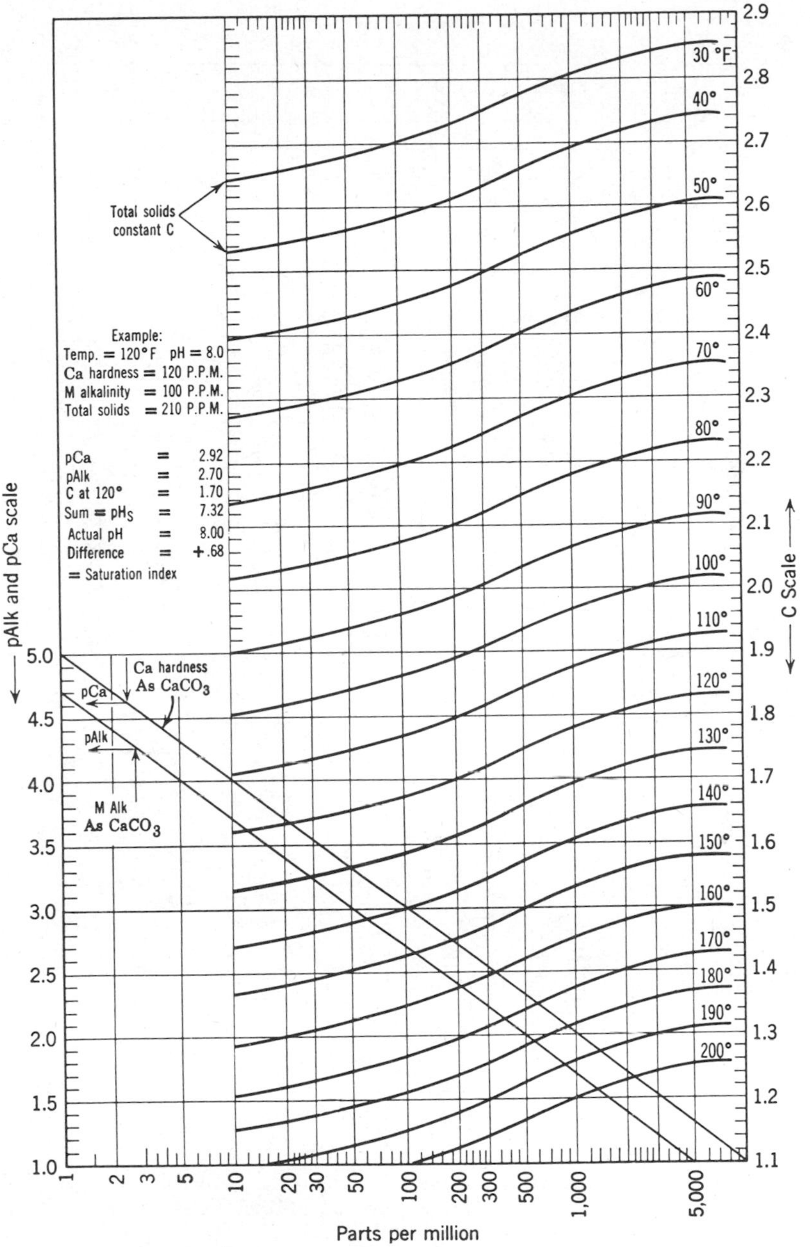
Total solids constant C
Example:
Temp. = 120°F pH = 8.0
Ca hardness = 120 P.P.M.
M alkalinity = 100 P.P.M.
Total solids = 210 P.P.M.
pCa = 2.92
pAlk = 2.70
C at 120° = 1.70
Sum = pHs = 7.32
Actual pH = 8.00
Difference = +.68
= Saturation index
Ca hardness As CaCO3
pCa
pAlk
M Alk As CaCO3
pAlk and pCa scale
C Scale
Parts per million
30 °F
40°
50°
60°
70°
80°
90°
100°
110°
120°
130°
140°
150°
160°
170°
180°
190°
200°
5.0
4.5
4.0
3.5
3.0
2.5
2.0
1.5
1.0
2.9
2.8
2.7
2.6
2.5
2.4
2.3
2.2
2.1
2.0
1.9
1.8
1.7
1.6
1.5
1.4
1.3
1.2
1.1
1
2
3
5
10
20
30
50
100
200
300
500
1,000
5,000

Fig. 4. Chart for calculating Saturation Index (Powell, Bacon, and Lill). ("Ca" and "alkalinity" expressed as ppm $CaCO_3$, temperature in degrees F.)

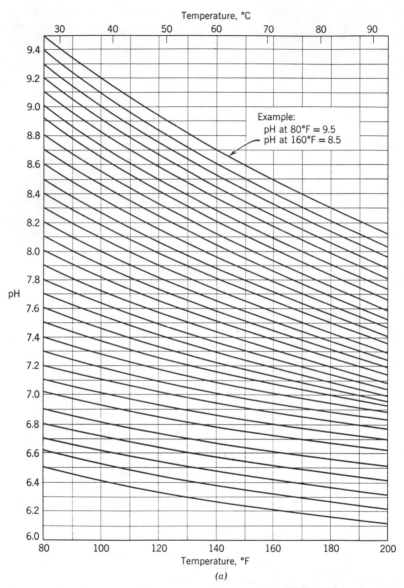

Fig. 5. Values of pH of water at elevated temperatures (Powell, Bacon, and Lill). (*a*) For water of 25-ppm alkalinity (methyl orange end point).

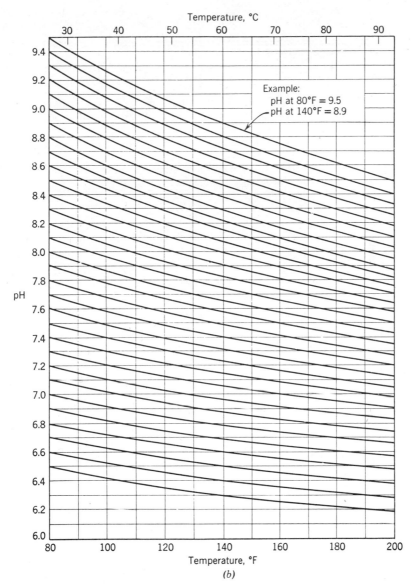

Temperature, °C

Example:
pH at 80°F = 9.5
pH at 140°F = 8.9

pH

Temperature, °F

(b)

Fig. 5. (*Continued*) (*b*) For water of 100 ppm alkalinity (methyl orange end point).

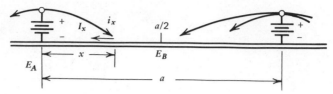

Fig. 6. Sketch of buried pipe cathodically protected by anodes distance a apart.
i_x = current density at pipe surface at distance x from point of bonding
I_x = total current in pipe at distance x
E = difference between measured and corrosion potentials of pipe
r = radius of pipe
R_L = resistance of metallic pipe per unit length
z = resistance of pipe coating per unit area

Then the change of current, I_x, in the pipe per unit length at x is equal to the total current entering the pipe at x, or

$$\frac{dI_x}{dx} = -2\pi r i_x \qquad (1)$$

By Ohm's law, the potential change along the pipe is given by

$$\frac{dE_x}{dx} = -R_L I_x \qquad (2)$$

Combining (1) and (2)

$$\frac{d^2E_x}{dx^2} = R_L(2\pi r i_x) \qquad (3)$$

At small values of i_x, polarization of the pipe surface is a linear function of the true current density, i_x', at the base of pores in the coating, or

$$E_x = k_1 i_x' \qquad (4)$$

Assuming that the resistance, z, per unit area of coating is inversely proportional to the total cross-sectional area of pores per unit area, true current density increases with z, or

$$i_x' = k_2 z i_x \qquad (5)$$

Combining (4) and (5)

$$E_x = kz i_x \qquad \text{where } k = k_1 k_2 \qquad (6)$$

Substituting (6) in (3)

$$\frac{d^2E_x}{dx^2} = \left(\frac{R_L 2\pi r}{kz}\right) E_x \qquad (7)$$

For an infinite pipeline, this differential equation has the solution

$$E_x = E_A \exp\left[-\left(\frac{2\pi r R_L}{kz}\right)^{\frac{1}{2}} x\right] \tag{8}$$

for the boundary conditions $E_x = 0$ at $x = \infty$ and $E_x = E_A$ at $x = 0$.

On the other hand, if the length of pipeline to be protected is $a/2$, or half the distance to the next point of bonding (see Fig. 6), and the potential E_x at $a/2 = E_B$, it follows that current in the pipeline at $a/2 = 0$, or $(dE_x/dx)_{x=a/2} = 0$. Introducing this boundary condition:

$$E_x = E_B \cosh\left[\left(\frac{2\pi r R_L}{kz}\right)^{\frac{1}{2}}\left(x - \frac{a}{2}\right)\right] \tag{9}$$

and

$$E_A = E_B \cosh\left[-\left(\frac{2\pi r R_L}{kz}\right)^{\frac{1}{2}}\frac{a}{2}\right] \tag{10}$$

The current in an infinite pipeline at any point, x, is given by substituting the derivative of (8) into (2), or

$$I_x = \left(\frac{2\pi r R_L}{kz}\right)^{\frac{1}{2}}\frac{E_A}{R_L}\exp\left[-\left(\frac{2\pi r R_L}{kz}\right)^{\frac{1}{2}}x\right] \tag{11}$$

The total current at $x = 0$ (sometimes called drainage current) multiplied by 2, to take care of current flowing from either side of pipeline to point of bonding, equals

$$I_A = \frac{2E_A}{R_L}\left(\frac{2\pi r R_L}{kz}\right)^{\frac{1}{2}} \tag{12}$$

Similarly, for a finite pipe

$$I_A = \left(\frac{2E_B}{R_L}\right)\left(\frac{2\pi r R_L}{kz}\right)^{\frac{1}{2}} \sinh\left[\frac{a}{2}\left(\frac{2\pi r R_L}{kz}\right)^{\frac{1}{2}}\right] \tag{13}$$

DERIVATION OF EQUATION FOR POTENTIAL DROP ALONG SOIL SURFACE CREATED BY CURRENT ENTERING OR LEAVING A BURIED PIPE

For a deeply buried pipe

$$i = \frac{j}{2\pi R} \tag{1}$$

But for a pipe buried h cm below the soil surface, less current flows in the h direction toward the soil surface than in other directions. The computation for current flow in this situation is approximated by assuming an image pipe located h cm above the soil surface which supplies an

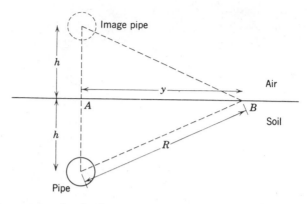

i = current density in soil at distance R from pipe
j = total current entering or leaving pipe per unit length
$\Delta\phi$ = potential difference between A and B at right angle to pipe
ρ = soil resistivity

Fig. 7. Sketch of buried pipe at which current enters or leaves, causing potential drop along soil surface.

equal current, j (Fig. 7). Then the component of current density along the soil surface equals

$$i_y = \frac{2j}{2\pi R}\frac{y}{R} \qquad (2)$$

(The same result can be obtained by assuming half-cylindrical current distribution in the soil multiplied by the factor y/R to account for zero current flowing to pipe in the h direction ($y = 0$) and an increasing current density, i_y, as distance, y, increases.)

The potential gradient along the soil surface by Ohm's law becomes

$$\frac{d\phi}{dy} = \frac{j}{\pi R}\frac{y\rho}{R} \qquad (3)$$

Since $R^2 = h^2 + y^2$

$$\Delta\phi = \int_0^y \frac{\rho j}{\pi R^2}\, y\, dy = \int_0^y \frac{\rho j}{\pi}\frac{y}{h^2+y^2}\, dy = \frac{\rho j}{2\pi}\ln\frac{h^2+y^2}{h^2} \qquad (4)$$

If we take $y = 10$ times distance h, or $y = 10h$

$$\Delta\phi = \frac{\rho j}{2\pi}\, 2.3\log 101 = 0.734\rho j \qquad (5)$$

where $\Delta\phi$ is in V, ρ in Ω-cm and j in A/cm. [See also, R. Howell, *Corrosion*, **8,** 300 (1952).]

DERIVATION OF EQUATION FOR DETERMINING
RESISTIVITY OF SOIL BY FOUR-ELECTRODE METHOD

Two metal electrodes, A and B, are at a distance $3a$ apart. Two reference electrodes, C and D, are located distance a apart and also a from A and B, respectively (see Fig. 8).

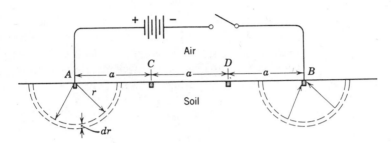

I = total current in battery circuit
r = distance from electrodes A or B
ρ = soil resistivity
$\Delta\phi$ = measured potential difference between electrodes C and D

Fig. 8. Sketch of four-electrode arrangement for determining soil resistivity.

Current density in the soil, assuming hemispherical symmetry at distance r from A or $B = I/2\pi r^2$. By Ohm's law

$$\frac{d\phi}{dr} = \frac{I}{2\pi r^2}\rho \tag{1}$$

$$\Delta\phi = \int_{r_1}^{r_2} \frac{I\rho}{2\pi r^2}\, dr = \frac{-I\rho}{2\pi}\left(\frac{1}{r_2} - \frac{1}{r_1}\right) \tag{2}$$

The potential difference between C and D by reason of current leaving $A = \Delta\phi_1$.

$$\Delta\phi_1 = \frac{-I\rho}{2\pi}\left(\frac{1}{2a} - \frac{1}{a}\right) = \frac{I\rho}{4\pi a} \tag{3}$$

Similarly, the potential difference, $\Delta\phi_2$, between C and D by reason of current entering electrode $B = I\rho/4\pi a$. Hence

$$\Delta\phi_{\text{total}} = \Delta\phi_1 + \Delta\phi_2 = \frac{I\rho}{2\pi a} \quad \text{or} \quad \rho = 2\pi a\frac{\Delta\phi}{I} \tag{4}$$

[See also S. Ewing and J. Hutchinson, *Corrosion*, **9**, 221 (1953).]

DERIVATION OF EQUATION EXPRESSING
WEIGHT LOSS BY FRETTING CORROSION

Assume n asperities or contact points per unit area of metal (or oxide) surface which, for mathematical convenience, are circular in shape. The average diameter of asperities is c and the average distance apart is s (Fig. 15, p. 160). In the fretting process, the asperities move over a plane surface of metal at linear velocity, v, each asperity plowing out a path of clean metal of width averaging c and of length depending on distance of travel. Behind each asperity on the track of clean metal, gas from the atmosphere adsorbs rapidly; followed in time by formation of a thin oxide film. The next asperity, moving in the same path as the first, wipes off the oxide film and leaves, in turn, a track of clean metal behind. The average time during which oxidation occurs is t. Then

$$t = \frac{s}{v} \tag{1}$$

The corresponding amount of oxide, W, removed by one asperity plowing out a path l long and c wide depends on the amount of oxide formed in time t. For thin film oxidation, the logarithmic equation is obeyed (see p. 188):

$$W = clk \ln\left(\frac{t}{\tau} + 1\right) \tag{2}$$

where τ and k are constants.

Preliminary to oxidation, we can also consider the situation of oxygen adsorbing rapidly as physically adsorbed gas, followed by conversion at a slower rate to chemisorbed oxygen atoms. The chemisorbed oxygen, in turn, reacts with underlying metal to form metal oxide, a reaction that is activated mechanically by asperities moving over the metal surface. Chemisorption limits the amount of oxide that is formed in such a process, the rate of chemisorption following an equation identical in form to that of (2).[4] Hence whichever process applies, the form of the final equation is essentially the same.*

* This argument is not compelling because the logarithmic term is eventually expanded and only the first term is used. This is equivalent to a linear rate of oxidation or of gas adsorption with time. A linear rate of gas adsorption suggests that the amount of oxygen reaching the clean metal surface as physically adsorbed gas may actually be controlling, rather than its conversion to chemisorbed oxygen atoms. This possibility is given support by the observed increase of fretting weight loss as the temperature is lowered, corresponding to increased rate and extent of physical adsorption at lower temperatures. The rate of chemisorption, on the other hand, usually decreases as the temperature is lowered.

[4] F. Stone, in *Chemistry of the Solid State*, p. 385, edited by W. Garner, Butterworths, London, 1955.

Substituting (1) into (2)

$$W = clk \ln \left(\frac{s}{v\tau} + 1 \right) \tag{3}$$

Assuming that relative motion of the two surfaces is sinusoidal, $2l$ is the total length of travel in any one cycle and x, the linear displacement from the midpoint of travel at time t, is given by

$$x = \frac{l}{2} \cos \theta \tag{4}$$

and

$$\frac{dx}{dt} = v = -\frac{l}{2} \sin \theta \frac{d\theta}{dt} \tag{5}$$

If f represents constant linear frequency, this is related to constant angular velocity by the expression

$$\frac{d\theta}{dt} = 2\pi f \tag{6}$$

Therefore, the average velocity is given by

$$\bar{V} = -\frac{\pi l f \int_0^\pi \sin \theta \, d\theta}{\pi} = 2lf \tag{7}$$

Hence, for n contacts or asperities per unit area of interface, weight loss, W, per cycle caused solely by oxidation is

$$W_{\text{corros}} = 2nlck \ln \left(\frac{s}{2lf\tau} + 1 \right) \tag{8}$$

To this must be added loss of metal by wear because each asperity, on the average, digs below the oxide layer and dissipates metal in an amount proportional to the area of contact of the asperities and the length of travel. The area of asperity, rather than the width, is important now because of the "tearing out" or welding action taking place during mechanical wear, in contrast to scraping off of chemical products from the surface, as discussed previously. A shearing off of asperities without welding also leads to wear dependent on total area of contact. For n circular asperities, weight loss per cycle is given by

$$W_{\text{mechanical}} = 2k'n \left(\frac{c}{2} \right)^2 \pi l \tag{9}$$

But $n\pi(c/2)^2$, the total area of contact, is equal[5] to the load, L, divided

[5] F. P. Bowden and D. Tabor, *The Friction and Lubrication of Solids*, Oxford University Press, New York, 1950.

by the yield pressure, p_m. The term p_m is approximated by 3 times the elastic limit; hence for mild steel, p_m equals 100 kg/mm² or 140,000 psi.
Therefore

$$W_{\text{mechanical}} = 2k'\frac{lL}{p_m} = k_2 lL \tag{10}$$

where k_2 is a constant equal to $2k'/p_m$. The total wear or metal loss per cycle is the sum of the oxidation or corrosion term, and the mechanical term

$$W_{\text{total}} = W_{\text{corros}} + W_{\text{mechanical}} \tag{11}$$

Returning to (8), the logarithmic term can be expanded according to

$$\ln(x+1) = x - \frac{x^2}{2} + \frac{x^3}{3} - \text{etc}$$

where x is equal to $s/(2lf\tau)$. When the latter expression is much smaller than unity, the square and higher terms can be omitted. This condition applies particularly to high loads (small values of s), high-frequency f, and large value of slip l. The constant, τ, for oxidation of metals, or adsorption of oxygen on metals, has not been determined empirically to any degree of precision. Empirical values for iron range from 0.06 to 3 sec. Assuming reasonable values for $\tau = 0.06$ sec., $f = 10$ cps, $l = 0.01$ cm, s (distance between asperities) $= 10^{-4}$ cm, then $s/2lf\tau = 0.008$. Therefore, whenever experimental conditions approximate those cited and higher terms of the logarithmic expansion can be neglected

$$W_{\text{corros}} = \frac{ncks}{f\tau} \tag{12}$$

This expression, it will be noted, is equivalent to assuming, from the very start, a linear rate of oxidation or of gas adsorption on clean iron where k/τ is the reaction-rate constant. The linear rate reasonably approximates the actual state of affairs for very short times of adsorption or oxidation.

Since the number of asperities along one edge of unit area is equal to $\sqrt{n}$, it follows that $s + c$ is approximated by $1/\sqrt{n}$. Also, recalling that $n\pi(c/2)^2 = L/p_m$, the terms n, c, and s can be eliminated from (12), or

$$W_{\text{corros}} = \frac{k_0 L^{1/2}}{f} - \frac{k_1 L}{f} \tag{13}$$

where

$$k_0 = \frac{2}{\sqrt{p_m\pi}}\frac{k}{\tau} \quad \text{and} \quad k_1 = \frac{4}{p_m\pi}\frac{k}{\tau}$$

Combining (10), (11), and (13), we have the final expression for fretting as measured by weight loss corresponding to a total of C cycles:

$$W_{\text{total}} = (k_0 L^{\frac{1}{2}} - k_1 L) \frac{C}{f} + k_2 l L C \tag{14}$$

CONVERSION FACTORS

Multiply inches penetration per year (*ipy*) by *696* × *density* to obtain mg per sq. decimeter per day (*mdd*).

Multiply *mdd* by *0.00144/density* to obtain *ipy*.

Metal	Density gm/cc	$\dfrac{0.00144}{\text{Density}}$	696 × Density
Aluminum	2.72	0.000529	1890
Brass (red)	8.75	0.000164	6100
Brass (yellow)	8.47	0.000170	5880
Cadmium	8.65	0.000167	6020
Columbium	8.4	0.000171	5850
Copper	8.92	0.000161	6210
Copper-Nickel (70–30)	8.95	0.000161	6210
Iron	7.87	0.000183	5480
Iron-silicon (Duriron) (84–14.5)	7.0	0.000205	4870
Lead (chemical)	11.35	0.000127	7900
Magnesium	1.74	0.000826	1210
Nickel	8.89	0.000162	6180
Nickel-copper(Monel)(70–30)	8.84	0.000163	6140
Silver	10.50	0.000137	7300
Tantalum	16.6	0.0000868	11550
Titanium	4.54	0.000317	3160
Tin	7.29	0.000198	5070
Zinc	7.14	0.000202	4970
Zirconium	6.45	0.000223	4490

CURRENT DENSITY EQUIVALENT TO A CORROSION RATE OF 1 MDD

$$1 \text{ mdd} = \frac{1.117 \, n}{\text{at. wt.}} \times 10^{-5} \text{ A/cm}^2$$

Reaction	A/cm²
$Al \rightarrow Al^{+++} + 3e^-$	12.4×10^{-7}
$Mg \rightarrow Mg^{++} + 2e^-$	9.19
$Fe \rightarrow Fe^{++} + 2e^-$	4.00
$Fe \rightarrow Fe^{+++} + 3e^-$	6.00
$Zn \rightarrow Zn^{++} + 2e^-$	3.42
$Cd \rightarrow Cd^{++} + 2e^-$	1.99
$Ni \rightarrow Ni^{++} + 2e^-$	3.81
$Sn \rightarrow Sn^{++} + 2e^-$	1.88
$Pb \rightarrow Pb^{++} + 2e^-$	1.08
$Cu \rightarrow Cu^{++} + 2e^-$	3.52

STANDARD OXIDATION POTENTIALS, 25°C
(See also p. 29)

$E°$ (V)

$Zn + 4NH_3 \rightarrow Zn(NH_3)_4^{++} + 2e^-$	1.03
$H_2 + 2OH^- \rightarrow 2H_2O + 2e^-$	0.828
$Pb + 3OH^- \rightarrow HPbO_2^- + H_2O + e^-$	0.54
$Pb + SO_4^{--} \rightarrow PbSO_4 + 2e^-$	0.356
$Ag + 2CN^- \rightarrow Ag(CN)_2^- + e^-$	0.31
$Cu + 2NH_3 \rightarrow Cu(NH_3)_2^+ + e^-$	0.12
$Ag + Br^- \rightarrow AgBr + e^-$	−0.095
$Sn^{++} \rightarrow Sn^{+4} + 2e^-$	−0.15
$Cu^+ \rightarrow Cu^{++} + e^-$	−0.153
$H_2SO_3 + H_2O \rightarrow SO_4^{--} + 4H^+ + 2e^-$	−0.17
$Ag + Cl^- \rightarrow AgCl + e^-$	−0.222
$2Hg + 2Cl^- \rightarrow Hg_2Cl_2 + 2e^-$	−0.2676
$4OH^- \rightarrow O_2 + 2H_2O + 4e^-$	−0.401
$2I^- \rightarrow I_2 + 2e^-$	−0.5355
$H_2O_2 \rightarrow O_2 + 2H^+ + 2e^-$	−0.682
$Fe^{++} \rightarrow Fe^{+3} + e^-$	−0.771
$2Br^- \rightarrow Br_2(l) + 2e^-$	−1.0652
$2H_2O \rightarrow O_2 + 4H^+ + 4e^-$	−1.229*
$2Cr^{+3} + 7H_2O \rightarrow Cr_2O_7^{--} + 14H^+ + 6e^-$	−1.33
$2Cl^- \rightarrow Cl_2 + 2e^-$	−1.3595
$Pb^{++} + 2H_2O \rightarrow PbO_2 + 4H^+ + 2e^-$	−1.455
$Mn^{++} \rightarrow Mn^{+3} + e^-$	−1.51
$Ni^{++} + 2H_2O \rightarrow NiO_2 + 4H^+ + 2e^-$	−1.68
$PbSO_4 + 2H_2O \rightarrow PbO_2 + SO_4^{--} + 4H^+ + 2e^-$	−1.685
$Co^{++} \rightarrow Co^{+3} + e^-$	−1.82
$Fe^{+3} + 4H_2O \rightarrow FeO_4^{--} + 8H^+ + 3e^-$	−1.9

Data from *Oxidation Potentials,* W. Latimer, Prentice-Hall, Englewood Cliffs, N. J., 1952 (with permission).

* At pH 7, 0.2 atm O_2, $E = -0.81$ V.

ABBREVIATIONS

atm	atmosphere
g	gram
lb	pound
at. %	atom percent
wt. %	weight percent
mg	milligram
kg	kilogram
eq	equivalent
in.	inch
ft	foot
μ	micron
cm	centimeter
d.	density
dm	decimeter
m	meter
ml	milliliter
oz	ounce
gal	gallon
km	kilometer
min	minute
hr	hour
A	ampere
mA	milliampere
μA	microampere
V	volt
mV	millivolt
Ω	ohm
s.h.e.	versus standard hydrogen electrode
C	coulomb
J	joule
ppm	parts per million
cal	gram calorie
mdd	$mg/(dm^2)(day)$
ipy	inch penetration per year
G	Gibbs free energy
F	Faraday
γ	mean ion activity coefficient
N	normal solution
M	molal solution

Index